DAIRY MICROBIOLOGY

Volume 2

The Microbiology of Milk Products

DAIRY MICROBIOLOGY

Volume 2

The Microbiology of Milk Products

Edited by

R. K. ROBINSON
M.A., D.Phil.

Department of Food Science,
University of Reading, UK

APPLIED SCIENCE PUBLISHERS
LONDON and NEW JERSEY

APPLIED SCIENCE PUBLISHERS LTD
Ripple Road, Barking, Essex, England
APPLIED SCIENCE PUBLISHERS, INC.
Englewood, New Jersey 07631, USA

British Library Cataloguing in Publication Data

Dairy microbiology.
   Vol. 2: The microbiology of milk products
   1. Dairy products   2. Food microbiology
   I. Robinson, R.K.
   630.2′76   QR121

   ISBN 0-85334-961-4

WITH 42 TABLES AND 31 ILLUSTRATIONS
© APPLIED SCIENCE PUBLISHERS LTD 1981

Printed in Great Britain by Galliard (Printers) Ltd, Great Yarmouth

Preface

The basic philosophy behind this text on dairy microbiology has already been discussed in the preface to Volume 1, but one particular aspect will bear repetition. Thus, it is important to reiterate that these two volumes form, in essence, one complete work, and hence that contributors to this volume will have anticipated an acquaintance with certain basic information. If this limitation is recognised, then the chapters that follow should provide a valuable insight into the microbiology of dairy products.

R. K. ROBINSON

Contents

List of Contributors

HELEN R. CHAPMAN, N.D.D., F.I.F.S.T.,
National Institute for Research in Dairying, Shinfield, Reading, UK.

J. G. DAVIS, O.B.E., D.Sc., Ph.D., C.Chem., F.R.I.C., F.I.F.S.T., F.R.S.H.,
J. G. Davis and Partners, 52 London Road, Reading, UK.

M. F. MURPHY, M.Sc., F.I.F.S.T.,
Department of Dairy and Food Technology, University College, Cork, Ireland.

H. LÜCK, D.Sc.,
Animal and Dairy Science Research Institute, Private Bag X2, Irene, Republic of South Africa.

R. K. ROBINSON, M.A., D.Phil.,
Department of Food Science, University of Reading, London Road, Reading, UK.

J. ROTHWELL, B.Sc., Ph.D., F.I.F.S.T.,
Department of Food Science, University of Reading, London Road, Reading, UK.

M. ELISABETH SHARPE, B.Sc., Ph.D., D.Sc., F.I.Biol.,
National Institute for Research in Dairying, Shinfield, Reading, UK.

A. Y. TAMIME, M.Sc., Ph.D.,
West of Scotland Agricultural College, Auchincruive, Ayr, UK.

1

Microbiology of Ice Cream and Related Products

J. ROTHWELL

Department of Food Science,
University of Reading, UK

It is probable that the Chinese some 3000 years ago were the first to mix snow and fruit juices together to form a dessert, and the Romans may also have used snow to 'ice' drinks during summer. Marco Polo is credited with bringing a recipe back from Peking to Venice in 1292 which included frozen milk, and was probably the original sherbet ice.

However, production of such frozen delicacies was generally on a very small scale, either in high class catering establishments or in the home, until the middle of the nineteenth century. In 1851, the first ice cream factory was opened in Baltimore by a Jacob Fussell, and from then onwards the industry grew enormously. In 1859 there was an annual production of 4000 US gallons in America, which had risen by 1899 to more than five million gallons, and is now over 1000 million gallons.

There are many types of frozen desserts. It is, therefore, necessary to classify them, and although descriptions vary from country to country, they generally follow the pattern given below. Legislation is usually based on similar nomenclature.

CLASSIFICATION OF FROZEN DESSERTS

Cream Ices

Creme glacée, eiskrem, cremadi gelato and roomijs are ice creams made with a statutory minimum content of butterfat (milk fat). Flavouring materials, including fruit, nuts and chocolate may be added to produce the corresponding flavoured cream ice. Many countries allow the use of milk fat only, and prohibit the use of any substitute fats.

Ice Cream

Ijs (The Netherlands), *mellorine* (USA), *glaces de consommation* (Belgium)
and *margarin-is* (Norway) are ice creams with a statutory minimum of fat,
some or all of which may be fat other than milk fat.

Milk Ices

Glace au lait, milcheis, gelato al latte and milkijs are ices based on milk,
normally without any extra added fat. Only milk fat must be used, and
minimum standards of 2·5 to 3 % are laid down in most countries.

Custards

The traditional milk ice, which is made by boiling 1 gal. of whole milk with
170 g of cornflour and 0·68 kg of sugar, is sometimes called a custard. The
true custard is probably made with eggs or egg yolk solids, and in the USA
must contain at least 1·4 % by weight of egg yolk solids. It is often called
French ice cream or French custard ice cream. A French product is called
glace aux oeufs, and this must contain at least 7 % egg yolk solids.

Ices or Water Ices

These are made from fruit juices diluted with water, and with additions that
may include sugar, citric, malic or tartaric acids, a stabiliser such as gelatin,
colour and flavour. They may be frozen with, or without, agitation and the
incorporation of air. If sold in a 'slushy' condition they may be called
'frappé'—if made with an alcoholic liquid in place of some of the water, they
may be known as a 'punch'. Water ices frozen without agitation, and
usually on a stick, are known as 'ice lollies' in the UK.

Sherbet

This is made from the same ingredients as a water ice, but incorporates
some ice cream, liquid milk, cream or milk powder. It is frozen with
agitation, and this incorporates about 25 % air.

   In addition, more complex products have been developed over the years.
Two of these are given below.

Cassata

This is made in a round mould. The confection is built up in layers of rich,
variously flavoured ice cream, some with fruit, some with liqueur, and
sometimes with nuts or chocolate. Fingers or slices of sponge cake,
sometimes soaked in liqueur may be added. The cassata is frozen for
several hours, then turned out of the mould for serving.

'Splits'
These are made on a stick, and consist of a central section of ice cream and an outer layer of fruit water ice. They may be dipped or coated in chocolate, broken nuts or biscuit crumbs.

ICE CREAM LEGISLATION

Composition
Ice cream is normally made from a liquid mix based on milk, cream, water, milk solids-not-fat, milk fat or other fat, sugar, emulsifying and stabilising agents, flavour and colour. This mix, with the ingredients in appropriate quantities, is heat treated, homogenised to reduce the size of fat globules, cooled and stored for up to about 24 h at a temperature usually around 3 °C. It is then subjected to the process of freezing in an ice cream freezer, where the temperature is rapidly lowered to about − 5 to − 7 °C by agitation, while air is incorporated to give the ice cream 'over-run'. The frozen ice cream, on emerging from the ice cream freezer, is usually in a stiff but plastic form, and may be dispensed for sale directly (as 'soft-serve') ice cream, or may be packaged in bulk or small retail packs, and then hardened at a very low temperature (− 20 to − 40 °C) before being sold as hard ice cream.

The wide variety of ingredients, and the possible variations in their quality, has given rise to legal quality and compositional requirements in many countries. In the UK there are two major items of legislation which refer respectively to the composition of ice cream, and to the methods to be used for heat treatment and storage of ice cream mix and ice cream. Ice cream in the UK has been largely based on milk products, although even before the Second World War, many manufacturers included a proportion of vegetable fat in the mix of at least some of their products. Then post-war shortages, particularly of milk fat, led to almost all the ice cream in the UK being made from fats other than milk fat, and the legislation which was introduced had to take this into account.

The various earlier regulations for ice cream composition were finally revoked and replaced by the Food Standards (Ice Cream) Regulations 1959 (SI 472) which, briefly, made the following requirements.

Ice cream or *ice* must contain not less than 5 % fat, and not less than 7·5 % milk solids-not-fat (SNF).

Ice cream or ice containing fruit pulp or fruit purée must either conform to the above standard, or must have a total fat and SNF content of not less than 12·5 %, which must include not less than 7·5 % fat and 2 % SNF.

Dairy ice cream, dairy cream ice or cream ice must conform to which ever of the above standards is appropriate, and must not contain any fat other than milk fat (except fat introduced by the use of egg, flavouring substances or emulsifying or stabilising agents as ingredients).

Dairy ice cream, dairy cream ice or cream ice containing fruit pulp or fruit purée must either conform to the above standard, or must have a total milk fat and SNF content of not less than 12·5%, which must include not less than 7·5% milk fat and 2% SNF.

Milk ice (including milk ice containing fruit, fruit pulp or fruit purée) must contain not less than 2·5% milk fat, and no other fat (except fat introduced by the use of egg, flavouring substances or emulsifying or stabilising agents), and not less than 7% SNF.

'*Parev*' (*Kosher*) *Ice* must contain not less than 10% fat, and no milk fat or derivative of milk.

The use of any artificial sweetener is forbidden in all these products.

Labelling regulations permit ice cream which contains non-milk fat to be sold as 'ice cream', and to bear a declaration that it contains skimmed-milk solids (SNF). All pre-packed ice cream for retail must be labelled to the effect that it contains vegetable fat or non-milk fat, whichever is appropriate. In addition, any wording or pictures on such packets which suggest butter, cream, milk, or anything connected with the dairy industry are prohibited, unless these are part of the trade name of the manufacturer.

Heat Treatment

After the severe outbreak of typhoid in Aberystwyth in 1947 caused by ice cream made by a carrier of the disease (quoted later), heat treatment regulations were introduced, and after some modification there is now a very complete set of regulations (Ice cream Heat Treatment, Etc., Regulation 1959 (SI 734) and amendment 1963 (SI 1083)), which may be briefly summarised as follows.

Ice cream mixture must not be kept for more than 1 h at any temperature which exceeds 7·2 °C (45 °F), before being pasteurised or sterilised as follows:

Pasteurisation
1. Temperature not less than 65·6 °C (150 °F) for at least 30 min.
2. Temperature not less than 71·1 °C (160 °F) for at least 10 min.
3. Temperature not less than 79·4 °C (175 °F) for at least 15 s.

Sterilisation
Temperature not less than 148·8 °C (300 °F) for at least 2 s.

Then it must be cooled to not more than 7·2 °C (45 °F) within 1·5 h of being heated, and must be kept at such temperature until it is frozen.

Exceptions to these conditions are:

1. If the mix has been sterilised and filled into sterile containers under sterile conditions, cold storage before freezing is not required.

2. Heat treatment is not required in the case of any mix which has a pH of 4·5, or less (e.g. ice lolly mix).

3. Mixes made by adding a prepared powdered mix (heat treated before it was dried) to cold drinking water need not be heat treated. These mixes must be frozen within 1 h of reconstitution. Sugar may be added as well as water.

In the case of HTST pasteurisation or sterilisation, full thermostatic control is required, and a positive drive pump must be used to ensure the correct holding time, together with a device for automatically diverting mix, or stopping the sterilisation process, if the mixture is not heated to the correct temperature.

After such heat treatment and careful storage of the liquid mix, it is to be expected that the mixture should have a very low bacterial content. It must be noted that composition and heat treatment requirements vary considerably from country to country. In the USA, there is legislation which applies to individual states, and there are also Federal regulations. An indication of the complexity of this situation is given in Arbuckle (1972), while Hyde and Rothwell (1973) give tables of the regulations existing at that time in respect of composition, heat treatment and, in certain countries, bacterial standards.

Mix Composition and Ingredients

In addition to these overall legal requirements, the composition of ice cream varies to a considerable extent depending on the type of ice cream required and, in particular, on the type of freezer which is used.

A mix for a large continuous freezer will contain about 10 % fat; 11 % milk solids-not-fat (SNF), 14 % sugar or sugar/corn syrup mixture, 0·5 to 0·75 % stabiliser and emulsifier, colour and flavouring. Such a mix will require about 90 to 100 % over-run. By comparison, a mix for a vertical freezer will contain about 8 % fat, 11·5 % SNF, 12 to 13 % sugar, 0·5 to 0·75 % stabiliser and emulsifier, colour and flavouring, and will require about 30 to 50 % over-run.

From this it will be seen that the SNF component is a most important part of ice cream. If milk fat is used as well, the total milk solids will approach 21 or 22 %, but even if milk fat is not used, ice cream will normally contain more milk solids than liquid milk.

Ingredients
Milk Based
Although many small-scale manufacturers use fresh, liquid whole milk as the basic ingredient, the major manufacturers mainly employ fresh concentrated skim-milk at a SNF content of about 30 %. Considerable use is also made of spray-dried skim-milk powder.

Fat
Milk fat may be supplied either as cream (which is very costly but produces an excellent 'cream ice'), fresh unsalted butter, or butter oil.

Other Fats
Ice cream fats are generally based on mixtures of partly hydrogenated vegetable fats, compounded to give a fairly sharp major melting peak around 30 °C. However, the whole of the fat must be melted below about 37 °C in order to avoid a clinging 'fatty sensation' in the mouth. Vegetable fats are completely bland and can take any flavour required. Increasing use is now being made of whey powders and modified whey products up to the level allowed by legal and technical requirements.

Sugar
The major sugar used is sucrose (cane or beet sugar), because of its solubility and its high sweetening power, but other sugars, notably glucose syrups produced from corn flour by hydrolysis of the starch, are used. These rather viscous syrups, with increasing dextrose contents depending on the extent of hydrolysis, give an improvement in texture, but careful selection is necessary otherwise the resulting ice cream may be too soft. Dextrose is also used to a limited extent.

Stabilising Agents
These, used at amounts up to about 0·5 % depending on the type, are intended to absorb any free moisture in the frozen ice cream, thereby making the ice cream resistant to 'heat shock'. They should also give body to the ice cream without making it too heavy or gummy, and give an attractive 'melt down'.

To a limited extent, milk proteins act in the same way, but additional stabilisation is usually required. The first substance used was gelatin, which, on hydration, produces a gel network during a period of about 4 h at 5 °C ('ageing')—other stabilising agents do not require any time for hydration. However, milk proteins from skim-milk powder require time to rehydrate, and time is necessary for correct and sufficient crystallisation of the fat to occur after cooling. A storage or 'ageing' time is, thus, still necessary for the best results on freezing an ice cream mix.

Many other substances are used now for the stabilisation of ice cream. These include sodium alginates, locust bean and guar gums, sodium carboxymethyl cellulose (CMC) and carrageen, and other seaweed derivatives. Often combinations of these substances are used, together with small quantities of phosphates or citrates. One problem, 'wheying off', caused by a precipitation of casein and the subsequent release of whey, may be brought about by locust bean and guar gums, and by some grades of CMC; the use of a small quantity of carrageen is very helpful in reducing this effect considerably.

Most of these stabilisers will dissolve reasonably well, but locust bean gum requires a temperature of about 70 °C for up to 15 min. It is thus not really suitable for HTST or UHT processing. However, guar gum, which is very similar in other respects, dissolves in the cold quite rapidly and may, therefore, be used in its place.

Suppliers of stabilisers often also make mixtures in which both stabilising and emulsifying agents are present, and these are most convenient for the ice cream manufacturer to use.

Emulsifying Agents

It is necessary to add an emulsifying agent to almost all ice creams. They are essential, if butter oil, butter or vegetable fats are used, to provide a surface active material in place of the natural fat globule membranes which surround milk fat globules in liquid milk and cream. In addition, homogenisation of a mix reduces the size of the fat globules, but increases very considerably the surface area to be 'protected'. An emulsifying agent is really a surface active agent, which acts by reducing the energy required to maintain the integrity of the fat globules. It will also materially assist in obtaining larger numbers of smaller, more uniform air cells, which, in turn, help to produce a smooth ice cream. The emulsifying agent, therefore, is needed to maintain the fat globules as individual separate units.

However, fairly recent investigations have shown that to obtain a reasonably 'dry' ice cream which will not melt too quickly, it is necessary for some destabilisation of the fat to occur. This effect may be brought about by certain types of emulsifying agent, but not too much of this type of emulsifier can be used or destabilisation may be so great that the fat forms large lumps ('butters'), instead of the optimum amount of clustering.

The first emulsifying agent used was egg yolk (the active ingredient being lecithin), and this was followed by glyceryl monostearate (GMS). Normal GMS has a monoglyceride content of 30–40%, but recent developments have shown that the 'molecular distilled' types, containing 60 to 80% monoglyceride, are very satisfactory.

The latest developments include the use of some of the 'Tweens' and 'Spans'—polyoxyethylene glycol and sorbitol esters. These are much more efficient in producing 'dry' ice cream, as they increase the amount of destabilised fat. The health regulations in some countries, however, still prohibit the use of some of these agents.

Other Ingredients

Other materials used in ice cream include glycerine in small quantities; this will add a little to the sweetness, and also produces a softer ice cream.

Flavourings

Most ice cream is flavoured by the addition of natural or synthetic flavours. A very large amount of vanilla (either bean, vanillin or a mixture of these) is made, and the second most popular flavours appear to be strawberry or raspberry, with chocolate a close third. Fruit and/or nuts or toffee pieces may be fed into the ice cream as it extrudes from the continuous freezer by special proportioning machines, and ripple ice cream is produced in a very similar manner by injecting a heavy fruit or chocolate syrup.

Manufacture

The manufacture of ice cream is a relatively complex operation, with a series of steps which, in both compositional and microbiological terms, contribute to the overall quality of the ice cream. It is not possible to give details here, but full information will be found in Arbuckle (1972) and Hyde and Rothwell (1973). Briefly, the steps are as follows.

1. After deciding the type of ice cream required, a mix formulation will be drawn up, and the ingredients weighed or measured into mixing vessels. The size of these, as well as method of mixing depends on the size of the enterprise.

2. After mixing, the ingredients are heat treated in either vats (Fig. 1) if on a small scale, or using a continuous plate heat exchanger (Fig. 2). UHT processing is normally only carried out by steam injection (Fig. 3) or by scraped-surface heat exchangers, as ice cream mix very quickly fouls the

FIG. 1. A Giusti Plastomix complete heat treatment vat suitable for small-scale operations. Reproduced by courtesy of T. Giusti and Sons Ltd, London.

FIG. 2. A large-scale mix processing area showing an Autobatch (automatic mixing) (right), the control room (centre), and the mix pasteuriser and homogeniser (left). Reproduced by courtesy of the APV Co. Ltd, Crawley, UK.

Fig. 3. An uperiser installation for sterilising the mix by direct injection of steam.
Reproduced by courtesy of the APV Co. Ltd, Crawley, UK.

plates of heat exchangers operating at temperatures in excess of about
110 °C. The actual operating conditions will depend on the legal
requirements of individual countries.

3. During the heat treatment, the mix will be homogenised. The
pressures used vary according to the fats; usually vegetable fats are treated
at lower pressures around $1600 \, lb \, in^{-2}$, while butter fat mixes are
homogenised at about $2200 \, lb \, in^{-2}$. If two-stage homogenisation is carried
out, the pressures for both types of fat are usually $2000 \, lb \, in^{-2}$ followed by
about $450 \, lb \, in^{-2}$. The object of homogenisation is to reduce the fat globule
size, so that fat globules do not rise to the surface. It also enables thorough
mixing to occur, and improves the whipping ability and air incorporation
properties of a mix. The homogeniser is, however, a complex piece of
equipment requiring thorough cleaning and sanitisation, and may cause
contamination of the mix if this has not been carried out properly. It is,
therefore, to be recommended that homogenisation of a mix is carried out
before the final heat treatment.

4. Cooling and ageing of the mix follows. The mix must be cooled

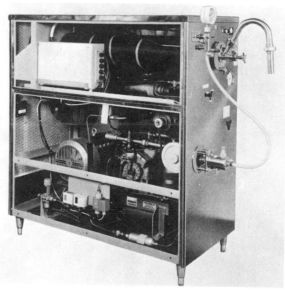

FIG. 4. A small-scale continuous freezer (with side panel removed). The ice cream mix is gravity-fed into the pump (right side of front panel), air is bled in, and the mix is pumped under pressure into the rear of the 'freezing' barrel. After freezing, the semi-solid ice cream is forced through the exit 'tap' at the front. Reproduced by courtesy of T. Giusti and Sons Ltd, London.

rapidly to about 4 °C and kept at that temperature until it is frozen, to avoid proliferation of any remaining organisms. During the storage of the mix at this temperature, milk proteins hydrate, fats begin to crystallise, and any added hydrocolloids absorb quantities of water. This is the process known as 'ageing'. Normally the mix should be frozen into ice cream within 24 h of heat treatment, and undue prolongation of storage can lead to the proliferation of psychrotrophic organisms with serious risk of mix spoilage.

5. The mix is next submitted to the *'freezing'* process where its temperature is reduced, under considerable agitation, in the ice cream freezer (Fig. 4). At the same time air is incorporated to give an aerated product with over-run (the amount of expansion caused by this incorporation of air) of up to 100 and even 120 %, depending on the type of freezer and the ice cream required. Small vertical freezers normally only give over-run of about 50 %, while large continuous freezers (Fig. 5) can produce ice cream with over-runs of up to 120 % by the incorporation of air under pressure. On a large scale, this air will be admitted through filters to prevent the ingress of airborne organisms.

Fig. 5. A large-scale continuous freezer installation at Ashford Creameries, Stourbridge, UK. Reproduced by courtesy of the APV Co. Ltd, Crawley, UK.

6. On leaving the freezer, the ice cream will normally be packaged, either in bulk, in smaller sized 'family' packs of 1 litre or less, or in 'individual' retail packs. In some cases, the ice cream is blended with water ice, and may be frozen on a stick or in a cone (single portions), and covered with a chocolate or flavoured couverture, broken biscuits or nuts. All these are frozen hard in wind tunnels at $-40\,°C$ or in hardening rooms, and then kept at a temperature of about $-30\,°C$ until, and during, distribution.

Some ice cream is sold direct from a dispensing freezer as 'soft-serve' ice cream (Fig. 6), either on cones, or in various types of made up sweets in cafés and restaurants, or from vehicles complete with their own electricity generation equipment.

It is essential that the bacteriological standard of the finished product should be as high as possible. In particular there must be an absence of all pathogenic organisms, as 1 gal. of ice cream can serve at least forty to fifty good sized portions. Often the recipient is a small child, least able to resist the onset of disease. In the UK there were several cases of food poisoning before the Second World War, but, as there were no heat treatment requirements at that time, the significant cases are those occurring later.

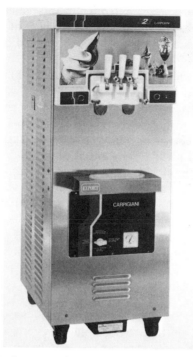

FIG. 6. A Carpigiani 'soft-serve' freezer with twin barrels. Reproduced by courtesy of Morrison Equipments Ltd, Southampton, UK.

Examples of Outbreaks of Food Poisoning Caused by Ice Cream

Hobbs and Gilbert (1978) quote two outbreaks of food poisoning in the UK since the Second World War, which have been directly attributed to the consumption of ice cream. One, in 1945, was due to staphylococci, carried by a worker in the cookhouse of an army hospital, which were introduced into batches of ice cream mix after the ingredients had been cooked. The mix cooled slowly overnight and was frozen 20–30 h later. Around 700 people were affected by a staphylococcal toxin which developed during this period.

A major outbreak of typhoid fever occurred during the summer of 1947 in Aberystwyth, soon after ice cream was allowed to be manufactured after several years of prohibition during the war (Evans, 1947). About 210 cases were reported including four deaths, and ice cream from one particular source appeared to be a common factor in all cases. It was found that this ice cream manufacturer was an active urinary excretor of typhoid bacteria

(*Salmonella typhi*), and specimens from him and patients were shown to be of the same phage type. The manufacturer had been ill with typhoid fever in 1938 but, at the termination of his recovery, he had been declared free of typhoid.

Other cases reported by Hobbs and Gilbert (1978) include paratyphoid at a North Devon holiday resort, where the cause was found to be small doses of bacteria (*S. paratyphi*) on the hands of the vendor of the ice cream and his wife. A case of shigella dysentery is also mentioned, caused by a monkey in the pets' corner of a large store which touched an ice cream cornet being eaten by a child. *Shigella flexneri* 103Z was isolated from a stool specimen after the child became ill, and was also found in specimens from the monkey. The infecting dose in this case was probably that transferred by the animal's paw, and thus there had been little or no time for an increase in the number of organisms.

Between 1950 and 1955 there were 11 outbreaks of food poisoning in the UK due to ice cream and ice lollies involving salmonellae, two involving staphylococci, and six of unknown cause. Since then, there do not appear to have been any further outbreaks from ice cream reported. This is almost certainly due to the fact that the heat treatment regulations, first introduced in 1947, became really effective from about 1950. In addition, the introduction of a modified methylene blue test, suggested in 1947 (Public Health Laboratory Service), and inspections by local health authorities have played a major role in improving the hygienic quality of ice cream. The work of the trade organisations (Ice Cream Federation and Ice Cream Alliance) must not be overlooked in helping to bring about this highly satisfactory state of affairs.

There are still cases of food poisoning caused by ice cream reported from overseas, although most countries, in which there is any appreciable production and sale of ice cream, have relatively strict standards and heat treatment requirements. These vary widely between countries as will be seen from Table I. However, in most countries some form of heat treatment of the mix is required, usually at higher temperatures than for milk pasteurisation, followed by cooling, usually to below $7.2\,°C$ and freezing within 72 h of heat treatment.

Standards
It is in the area of bacteriological requirements that even greater variations are to be found. Total counts vary from $300\,000\,cm^{-3}$ (or per gram—the use of weighed quantities is definitely to be recommended as the air content of ice cream may prove troublesome when measuring by volume) to

TABLE I

Country	Minimum heat treatment standards	Bacteriological standards
Australia	68 °C for 30 min	Total count 50 000 g^{-1} Coliforms absent in 0·1 g No pathogenic organisms
Belgium	Must be made from pasteurised milk	Total count 100 000 g^{-1} No coliforms in 0·1 g
Czechoslovakia	68 to 70 °C for 30 min	Total count 100 000 cm^{-3} Coliforms 100 cm^{-3}
Eire	65·5 °C for 30 min or 71 °C for 15 min, must be cooled to 7·2 °C or lower within 1·5 h of heating	No standards
Fiji	Must be boiled, or pasteurised: 68·5 °C for 20 min or 74·4 °C for 10 min	Total count 50 000 cm^{-3} No coliforms in 0·1 cm^3
France	Reconstituted or fresh milk must be boiled for 1 min or heated for 2 to 3 min at 80 to 85 °C, whole mix 60 to 63 °C for 30 min	Total count 300 000 cm^{-3} No pathogenic organisms
Israel	Pasteurised under approved conditions	Total count 100 000 cm^{-3} Coliforms 20 cm^{-3}
Japan	68 °C for 30 min or equivalent	Total count 50 000 cm^{-3} (10 000 cm^{-3} if 3 % milk fat)
Netherlands	None specified	Total count 100 000 cm^{-3} No coliforms in 0·1 cm^3 No pathogenic organisms
New Zealand	Boiling, or 74 °C for 10 min or 69 °C for 20 min	No coliforms in 0·1 cm^3
Switzerland	None, but the usual practice is 75 °C for 15 to 40 s	Total count 25 000 cm^{-3} Coliforms absent from 0·1 cm^3 No pathogenic organisms
United Kingdom	a 65·6 °C for 30 min b 71·1 °C for 10 min c 79·4 °C for 15 s d 148·8 °C for 2 s All followed by cooling to below 7·2 °C within 1·5 h of heating	None; a methylene blue test is used by public health authorities as a guide to hygienic quality

(continued)

TABLE I—contd.

Country	Minimum heat treatment standards	Bacteriological standards
USA	Individual states provide requirements which vary, examples are: 68·3 °C for 30 min 79·4 °C for 25 s	Total count 50 000 to 100 000 g^{-1}
Proposed International Dairy Federation Standards	Pasteurised according to an officially recognised method; mix should be kept at 5°C or lower for a maximum of 72 h	Total count 100 000 g^{-1} Coliforms 100 g^{-1} No pathogenic organisms which pasteurisation should destroy
Proposed EEC Standards	Pasteurised according to an officially recognised method; mix must be frozen within 72 h of pasteurisation	Ice cream: Total count 100 000 g^{-1} Coliforms 100 g^{-1} Ice cream with discernible addition of unpasteurisable substances: Total count 200 000 g^{-1} Coliforms 200 g^{-1}

After: Hyde and Rothwell, 1973.

25 000 g^{-1}, coliforms from 100 g^{-1} (or cm^{-3}) to absent from 0·1 g, and some, but not all, countries stipulate an absence of pathogenic organisms.

In the UK, the modified methylene blue test has been used by local health authorities for many years. Despite some drawbacks (e.g. it is not possible to estimate numbers), its use appears to have been valuable in helping to raise the hygienic quality of ice cream to a high level. This has been done in conjunction with a system of frequent inspection by the authorities.

In countries where a count is applied, problems can arise if a sample contains more than the 'permitted' number of organisms, particularly in borderline cases, as to whether the sample should be condemned or not. Normally such ice cream should be subjected to further investigation, and this is suggested by the International Dairy Federation standard which states that 'all test results obtained by central authorities should be reported to the responsible sources for immediate investigation where appropriate'.

The International Commission on Microbiological Specifications for Foods (ICMSF) suggests the limits for ice cream shown in Table II.

TABLE II
LIMITS FOR ICE CREAM SUGGESTED BY THE ICMSF

Test	Limit per gram	
	m	M
Simple ice cream		
Standard colony count	10^4	2.5×10^5
Coliforms	10	10^3
Staph. aureus	10	10^2
Salmonella spp.	0	0
Complex, i.e. with added ingredients		
Standard colony count	2.5×10^4	2.5×10^5
Coliforms	10^2	10^3
Staph. aureus	10	10^2
Salmonella spp.	0	0

m = Values equal to or less than are acceptable; above m is defective or marginally acceptable.
M = A count above M is not accepted.
From: ICMSF, 1974.

It is suggested that the ICMSF values are standards which could well be applied in many, if not all, countries.

Factors Which Affect the Bacteriological Content of Ice Cream
There are many factors which affect the bacteriological content of ice cream, not least the fairly complex production process. These factors include those described in the following three sections.

The Raw Materials
Provided that the conditions of storage and handling are satisfactory, there should be very few problems of high numbers of bacteria being introduced to the mix by the raw materials. Skim-milk concentrate, liquid milk and cream should have been subjected to adequate heat treatment by the supplier, and if kept under proper refrigeration and used promptly should be quite satisfactory. Skim-milk powder may, on occasions, contain numbers of *Bacillus cereus* and, although this is not often a health hazard, it is preferable for the numbers to be kept as low as possible. Bacilli are very active reducers of methylene blue and can produce clotting in milk, and as they will grow at relatively low temperatures, they may, in extreme cases, cause spoilage of the mix.

Granulated sugar should be almost sterile, and the only organisms which may normally be present are small numbers of yeasts. Sugar syrups, whether sucrose or mixtures of sucrose and corn syrups, again should contain only a few yeasts, but it should be remembered that osmophilic yeasts may be able to grow in these syrups. Tests for yeasts should, therefore, be made on all bulk deliveries of sugar and sugar syrups.

Butter and butter oil are also products made under careful control, and from highly heat treated cream. A very high microbiological standard is to be expected, and spoilage is usually caused by rancid and other off-flavours developing as the result of either chemical or enzymically induced changes. Nevertheless, tests for yeasts, moulds, mesophilic bacteria, coliforms and the presence of lipolytic organisms should be carried out. In particular, the presence or absence of *Pseudomonas fragi* should be noted, as this can cause unpleasant taints in butter. Butter should be stored at temperatures around − 18 to − 20 °C (about 0 °F) for the best keeping quality, and as for all other ingredients, stock control should ensure strict rotation.

Vegetable fats are normally produced by very high temperature refining and de-odorising processes. They contain almost no moisture, and should normally have very few contaminating organisms.

Stabilisers and emulsifiers rarely present problems, but gelatin, as an animal product, may be a hazard, and should be obtained from reputable suppliers and kept quite cool and dry, as indeed should all the dry materials which are used in ice cream manufacture.

Many foodstuffs are added to ice cream now, either mixed with it or as coatings. These include fruits, both canned and fresh or frozen (in concentrated sugar syrup), nuts, chocolate, and colours and flavours. Most of these, particularly canned fruits should be satisfactory, but fresh and frozen fruit may contain yeasts; nuts may be infected with moulds; and coconuts must be heat treated, when desiccated, to avoid contamination by salmonellae. The examination of these ingredients should include a visual inspection, and the enumeration of mesophilic bacteria, coliforms, yeasts and moulds.

Colours may become infected by careless handling, and this must obviously be avoided by good management practice. Flavours, normally added to the mix after it has been heat treated, must also be handled with great care to avoid contamination, and in common with all the other ingredients, stored in cool, dry conditions.

The Efficiency of Processing

It will be obvious that, as an ice cream mix is a heat treated product, its quality will depend on the manner in which the processing equipment is operated and maintained.

As stated earlier, most countries specify heat treatment conditions, and these must be adhered to. Holder-processing in vats does have a greater safety margin than short-time processing, but it is essential that the holding time should be strictly observed. This is usually under manual control, so that management must ensure that the time is being adhered to. In the case of continuous operating plate heat exchangers, the time depends on the design of the equipment and rate of mix flow through the plant. This should be set on installation and must not be altered.

The whole of the process, which includes cooling and keeping the mix under 7·2 °C until it is used, must be carefully monitored by management. On a large scale, recording thermometers should be installed not only on the heat treatment equipment, but also on the storage vats.

Ice cream mix has been shown (e.g. Singh and Ranganathan, 1978) to give greater protection to micro-organisms than milk. These authors found that D-values at 50 and 55 °C of three strains of *Escherichia coli* were approximately double those obtained with cow's milk. This further stresses the importance of following the legal requirements for time and temperature of pasteurisation.

Hygiene

As well as careful operation of the equipment, according to the laid down regulations, it is obviously most important that the plant must be properly cleaned and sanitised at the end of each day's run, and before starting to process. Not only is the processing equipment important, but ancillary equipment, particularly at the final selling point, must be kept in a properly hygienic condition.

Contamination arising from the handling of ingredients, packaging materials and the product during processing and distribution must be eliminated, or kept to the very minimum. Occasional sources of trouble are the packaging materials themselves, particularly where the ice cream is to be directly in contact with them, as in the 2 or 4 litre boxes, or the smaller waxed board or paper containers. Storage of these materials must be carefully managed to avoid any trouble.

Probably the most important and dangerous source of contamination is the operative. Thus, the ice cream mix will be heat processed in enclosed equipment and then frozen with the very minimum of handling, but during the final stages of packaging, and at the retail sales point, it will be subject to many more opportunities for human contamination. The personal hygiene and habits of all who are in the factory, whether it is large or small, must be above reproach. Education, as well as medical inspection, is necessary, and if an operator is unwell for any reason, he or she must not be allowed to work without a full medical clearance. Many major food poisoning outbreaks have been caused by human contamination.

The method of sale adopted has a major bearing on the amount of contamination to which the product is subjected. Ice cream sold pre-packed in retail portions, which only has to be handed to the consumer in its wrapping, should be subject to the least contamination of all. At the other end of the scale is the sale of cones, and other individual portions dipped from bulk ice cream. Here there is the possibility of considerable contamination, unless the equipment used (servers, etc.), the method of dispensing, and the personal hygiene of the person serving are all of a very high standard. Scoops should, wherever possible, be allowed to stand in a running stream of cold, pure water, so that ice cream residues are rapidly removed (see Stenzel, 1978).

Soft-serve ice cream, sold direct from a dispensing freezer, may also prove a hazard unless similar stringent precautions are taken. In addition, the freezer itself has to be carefully cleaned and sanitised each day.

The freezing process also has some effects on the bacterial content of the product. In the freezer, as already mentioned, the temperature of the mix is rapidly reduced, while air is incorporated, accompaned by vigorous whipping of the ice cream. In 1970, Alexander and Rothwell reported that considerable destruction of bacteria occurred during freezing in vertical and soft-serve freezers after a critical point of ice crystal formation had been reached. They attributed this loss to mechanical damage caused to the bacteria by ice crystals. In a continuous freezer, however, no doubt due to the rapid freezing during a very short period of agitation, numbers of survivors fluctuated considerably, but were generally greater than in other types of freezer.

Once the ice cream leaves the freezer, any further reduction of temperature occurs in the hardening room or tunnel under quiescent conditions. The ice cream will then be held at temperatures around $-20\,^{\circ}C$ until it is sold, and this period has been shown to cause very little change in the numbers of bacteria even over long periods. Pathogenic organisms, amongst many others artificially inoculated before freezing the mix, have been shown to survive for even years. The literature is extensive, ranging from Wallace and Crouch (1933) to Abd-El Bakery and Zahra (1978), and it must be stressed that there are many reports of the survival of organisms, such as *Mycobacterium* and *Salmonella*, as well as other less harmful, but often more resistant, types. It is, therefore, essential that bacteriological content of ice cream from the freezer should be as low as possible, as neither the final hardening process nor low-temperature storage can be relied upon to reduce the numbers appreciably; pathogenic organisms should be absent.

Bacteriological Control

Table I gives an indication of the variation in bacteriological standards for ice cream and frozen desserts from country to country. Indeed some countries, among them England and Wales, have no standards at present, and rely on modified methylene blue tests together with a system of inspection of factories by local health authorities.

As in all systems of bacteriological control, it is obvious that it is the intention of the system to ensure that products reach the consumer in a satisfactory condition. As Lloyd (1969) points out, large production units which operate at high speeds and where the products are becoming more complicated, e.g. with coatings which often contain nuts and confectionery items, obviously raise major problems of control. However, it seems to be the experience in many countries that ice cream produced on a small scale often has a poorer bacteriological quality than the output of a large-scale factory (e.g. Rondi *et al.*, 1977; Hamann and Weber, 1978), for often, these small factories have no system of quality control at all. This point cannot, however, be validated in the UK where, despite the fact that the Ice Cream Alliance, representing small firms who are concerned with all aspects of the trade including manufacturers of ice cream and retailers, has about 1500 members, standards of hygiene are generally satisfactory, and there has been no outbreak of food poisoning attributable to ice cream since 1955.

In those countries where standards do exist, bacteriological control will involve the use of tests to check on the ice cream as required by the regulations. In this respect, it is probably useful to refer to the International Dairy Federation Standard (No. 46: 1969), where the maximum per gram of product is a total aerobic count of 100 000 colonies and 100 colonies of coliforms. The methylene blue test used for ice cream testing in England and Wales is described in the Supplement No. 1 (1970) to British Standard 4286: 1968. The results of the testing are interpreted as shown in Table III.

The time taken to reduce the methylene blue is that period of time required, using incubation at 37 °C (in a water bath), to decolourise the dye after the methylene blue tubes (2 ml sample, 7 ml Ringer's solution and 1 ml methylene blue) have been previously incubated for 17 h at 20 °C. Throughout a year of sampling, it is expected that 50 % of all samples would be Grade 1, 80 % Grade 1 and 2, not more than 20 % Grade 3 and none

TABLE III
INTERPRETATION OF RESULTS OF ICE CREAM METHYLENE BLUE
TEST

Provisional grade	Time taken to reduce methylene blue
1	4·5 h or more
2	2·5–4 h
3	0·5–2 h
4	0 h

Grade 4; and well produced ice cream should have little difficulty in achieving these results.

In an investigation in 1970, Alexander and Rothwell showed that, if ice cream was inoculated with *E. coli* type I, approximately 100 000 organisms per ml were required to give a Grade 4 result, while less than 50 000 cells of *B. cereus* produced the same reduction; to obtain Grade I necessitated less than 10 000 *E. coli* or below 500 *B. cereus* organisms per ml. This type of variation was commented on by Hobbs (1967) who stated 'the methylene blue test is quick, easy to perform and requires little apparatus. As a test it cannot be regarded as biologically accurate, and it is subject to variations due to the microbiological flora, but its value has been proved in practice.... It helps to keep pasteurisation techniques and cleaning procedures at an efficient level'.

Harrigan and McCance (1976), amongst others, query the use of the test, 'as it measures the ability of micro-organisms present in the ice cream to grow in a dilution of the ice cream when kept at temperatures of 20–37 °C. Yet ice cream will be kept frozen until consumption, and many organisms which could be present, especially certain pathogens, would not affect the test result. In these circumstances it is difficult to see how the methylene blue test assesses either the keeping quality of the ice cream or its potential health hazard'.

It must be pointed out, however, that very few, if any, general tests would distinguish pathogenic organisms—certainly the standard colony count does not, and in order to check for these, it would be necessary to carry out specific tests.

It is of interest to note that the UK is the only country using the test in this way, and that EEC countries, and most other countries specify a standard colony count. Even in the UK there are strong moves to replace the methylene blue test by a colony count, and a possible standard has been suggested, viz.—a sample should be considered as unsatisfactory if the colony count exceeds $50\,000\,g^{-1}$ (without, at present, stating what is meant by 'unsatisfactory').

The IDF Standard 61 (1971) gives a reference method for the colony count which involves incubation at 30 °C for 72 h, as also does the British Standard (4285: 1968).

The use of a coliform count is probably much less controversial as ice cream is a heat treated product, and if the processing has been carried out satisfactorily in equipment of a high hygienic standard, the count should be very low. The IDF requirements of less than 100 per gram, and the suggested UK standard where presence of coliforms in $0.01\,g$ is regarded as positive, should be quite capable of being met.

The IDF Standard 62 (1971) gives both a reference and a routine method for counts of coliform bacteria. The routine method involves the inoculation of a series of dilutions of the sample into violet red bile agar, and the characteristic red colonies are counted after 22 h at 30 °C. It should be noted that the use of this routine test, together with the methylene blue test, gives results which indicate a great deal about the hygienic standard of the sample in around 24 h, during which time, the relevant day's production would be in the hardening tunnels or rooms, and if satisfactory results are obtained, it can then be dispatched without any undue delay.

The reference method for coliforms involves the inoculation of serial dilutions of the sample into tubes of brilliant green lactose bile broth, incubating the tubes at 30 °C for 48 h, confirmation of positive results (production of gas in the Durham tubes), and then assessing the most probable number (MPN) of organisms per gram of original sample using the appropriate tables.

Whatever method of testing is adopted, it is essential that sampling should be carried out in a proper manner. The procedure is given in the IDF standards for colony counts and coliform counts, and these instructions should be followed carefully. For example, it is essential that samples are taken under aseptic conditions (unless an assessment of the method of dispensing is also required), placed into sterile, wide-mouth jars, and the sample kept frozen (− 15 °C maximum temperature) until the ice cream is to be tested. Problems arise with composite ice creams, including various flavours, nuts, chocolate and so on, and the solution of these can test the ingenuity of the bacteriologist.

The planning of the system of bacteriological control is most important (Lloyd, 1969). He felt that it was unwise to rely on a system based only on examining the final product, and that a system based on monitoring not only the final product, but also the raw materials and the product at various

TABLE IV

SUGGESTED TESTS FOR RAW MATERIALS

Raw material	Bacteriological tests
Milk	Total count, coliforms
Milk powder	Total count, spore-forming organisms
Butter	Total count, coliforms, *Pseudomonas fragi* and other lipolytic organisms, yeasts and moulds
Cream	Total count, coliforms
Vegetable oil	Yeasts and moulds, coliforms, lipolytic organisms
Sugar	Total count, yeasts and moulds, coliforms
Stabilisers and emulsifiers	Total count, coliforms
Fruit	Total count, coliforms, yeasts and moulds
Nuts	Total count, yeasts and moulds, coliforms
Confectionery items	Total count, coliforms, yeasts and moulds, staphylococci

After: Lloyd (1969).

stages during the manufacture, was probably much more efficient and less costly.

The importance of the quality of the raw materials has already been stressed. Regular samples of these for bacteriological examination are required, as well as a strict control of their specifications, and Table IV provides some suggestions for the routine testing of raw materials.

It is particularly important that the bacteriological quality of items such as fruit, nuts and confectionery, which will not be heat treated before adding to the ice cream, be checked very carefully indeed before they are used.

An assessment of the efficiency of cleaning and sanitisation of a plant may be carried out, in part, by an examination of the finished product over a period of time. Any variation from the normal, regularly obtained results for total colony count and absence of coliform organisms could indicate a lowering in the standard of cleaning. However, it is suggested also that swabs of particular areas of the equipment should be carried out at intervals, and unless trouble areas are indicated, the swabs should be taken at different points on each successive occasion.

Wherever possible the use of rinses instead of, or in addition to, swabs can be very helpful in the maintenance of good hygiene.

The final test is, of course, of the products themselves. The tests usually carried out for routine control, as has already been indicated, include (in England and Wales) a modified methylene blue test, coliform test and total colony count. As in all examples of bacteriological control work, a great deal depends on how the test results are interpreted. All these tests have their advantages and their disadvantages.

For example, the wide variation which is normally accepted with the total colony count is well known, and the modified methylene blue test is also known to produce some anomalous results. Thus, the simultaneous occurrence of high counts including coliforms and a high methylene blue grading (and occasionally the reverse—particularly if bacilli are present) is not unknown, while the true significance of the presence of coliform organisms in ice cream is often not fully appreciated.

Coliforms, particularly *E. coli*, are regarded as an indication of recent faecal pollution if they are found in water (they die out rapidly in water) but, in ice cream, as in many other foods, they do not die out, and conditions, except in frozen ice cream, are, in fact, favourable for their growth. The presence of coliform organisms in ice cream is almost invariably due to contamination from equipment which has not been properly cleaned and sanitised, and although such an infection may have initially consisted of only a few organisms, the cells will have multiplied due to conditions being favourable (they will grow, for example, at temperatures above about 7 °C). It should be remembered, however, that if coliform organisms are detected in a heat treated product, such as ice cream, this could also indicate either faulty or inadequate heat treatment, and as such is obviously unsatisfactory.

Even with the wide variations reported to be obtained from total counts, it is usually quite possible to state whether an ice cream sample is satisfactory, doubtful or unsatisfactory, and the intelligent use and interpretation of the three tests mentioned should enable proper bacteriological control to be carried out.

Plant Hygiene

In order to obtain the best bacterial standards which are required, the utmost attention must be given to the cleaning and sanitisation of all the plant and equipment used, to the personal hygiene of all the operators, and to the general cleanliness of the factory. It must be remembered that until ice cream is sold, and subject to contamination at that stage, the bacteriological content will change only little from that at the freezing stage.

Factory and plant hygiene in the ice cream industry is very similar to that in most parts of the dairy industry, and depends on the satisfactory use of up-to-date cleaning procedures. These are, however, complicated by the fact that some equipment employs refrigeration, notably the ice cream freezer itself and the mix storage vats, in both of which the ice cream, and hence the cleaning and sanitising media, are in contact, through the stainless steel of the equipment, with refrigerant liquid or gas. High-temperature washes should preferably not be used in these equipments, as this may cause a build up of high-pressure refrigerant gas.

One other item of equipment is also a potential source of contamination. This is the 'soft-serve' or dispensing freezer, and manufacturers of these lay down strict cleaning routines which must be carefully followed. In particular, ice cream must not be allowed to remain in the freezer overnight, and the personal hygiene of the operator must also be of a very high standard. A typical cleaning routine for such a freezer is as follows.

1. The surplus ice cream must be removed from the freezer.
2. Cool or warm (50 °C) water is used to rinse as much of the remaining ice cream from the freezer as possible.
3. The freezer should then be dismantled.
4. Each part should be cleaned in a detergent or a detergent/sanitiser solution by hand scrubbing. Particular care has to be taken to remove rubber rings and gaskets, and to clean them and the space they occupied. These parts, still fully dismantled should remain in the detergent or detergent/sanitiser until the rest of the freezer is washed.
5. The chassis (cylinder and hopper, etc.) of the freezer should next be washed using a similar solution.
6. Before re-assembly all the parts should be rinsed in water and placed in a sanitising solution.
7. The equipment is then re-assembled and sanitised carefully prior to being used again.

Detergents and sanitising agents suitable for use in the dairy and ice cream industry have been described, along with recommendations for their use, in two British Standards—BS 2756, 1970, 'Recommendations for the use of detergents in the dairying industry' and BS 5305, 1977, 'Recommendations for sterilization of plant and equipment used in the dairying industry'.

The sequence of operations for cleaning and sanitising other equipment will differ in detail according to the type and size of the factory. The sequence which is normally adopted is as follows.

1. Rinsing of the plant with warm water.
2. Dismantling of the plant.
3. Washing with detergents and physical methods of cleaning.

4. Re-assembling.
5. Rinsing with clean water—the temperature will depend on the type of equipment.
6. Sanitising—in many cases using a chemical sanitising solution.
7. Final rinsing, draining and allowing of the equipment to dry.

It is advisable to carry out this routine at the end of each day's operation and then, before beginning to process the next day, to repeat the sanitising and rinse stages (stages 6 and 7, above).

The sequence given may vary somewhat if large-scale equipment is being used, and the whole plant has been designed for cleaning-in-place (CIP) procedures. In this case, dismantling and re-assembly will not be required, the whole operation may be automated, and in most cases, the results obtained are often better than when hand washing is carried out because of a reduction in the possibility of contamination.

The sections which usually prove most difficult to clean and sanitise are freezers, although the latest large-scale freezers are available for CIP; homogenisers, which always pose problems due to their intricate design and the numbers of individual parts; mix holding tanks, and tanks for the storage of raw materials.

In the case of raw material tanks, it is recommended that the tanks used for holding liquid milk or milk concentrate should be cleaned every day; sugar syrup tanks at least once every 14 days, or each time the tank is emptied; fat tanks as often as possible to avoid the mixing of old fat (which may rapidly oxidise due to the fat being kept at a relatively warm temperature so that it is liquid) with new fat—these tanks must be completely dry before the fresh fat is allowed in; and chocolate tanks, where only hand cleaning may be possible to give thorough removal of the chocolate. Here, again, considerable problems will arise if the tank is not properly dried after washing.

Batch heat treatment vats, HTST plant and UHT equipment are cleaned in exactly the same way as milk or cream treatment plants (Zall, 1981), and as described in the *Cream Processing Manual* (Society of Dairy Technology, 1975).

A very full discussion of cleaning and sterilisation in the ice cream industry was given by Herschdoerfer (1974), while other useful information is given in Arbuckle (1972) and in Hyde and Rothwell (1973).

There are also special machines used in the ice cream trade, such as ripple machines, cup fillers, fruit and nut feeders, chocolate enrobers, and rotary and in-line stick confection machines. These are very varied in the methods by which they need to be cleaned, but usually a scheme similar to that detailed for the cleaning and sanitising of a 'soft-serve' freezer will be satisfactory.

Hygiene at the Final Selling Point

This is one of the weakest links in the whole chain. 'Soft-serve' ice cream, and ice cream scooped from bulk containers is handled to a greater or lesser extent, of necessity. Personal hygiene is obviously of the utmost importance and this, in fact, applies even if prewrapped ice cream is being sold. Cleanliness of hands, clothing and habits are all important, and the operators must be trained in the best methods of maintaining this, and in the distribution of the individual portions of ice cream.

The equipment—servers, wafer holders and so on—has to be kept free of ice cream residues, which may melt and allow the growth of bacteria to recommence. These items of equipment should preferably be kept in a running stream of cold water. Often they are kept in a jug which once did contain fresh water, but more usually consists of a fairly milky solution which can contain large numbers of organisms on a hot day. If a jug has to be used, the water must be changed regularly before it can become a source of contaminating bacteria.

CONCLUSION

In order to produce ice cream which will not only be a pleasant and nutritious food, but also one which does not present a health hazard, it is necessary to pay attention to a wide range of details. These include careful selection and testing of the raw materials, the use of correct processing conditions in equipment properly cleaned and adequately sanitised, and finally satisfactory handling of the product at the sales point. Only a small quantity of ice cream has to be contaminated to produce a hazard for a large number of people, and to avoid this, it is essential that all operators are properly trained in every way, and that bacteriological control should be carried out carefully and the results acted upon.

REFERENCES

ABD-EL-BAKERY, M. A. and ZAHRA, M. (1978) *Research Bulletin*, Faculty of Agriculture, Ain Shamo University, Egypt.

ALEXANDER, J. and ROTHWELL, J. (1970) *J. Fd Technol.*, **5**(4), 387.

ARBUCKLE, W. S. (1972) *Ice Cream.* AVI, Westport, Connecticut.

EVANS, D. I. (1947) *The Medical Officer*, 25th Jan., 39.

HAMANN, R. and WEBER, H. (1978) *Archiv Fur Lebensmittel-Hygiene*, **29**(1), 5.

HARRIGAN, W. F. and McCANCE, M. E. (1976) *Laboratory Methods in Food and Dairy Microbiology*, revised editn. Academic Press, London and New York.

HERSCHDOERFER, S. M. (1974) *Inter Ice Seminar*, Solingen, Germany.

HOBBS, B. C. (1967) In: *Quality Control in the Food Industry*, Vol. 1, Ed. Herschdoerfer, S. M. Academic Press, London and New York.

HOBBS, B. C. and GILBERT, R. J. (1978) *Food Poisoning and Food Hygiene*, 4th editn. Edward Arnold, London.

HYDE, K. A. and ROTHWELL, J. (1973) *Ice Cream.* Churchill Livingstone, Edinburgh.

ICMSF (1974) *Microorganisms in Foods*, **2**. International Commission on Microbiological Specifications for Foods, University of Toronto Press.

LLOYD, T. P. (1969) *Dairy Industries*, **34**(4), 271; **34**(6), 363.

PUBLIC HEALTH LABORATORY SERVICE (1947) *The bacteriological examination and grading of ice cream.* Monthly bulletin of the Public Health Laboratory Service, London.

RONDI, M. G., PELLOGRINI, M. G. and CASERIO, G. (1977) *Latte*, **2**(5), 307.

SINGH, R. S. and RANGANATHAN, B. (1978) *XXth Internat. Dairy Congr.*, E, 855.

SOCIETY OF DAIRY TECHNOLOGY (1975) *Cream Processing Manual.* SDT, Wembley, UK.

STENZEL, W. (1978) *Offentliche Gesundheitswesen*, **40**(6), 381. Cited in: *Dairy Sci. Abstr.*, (1980) **42**(4), 2130.

WALLACE, G. I. and CROUCH, R. (1933) *J. Dairy Sci.*, **16**, 315.

ZALL, R. R. (1981) In: *Dairy Microbiology*, Vol. 1, Ed. Robinson, R. K. Applied Science Publishers Ltd, London.

2

Microbiology of Cream and Dairy Desserts

J. G. DAVIS

J. G. Davis and Partners, Reading, UK

These two products are generally regarded as luxury foods, and have shown a marked growth in sales in recent years. In England and Wales, over the past year, the quantities of milk used for manufacture were roughly as follows (million gallons): butter 656, cheese 417, fresh cream 196, condensed milk 82, milk powder 27, sterilised cream 2·8, and other products, including dairy desserts and yoghurt, 11; the 'other products' are largely based on separated milk, i.e. they are low fat products. Thus it will be seen that after the main products butter and cheese, cream is by far the most important dairy product in terms of milk utilisation.

Cream and dairy desserts have other things in common. They are relatively expensive and perishable, so that they have to be held at low temperatures unless sterilised. Unlike milk which is generally produced one day, pasteurised the next, delivered the next day and consumed by the following day, these luxury products tend to be bought for weekends, holidays and other special occasions, so that keeping quality problems often arise unless the technology is adequate.

The high cost of cream arises from the volume of milk required to produce it, and this varies according to the fat content of the milk, and the richness of the cream (see Table I). The higher the fat content the greater the cost, and those dairy desserts which contain cream are naturally also costly.

A further factor is the cost of modern packaging. This, together with the necessary refrigeration, inevitably results in a high cost in relation to the price of ingredients. Nevertheless, dairy desserts are extremely popular due to their attractiveness as convenience foods.

J. G. Davis

TABLE I

QUANTITY OF CREAM THAT CAN BE PRODUCED FROM 100 kg
OF LIQUID MILK

Fat content of milk (%)	Quantity of 18% fat cream (kg)	Quantity of 48% fat cream (kg)
3	16	6
3·5	18	7
4	20	8
4·5	23	9

CREAM

In the UK, cream is defined in The Food and Drugs Act (1955) and in the Cream Regulations (1970) as 'that part of milk rich in fat which has been separated by skimming or otherwise'. Clotted cream means 'cream which has been produced and separated by the scalding, cooling and skimming of milk or cream'. Pasteurised cream means 'cream which has been subjected to heat treatment so as to pasteurise it or has been produced from pasteurised milk'. Sterilised cream means 'cream which has been subjected to a process of sterilisation by heat treatment in the container in which it is to be supplied to the consumer'. UHT cream means 'cream which has been subjected in continuous flow to an appropriate heat treatment and has been packaged aseptically'. Untreated cream means 'cream which has not been treated by heat or in any manner likely to affect its nature and qualities and has been derived from milk which has not been so treated'.

There are eight types of cream, defined only in respect of their fat content, produced in the UK.

	Minimum fat content (%)
1. Half cream	12
2. Sterilised half cream	12
3. Single cream or cream	18
4. Sterilised (or canned) cream	23
5. Whipping cream	35
6. Whipped cream	35
7. Double cream	48
8. Clotted cream	55

Certain additives are permitted in some types of cream (Cream Regulations (1970)).

Whipping and Whipped Creams

The microbiology of these creams is identical to that of other creams, making allowance for a fat content of 35–40 %. All forms of processing, including whipping, provide a source of contamination, and so hygienic precautions are particularly important. It is essential to whip cream at about 5 °C, and maintain it at this or lower temperatures (see section on Bakers' Cream).

Half Cream and Single Cream

These are practically universal for coffee, and single cream may be used domestically if double cream is considered to be too rich.

Double Cream

This is generally used for fruit, trifles, etc., which richness and viscosity are required. The keeping quality of double cream appears to be somewhat better than that of lower fat creams, other things being equal.

Clotted Cream

This is the most popular holiday type, made in Devon and Cornwall and often sent away. In the traditional or farmhouse method, milk (Channel Island or South Devon) would be put in aluminium pans, 30 cm in diameter and 20 cm high, and held 12 h for the cream to rise. It was then put in a steamer until a ring of solidified cream 2–3 cm wide formed round the edge. The cream was then ladled off with a perforated dipper.

In the modern factory method, 60 % fat cream is put into rectangular aluminium pans (30 × 30 × 20 cm) and held in water at 95–100 °C for 1–2 h. It is then cooled in the refrigerator and carefully ladled into cans; rough treatment may lead to gritty particles.

The more severe heat treatment which clotted cream receives results in a different microflora, in which aerobic spore-formers of the *Bacillus subtilis* type are usually prominent.

Cream Powders

An interesting technical development, especially for caterers and food processors, is the manufacture of powdered cream. This is usually made from fresh cream, or anhydrous milk fat and skim-milk powder. Methods are described by Kieseker *et al.* (1979).

Other Types of Cream

In addition to the types mentioned above, there are other terms applied to cream used in the food industry. 'Artificial' or 'reconstituted' cream is defined in the Food and Drugs Act (1955) as a 'substance not being cream which has no non-milk ingredients (except water) and is usually made from butter and milk'. 'Imitation' cream is made by emulsifying edible fats or oils in water with other permitted substances. Both types are extensively used in the food industry, especially for cakes and sweets, in place of real cream, and their methods of processing and problems of hygiene, storage and quality control are virtually identical with those of real cream.

When real cream is used by bakers and others in the food industry it is commonly called *bakers' cream*, and this product is an important part of the cream trade as it constitutes about half the cream sold in the UK; it generally causes more problems than ordinary retail cream. The type usually made may contain from 32–40 % fat, although it can only legally be called whipping cream if it contains at least 35 % fat. The form of processing may be normal (for example, heating to 73 °C for 15 s, standardising, and heating to 87 °C without holding), but the cream may then be filled into large cans or churns, left in a warm atmosphere, and generally not receive the care in hygiene that is necessary. Manipulation in the bakery and utilisation in the manufacture of cakes, etc., may be most unhygienic.

Another factor is that sugar and certain additives may be put into whipping or whipped cream (Cream Regulations (1970)). In addition, the bakers' requirement for cream may be spasmodic. He may require some at short notice, and at other times he may over-order and return a bulk of cream which the dairy has to re-use or discard. Bakers and other users of cream may allege that a batch of cream is of poor quality or 'contains coli' to justify returning it to the dairy.

Thus from several points of view there may be problems with bakers' cream. It is important that control be exercised in manufacture, hygiene and temperature at all stages, and it is recommended that cream should always be bought and sold on both chemical and microbiological specification, bearing in mind the necessary precautions for the latter (see Tables VI and XI, later; see also Davis, 1963, 1969a).

Fresh Cream Produced on the Farm

Cream production is essentially a dairy operation, but cream may still be produced on farms and sold direct to the public, just as untreated milk is sold. Hygiene is very variable, and high counts of bacteria, yeasts and moulds may be found in 2 or 3 days, and the shelf-life is correspondingly short.

A simple farmhouse pasteurisation (heating at 65 °C for 30 min) will reduce the bacterial count to about 1 % or less of the original, but unless followed by rapid cooling to 5 °C, it will have little ultimate effect on keeping quality. The treatment may be repeated the following day with advantage, provided the cream is hygienically cooled immediately after heating.

Common taints found in farm-produced cream include sour, rancid, cheesy, stale, bitter, putrid and yeasty; a slight, ill-defined taint may be described as stale or unclean. These problems are always associated with high microbial counts, with a predominant organism, such as *Pseudomonas*, *Micrococcus* or yeast, being responsible for the dominant taint. Ropiness or sliminess may be caused by some coliforms or lactic streptococci. In bad cases of spoilage, gas may be formed, usually by lactose-fermenting yeasts, and mould growth may be visible on the surface of the cream, e.g. *Geotrichum candidum*. Souring by lactic acid bacteria may repress putrefactive organisms, but their activity will stimulate yeasts and moulds. Sweet curdling may be caused by rennet-type enzymes produced by aerobic spore-formers, which can also be responsible for bitterness.

However, as with other types of dairy processing, such as cheesemaking, the gap in technological proficiency between farms and dairies has been steadily diminishing, except in scale of operation. Hence, the microbiological quality of farm-produced cream should not differ significantly from that retailed by a creamery.

Manufacture on an Industrial Scale

Apart from cultured or soured cream, the entire process of manufacturing, packaging and distributing cream is a matter of keeping bacteria out, or preventing contamination, and keeping the growth of the few organisms that are present to a minimum.

The overall system includes the following stages:

(1) production of milk on the farm,
(2) transport to the dairy,
(3) storage in the dairy,
(4) separation,
(5) standardisation of cream to desired fat content,
(6) heat treatments of milk and cream,
(7) storage after heat treatment,
(8) packaging,
(9) storage and distribution of cartoned cream, and
(10) sale—possibly a multi-stage operation.

Thus cream technology is more complicated than milk technology, and there are more opportunities for problems to arise.

To produce the best possible cream, each of the above stages must be carried out as efficiently as possible; particularly important are adequate heat treatment, storage and distribution at 5°C, with excellent hygiene throughout. The old saying that the strength of a chain is the strength of the weakest link is particularly applicable to cream. The most important aspects are contamination after heat treatment, and temperature of storage and distribution.

Notes on the Stages in Cream Production

(1) The hygienic production of milk is of the greatest importance for cream, because although vegetative cells are easily killed by heat treatment, spores are not, and types, such as *B. cereus*, can be a major cause of failure in the methylene blue (MB) test. If the spore count of the milk is high (over $100\,ml^{-1}$) it may be worth while to reduce this by high-speed centrifugal methods for, as aerobic spore-formers tend to form chains, these are more easily removed than single cells.

(2) and (3) As psychrotrophs are of considerable importance in a long shelf-life product such as cream, the temperature of the milk should never be allowed to rise above 5°C in order to keep their growth to a minimum.

(4) and (5) Ideally the milk should be separated to give a cream of the desired fat content, but in practice this is rarely possible, and the standard method is to separate at a slightly higher fat content and then standardise with milk or separated milk. Delays at these stages can be dangerous, particularly for those dairies producing only small amounts of cream.

(6) Whether the milk is heat treated or not, it is always best, and a standard practice, to give an adequate heat treatment to the cream after standardisation; to omit this is to ask for trouble.

(7) The complexity of a modern large dairy may necessitate storage after heat treatment and before packaging. This should be avoided if possible. Every item of equipment with which the cream comes into contact represents an additional source of contamination, and such aspects are often overlooked.

(8) The packaging of cream is almost invariably done by automatic fillers, and cartons are practically universal. These should be almost sterile from their method of manufacture, and if contamination is found at this stage, the trouble is almost certain to be the filler. The use of fillers for more than one product (e.g. yoghurt, etc., and cream) is fraught with danger. If

this procedure is necessary, the whole section of equipment should be cleaned and 'sterilised' before being used for cream.

(9) Refrigeration must always be positive (i.e. maintained) for perishable products such as cream. The larger operators have a double advantage; the greater the bulk of product being processed the less is the relative contamination, and the slower the rise in temperature of refrigerated products whether in bulk (tanks and churns) or in cartons.

(10) The best sales system is for the processor to deliver direct to the customer, which is generally the standard procedure for milk. Unfortunately the distribution and sale of cream may be complicated, especially in small urban units. Cream may pass from wholesale-processor to a retail dairy and thence to a shop. It is then of the greatest importance that the temperature of the cream should be maintained at 5 °C *at all times*. The cabinets in supermarkets are usually efficient in this respect, but small shops may not be adequately equipped, and shop keepers may not realise the importance of storage temperature.

The Storage of Raw Milk
Bulk collection of milk in Britain became universal in 1979, so that all milk now arrives at distributors' and manufacturers' premises at 5 °C. It is common practice for milk to be stored in creameries at 5 °C for up to 48 h and sometimes longer. There is normally no problem with milk quality for milk so held, although psychrophiles and psychrotrophs can grow very slowly at below 5 °C. These are generally biochemically active against fat and protein, but do not usually ferment lactose to lactic acid. Thus cold-stored milk does not sour but can develop taints, and although the organisms are nearly all killed by pasteurisation, they produce enzymes which can survive pasteurisation and continue to produce changes in the pasteurised product.

Separation
Cream is usually made by separating milk in a mechanical centrifugal separator at a temperature between 35 and 45 °C. This is an ideal temperature for bacterial growth, and so it is advantageous to use modern equipment separating milk at lower temperatures, e.g. 25–30 °C; this has the additional advantage of causing less physical damage to the fat globules. Separation also helps to purify the milk by removing dirt, somatic or body cells, and any other foreign matter in the form of slime which is removed mechanically from the separator. Some bacteria, especially clumps of large organisms, including spore-formers, are also removed in

this slime. The times between separation, homogenisation (if performed), standardisation and heat treatment should be as short as possible.

Microbiological Problems of Standardisation

Cream is normally standardised by separating the milk at a slightly higher fat content than required, and then standardising it by adding the calculated quantity of skim-milk. This procedure constitutes a microbiological hazard because milk is usually separated at about 40°C (ideal for microbial growth), the separator may contaminate the cream, the cream has to be held for a short time, and the skim-milk may not be of good quality.

In practice it is a question of balancing the disadvantages of various methods, but irrespective of the method adopted, it is always preferable to heat treat the final cream and cool and package it immediately. As far as practicable, milk and cream should be held, during processing, at about 5° or above 60 °C. It is expensive to cool warm milk or cream to 5 °C, but in hot weather, failure to do so may be disastrous.

Homogenisation

Half cream (or coffee cream) and single cream are normally homogenised to increase viscosity, and this treatment may have two effects.

1. Homogenisation necessitates an extra treatment and so extra contamination of the cream. It may involve holding the cream warm for a short time, and is always results in breaking up clumps of bacteria. These three factors are usually held to be responsible for any fall in keeping quality which may occur. Homogenisation should, therefore, take place immediately before the final stage of heat treatment.

2. The splitting of the fat globules removes the fat globule membrane and greatly increases the surface area of the fat, thus favouring the action of any lipase enzymes which may be present.

There is a general belief in the dairy trade that homogenised milk (and perhaps cream) do not keep as well as the unhomogenised products. If true, this may be because of the extra contamination, the splitting of bacterial clumps, or the oxidised flavour and other taints (lipolytic) brought about by homogenisation.

Heat Treatment

Some form of heat treatment has been used for cream for many years, and cream behaves similarly to milk in all respects microbiologically. Heat penetration is slightly slower, but the chief difference in microbiological behaviour is caused by the special demands of retail distribution and keeping quality.

In the early years of this century, a crude form of holder pasteurisation was practised, but since 1940 the HTST continuous flow method at 80–90 °C, with or without a holding section of 15 s nominal, has been almost universally used.

Unlike liquid milk and ice cream, there are no legal requirements for the heat treatment of cream, but recommendations have been made for minimum temperature–time treatments, and the conditions used by processors are controlled by four requirements:

(1) destruction of all pathogens;
(2) achievement of desired shelf-life;
(3) avoidance of 'cooked taints' which result from the production of volatile sulphur compounds when milk and cream are heated above 80 °C; these usually disappear in 1 or 2 days;
(4) destruction of milk enzymes, particularly lipases which may cause rancidity.

The normal HTST method used for cream achieves all these objectives, provided that hygiene and subsequent refrigeration are adequate.

The optimum heat treatment is useless unless the cream is cooled rapidly, otherwise the surviving bacteria, and particularly the spores, will grow and produce taints and physical defects. An old 'trick of the trade' was the technique known as 'rebodying', which was to cool the cream quickly to 30 °C or less, and then to continue the cooling slowly down to 5 °C. This procedure can give a cream of much higher viscosity, but there is obviously a source of danger in such slow cooling. This device should be used, if at all, with caution.

Holder Pasteurisation

This method, involving heating cream at about 65 °C for 30 min, is obsolete as far as ordinary dairies are concerned, but has advantages for farms and very small dairies processing up to 200 litres a day. Equipment and control are very simple. However, all sources of contamination, particularly filling into cartons, become potentially more dangerous, and cooling is inevitably slower; although heating and cooling in the same vessel eliminates one cause of contamination.

The High Temperature–Short Time Method

The HTST method is now universal for cream in virtually all dairies. Its

FIG. 1.   A typical processing area for cream showing the Paraflow heat exchangers now used for pasteurising cream. It is also apparent that the layout allows for a high standard of cleanliness. Reproduced by courtesy of the APV Co. Ltd, Crawley, UK.

main advantages are in compactness, thermodynamic efficiency, speed of operation, total enclosure and overall cost, and it can give a cream of very good keeping quality, provided it is properly operated. The entire sequence of operations—reception of raw milk, warming to separation temperature (30–40 °C), homogenisation (if done), pasteurisation, storage and filling—can be carried out in one continuous process in an entirely enclosed system, with obvious advantages (see Fig. 1). The equipment used for the cream should be designed to treat the cream as gently as is practicable.

The officially recommended (but not legally compulsory) conditions are:

(1)   not less than 80 °C without holding;
(2)   not less than 74 °C with at least 15 s holding.

The 'holding time' is nominal, and when stated to be 15 s, the actual time is usually 18–20 s because of the inter-connecting pipe work.

When the temperature of pasteurisation is raised or the holding time increased, there is an increased kill of non-spore-forming organisms, but with spores there is a complicating factor which may result in an increase in the colony count. The germination of spores is a complex subject, but one of the most important factors is the stimulus given to germination by heat. In the present context, because of the requirement for a long shelf-life, the protective effect of fat on bacteria, and the slower heat transfer in cream, it is usual to heat cream appreciably above the statutory minimum for milk (71·1 °C for 15 s nominal), either by flash heating or with a holding period. However, this higher heat treatment may, in fact, have an adverse effect on keeping quality because this more drastic exposure may increase the germination rate of spores which survive.

This anomalous effect is well shown by the results of Brown *et al.* (1980). Little difference was found in creams held for up to 9 days at 7 °C, but after this, appreciably higher counts were found when temperatures of 76·5, 79·0 and 81·5 °C (with a holding time of 15 s) were used rather than 74 °C. Similarly with a holding time of 1 s, a temperature of 80 °C gave lower counts after 6 days storage than 82·5, 85·0, 87·5 and 90 °C.

Shelf-life tests on these creams confirmed these results, the lowest temperatures giving values of 20 days, and 87·5 °C for 1 s only 9 days.

'*In-bottle*' *Pasteurisation*

It was established many years ago that if milk was pasteurised in the bottle, an excellent keeping quality could be obtained because subsequent contamination became impossible. Unfortunately the cost of the process rendered it uneconomic. However, a higher profit margin is obtainable with cream, and some dairies have utilised this principle to obtain nearly sterile cream. Thus the milk would be pasteurised and separated, and the bottled cream heated to 65 °C or higher for 30 min. Some dairies have employed the principle of 'Tyndallisation' in which the liquid is heated for three successive periods of 0·5 h to secure virtual sterility.

Tyndallisation

In the early days of bacteriology, considerable trouble was experienced through the inability of the available processes to kill spores in a medium. Tyndall in 1877 suggested that if a medium was heated at 100 °C for 30 min on three successive days, first the vegetative cells would be killed, and then the spores would germinate and be killed on the second and third days. The idea has been resurrected from time to time for foods which would be affected organoleptically by autoclaving. The method fails if the medium does not permit the spores to germinate, and is unreliable for anaerobic and thermophilic aerobic spores. Various forms of the method have been suggested, a recent one by Pien (1977) being based on double HTST pasteurisation. The main reason for the failure of all these methods is the unpredictability of the germination of spores (Franklin, 1969; Jayne-Williams and Franklin, 1960).

Brown *et al.* (1979) investigated a double pasteurisation treatment separated by periods of aerobic and anaerobic incubation at 30 °C. None of the treatments had any effect on the spore load or the storage life of the creams. The organoleptic scores were not influenced by the different types of treatment.

A method formerly used for making 'long life' cream was to heat the cream in bottles for 30 min at 105 °C on the first, second and fourth days, leaving the bottles at room temperature for the third day. This period allowed surviving spores to germinate, and the cells to be killed on the fourth day. This method was claimed to give a very long keeping quality, but the serum began to separate after a month.

The Ultra-High Temperature (UHT) Process

Milk which has been subjected to this treatment is known as 'ultra-heat treated milk' (HMSO, 1965) and must have been heated to at least 132 °C for at least 1 s. It must be immediately put into a sterile container with aseptic precautions. There are three important advantages of this process over the traditional sterilisation.

1. The milk is sterilised, i.e. all forms of life are destroyed.
2. The process is very rapid.
3. The treatment induces very little cooked flavour in the milk.

The method has obvious advantages where a long keeping quality is required, as microbiologically, the milk will keep indefinitely without refrigeration. The milk finally turns into a gel as the result of the action of enzymes which survive the heating process, but this may require 6 months or longer according to circumstances. The original slightly cooked flavour soon passes off, and is usually replaced ultimately by a stale 'cardboardy' or oxidised flavour.

The UHT method has been used extensively for retail cream, especially for small units in the catering industry. Single or half cream is generally used, and packed in very small cartons. The UHT method becomes progressively more difficult to control as the fat content of the cream rises. Methods for avoiding recontamination of UHT-treated milk and milk products have been described by Flückiger (1979). Hydrogen peroxide, ethanol and ethylene oxide are used commercially to sterilise the paper and plastics used in aseptic cartoning.

An interesting method for the rapid sterilisation of cream and other liquids is the ATAD friction process. In this process, the liquid is preheated to 70 °C and then heated at 140 °C for 0·54 s, and it can be applied successfully to 12 and 33 % fat creams (Alais *et al.*, 1978).

Packaging of Cream

The packaging of any perishable food is technically demanding, because the greatest efficiency in processing is invalidated if the product is unhygienically filled into unclean containers. Until recently, retail cream was filled into bottles or jars and sold like milk, but cartons are now used almost universally. The growth of supermarkets is largely responsible for this, and in addition, the greater profit margin for cream enables dairies to use cartons.

In the modern dairy, the carton is the last item coming under control in the sequence of operations. Cartons, whether of paper or plastics, are virtually always of good hygienic quality because of their method of manufacture, provided they are stored in a clean atmosphere.

Bulk quantities of cream (2000–15 000 litre) are now transported in stainless steel tankers. Intermediate quantities, such as 50 litres, may be distributed in aluminium or steel cans or churns. The same principle of efficient sanitation applies to all metal containers. They must be as effectively cleaned and 'sterilised' as the equipment in the dairy.

Increasing use is now being made of plastic bags holding from about 5–10 litre. As with cartons, these are virtually sterile from their method of manufacture. Bacteriologically, all that is required is to store them under clean conditions and prevent contamination in filling and sealing.

Sterilised (or Canned) Cream and Half Cream

Sterilised cream must contain not less than 23 % fat, and sterilised half cream not less than 12 %, but otherwise, and from the microbiological point of view, they are identical.

After standardisation, the cream is heated to homogenising temperature, usually about 65 °C. After homogenisation at about 2500 lb in^{-2}, the cream is filled into cans at a temperature of 30–50 °C. As filling is a well recognised source of contamination, some makers fill at 75–80 °C, but viscosity may be affected. The most efficient method, and one generally adopted now, is to use an initial UHT treatment at 140 °C for 2 s to kill the spores prior to canning. A much milder treatment of the cans can then be used, as it is only necessary to kill adventitious contaminants.

The sterilisation process may be by the batch or the continuous method; the latter is much to be preferred. A stationary retort is unsuitable, but a rocking autoclave or, still better, a rotary steriliser can give a good product. Temperature–time conditions vary according to size of the can (SDT, 1975; Crossley, 1955).

Continuous systems are being increasingly used for large-scale production. In the three-stage cooker, the cans are transferred by a conveyor-valve system through a preheating, sterilising and cooling process of the type long used for evaporated milk. The trend is now towards the hydrostatic system in which the pressure required to give the necessary temperature is maintained by columns of water at the entrance and the exit of the equipment. Cans are heated at 116–121 °C for 30 min, but if the UHT treatment is applied first to the cream, a lower temperature and shorter time can be used.

The microbiological problems of sterilised cream are somewhat different from those of pasteurised cream. The cream is theoretically sterile, but problems similar to those of evaporated and sterilised milk can occur. If the cream becomes contaminated with very resistant spores of B. subtilis, these can germinate and produce a bitter taint and thinning of the cream by the production of lipolytic and proteolytic enzymes. This fault is usually associated with poor quality raw milk high in spores, and dirty equipment. Such spores can sometime survive heating at 120 °C for 40 min (Nichols, 1939). Pending improvement in the hygienic production of the raw milk, it may be necessary to raise the temperature of sterilisation.

If any microbiological defect does arise, then the causative organism should be identified, as this information can provide an important clue to the source of infection. Thus, if a non-spore-forming organism is responsible in a sterilised product, this indicates contamination after sterilisation, and in a canned product, a defective can or a 'leaker'. In the latter case, bacteria can then enter the cream from the cooling water, or elsewhere, and cause various faults. For example, common water-borne organisms, such as Proteus, can cause bitterness and thinning, coliforms can form gas, and cocci give rise to acid curdling. It is important to remember that once the cream has been sterilised, it is just as good a medium for microbiological growth as raw milk or cream, if it becomes contaminated.

Canning is a technology in its own right. A very useful book is that by Hersom and Hulland (1980), and FDA (1978) gives not only detailed descriptions of laboratory control methods, but includes very helpful sections on the control of canning operations.

Konietzko and Reuter (1980) working with spores of B. stearothermophilus found a linear thermal death curve in the range 130–145 °C, but for homogenised 25 % fat cream, the best temperature–time combination for sterilisation is 15 min at 121 °C or 10 min at 122 °C (Dhamangaonkar and Brave, 1978).

Careful control is essential at all stages in the manufacture of canned foods. Ordinary microbiological control methods are of little use because one surviving spore in a can may germinate weeks or months after manufacture and so produce a defect. A few cans from each batch should be subjected to accelerated storage tests by incubation at 37 and 55 °C for at least 7 days. Growth can often be detected by external inspection (swelling of cans or shaking them), and by internal examination of others. Retention of the whole batch at ambient factory temperature for a month followed by inspection before despatch is a wise precaution.

In order to obtain a satisfactory sterilisation in the manufacture of canned cream, the heating process must be evaluated by a suitable method (Hersom and Hulland, 1980). Using heat penetration data, Board and Steel (1978) have compared sterilising values (F_0) obtained using an automated version of Gillespy's method with those obtained by the general method. The former usually gave lower F_0 values, but the differences were not significant.

Frozen Cream

There has recently been an increasing interest in frozen cream (Ramuz, 1979; Tressler *et al.*, 1968). Freezing is the least objectionable method for preserving perishable foods. Organisms cannot grow, and nutritional value is maintained virtually unaltered. Textural and flavour problems may occur, depending on the method of freezing and type of product.

The cream should be pasteurised at 75–88 °C for 15 s, cooled to 1 °C as quickly as possible and filled into stainless steel containers, best quality tinned cans *in perfect condition*, or plastic or paper containers. It must be frozen as quickly as possible and stored at −18 to −26 °C, the lower the better. A keeping quality of 2–18 months with an average of about 6 months can be expected.

Freezing cannot be used to reduce bacterial numbers, although a few organisms may be killed mechanically by the formation of ice crystals. Normally pasteurised cream would be used, and the microbiology would be the same as for the non-frozen product.

The General Assessment of the Hygienic Condition of a Dairy Plant

With all perishable heat-treated foods, the keeping quality or shelf-life is largely determined by the extent of post heat treatment contamination. The control of the sanitation (cleaning and 'sterilising') of the equipment is thus second to none in its importance in the operation of any dairy. Control can be considered in two parts—the methods of cleaning and 'sterilising' (SDT, 1959; Davis, 1955, 1956) and the laboratory examination of the sanitised plant to check for 'sterility'.

Methods for checking the efficiency of cleaning (removal of all soil) include the lipstick method, removal of available chlorine, physical examination, and staining by iodophors (Davis, 1956). Methods for checking 'sterility' include swab tests (still the most popular), rinse tests, the agar sausage or imprint tests, and comparative counts or other tests on the first, and later, containers from the filler.

'Good manufacturing practice' is a term used to embrace all operations in the dairy, including hygiene; and 'quality control' refers to laboratory and ancillary tests. The two together assure the manufacturer and the customer of a good product. Thus, quality assurance = good manufacturing practice + quality control. Recommendations on hygienic precautions in a processing dairy have been given by Davis (1972).

Hygienic Control in Cream Processing
Ideally this should extend from the production of milk on the farm to the delivery to the customer, and, essentially, hygiene means the prevention of contamination of the cream at all stages. The greatest hazard is usually from dirty equipment. All items which come into contact with the cream in any way at any stage must be sanitised by removing all soil, and killing residual organisms by heat or chemical disinfectants, such as chlorine compounds. The requirements are exactly the same as for milk (Davis, 1956, 1972; SDT, 1959, 1966). The two most important factors controlling shelf-life are cleanliness or 'sterility' of equipment, and temperature of the cream. Rothwell (1969) has made recommendations on precautions to be taken in the processing, packaging and distribution of cream.

Although there are no legal requirements in the UK for the processing and distribution of cream (as there are for milk), the relevant government departments have issued a Code of Hygienic Practice for Cream (this is reproduced in SDT (1975), p. 114).

One of the most important aspects in a dairy is complete and constant liaison between the bacteriological laboratory, the management and the engineering or processing staff. Fundamental precautions may be overlooked if this liaison is lacking. The reason for very bad bacteriological results is usually that an operative has not done his job properly, which implies poor supervision by the managerial staff. The bacteriological staff should always be available for consultation. Instructions to the processing staff should be via the manager on the advice of the bacteriological staff, and all quality control tests should be surprise tests as far as is practicable.

At least one responsible member of the laboratory should be available as soon as processing begins in the dairy, even if the full laboratory routine has not started. If a technical problem occurs, a knowledgeable person is then able to advise.

Some practical and legal aspects of cream manufacture have been discussed by Jackson (1978).

Control of 'Sterility' of Air Near Fillers

Contamination of the air around the fillers is often overlooked or assumed to be negligible. This may be a dangerous assumption. If the dairy is surrounded by vegetation (especially fruit trees), the air may be rich in yeasts and moulds. Check tests should always be made near lines handling cream and/or dairy desserts. Special apparatus may be used for this purpose, but it is usually expensive, and it is quite adequate to expose plates of malt or unhopped beerwort agar at pH 6·6 in the vicinity of the fillers for a specified time, e.g. 60 min. The plates are then incubated at 22–27 °C for 3 days and counted. Yeast and mould colonies grow quickly on these media and are easily recognised. This medium is also excellent for growing bacteria, and so can be used for a general assessment of the 'sterility' of the air. If the total exceeds 60, or if the number of yeasts *or* moulds exceeds 10, special measures should be adopted to reduce the number.

Windows should be closed, rubber air-lock doors fitted, and a suitable air conditioning/filtering system installed.

Quality of Water in Dairies

The bacteriological quality of water as used in a dairy is not necessarily as good as that of water entering the dairy, especially if fed through a storage tank. This point is often overlooked. Moreover a water of potable quality is not necessarily good enough for cream processing because pseudomonads, aerobic spores, etc., which are dangerous for cream, are not considered in respect of public health control work. All water in a dairy should be chlorinated, e.g. at 5 ppm available chlorine, and regularly checked (Davis, 1956, 1959b).

In-Line Testing of Cream Equipment

The modern trend to ever larger processing units has resulted in very complicated systems, whose operations are often not only mechanised but controlled by computers. The sanitising (cleaning and 'sterilisation' of the equipment) is carried out by a cleaning-in-place (CIP) system involving typically a cold water rinse, hot caustic alkaline wash, a rinse, hot water or cold hypochlorite treatment and a final rinse. With a properly planned, well constructed system with equipment in good condition and properly operated, there is normally no microbiological problem, but if there is a defect anywhere, faults can arise and these can sometimes be very elusive. Keeping quality can fall drastically, and even if taints do not develop, the cream may fail the methylene blue test used by the Public Health Laboratory Service (PHLS). A follow-up test may reveal a high count and sometimes the presence of coliforms.

The first step is then to check all the operations from the initial treatment of the raw milk onwards, and to examine the condition of all items in the processing line. If no apparent fault can be found, the next procedure is to take samples at various points in the processing line and examine them by a suitable microbiological test, a method commonly described as in-line testing. It is essential that the sampling be above reproach. A satisfactory way is to take the samples by hypodermic needle through an Astell seal using the usual precautions. When a severe heat treatment is used, e.g. 85–90 °C with 15 s holding, the surviving colony count is, or should be, very low, for example about 10 cfu ml^{-1}. However, as normally only 0.1 ml is plated, a reliable count cannot be obtained by this method. Alternative approaches may, therefore, have to be considered, and in one system, three aliquots of 10, 1 and 0·1 (or even 100, 10 and 1) ml are held at 30 °C for 24 h, and then three loopfuls from each are streaked onto milk agar or plate count agar; the plates are incubated at 30 °C for 24 or 48 h. The most probable number (MPN) of viable cells per ml is then calculated from the numbers of plates showing growth corresponding to the volumes of cream tested, using the McCrady tables as given in Thatcher and Clark (1968) or other reference works. Alternatively a small amount of a solution containing glucose, yeast autolysate and litmus may be added to each aliquot which is then incubated at 30 °C for 24 or 48 h. Any surviving organisms will grow in this enriched cream, and they can then be detected by biochemical changes and/or a reduction of the litmus (Davis, 1935).

The resulting counts are recorded against each sampling point, and this operation should be repeated twice before any conclusions are drawn. Some typical results are given in Table II.

A third method is simply to incubate the samples for 1 or 2 days at 30 °C and then plate them.

TABLE II
SOME TYPICAL RESULTS OBTAINED BY TAKING
IN-LINE SAMPLES OF CREAM DURING A SERIES OF
PROCESSING RUNS

Sampling point	Counts (*MPN per ml*)		
	No. 1	*No. 2*	*No. 3*
Exit from pasteuriser	4	3	<3
Entry to storage tank	7	4	11
Exit from storage tank	9	7	93
Entry to filler	39	9	120
Exit from filler	43	120	150
Filled carton	64	120	210

Sources of infection are shown by marked differences in counts. For example in test No. 1, the pipework was clearly responsible for contamination, in No. 2 the filler, and in No. 3 the storage tank. All samples must be tested under the same conditions.

Keeping Quality or Shelf-Life
The keeping quality (KQ) of retail cream is more critical than that of milk, for although cream usually receives a more severe heat treatment than milk the distribution system is different. Thus, milk is usually pasteurised one day, delivered the next, and consumed a day or two later, while cream has to be separated, standardised, cartoned and distributed, often through a retail dairy organisation, and then sold through supermarkets or shops. There may be no temperature control during this procedure, and as cream sales tend to be concentrated at weekends and on special occasions, and the carton may be opened and used more than once, the KQ requirements are severe. Some dairies now achieve, or at least aim at, a KQ of 14 days with the cream held at temperatures not exceeding 5 °C.

Causes of Poor Keeping Quality in Cream
The following are the main factors responsible for a short shelf-life in retail cream:

(1) poor quality raw milk, particularly with high spore and thermoduric counts;
(2) poor hygiene in separation;
(3) wrong choice of temperature for heat treatment;
(4) poor hygiene in processing;

(5)   poor hygiene in packaging; and
(6)   too high a temperature for storage and distribution.

Of these factors, the quality of the raw milk and the storage temperature of the end-product are, perhaps, the most important.

Quality of the Raw Milk

Bacteria found in milk may be placed in three groups from the point of view of keeping quality in cream:

(1)   thermolabile types which are killed by ordinary pasteurisation (71 °C with 15 s holding);
(2)   thermoduric types which do not form spores but survive ordinary pasteurisation;
(3)   aerobic spore-forming bacteria.

Groups (2) and (3) are of the greatest importance for cream.

It has been shown that an initially good (low count) cream will last for 6 days at 5 °C, but develops a high count (but no coli) and fails the methylene blue (MB) test after 2 days at 15 °C. An initially poor (high count) cream becomes sour, develops coli and fails the MB test after 6 days at 5 °C, and develops an unclean odour and taste and coli, and fails the MB test after 2 days at 15 °C (Davis 1969b) (Table III).

The bacterial population (P) of cream, after holding for a known time at a specified temperature, may be estimated by using the formula:

$$P = \text{initial number} \times 2^N$$

where

$$N = \frac{\text{Age in hours}}{\text{Generation time in hours}}$$

As the generation time in cream for any one organism is roughly constant for a considerable time at a low temperature, this emphasises the importance of obtaining a low count cream at the end of processing/filling, or at the start of the shelf-life of the cream.

The Effect of Temperature

From the microbiological point of view, cream is a remarkably standard product, in that the pH value, water activity, etc., are virtually constant. It follows, therefore, that for a given degree of microbial contamination, the

TABLE III
COMPARISON OF MICROBIAL GROWTH IN GOOD AND POOR CREAM HELD
AT 5 AND 15 °C

Age (days)	Held at 5°C			Held at 15°C		
	Total count at 30°C	Coliforms	MB (h)	Total count at 30°C	Coliforms	MB (h)
			Good cream			
0	1 900	− − −	>4·5	1 900	− − −	>4·5
1	2 100	− − −	4·5	12 000	− − −	1
2	3 200	− − −	4·5	>1 600 000	− − −	0
3	2 900	− − −	4·5			
4	2 100	− − −	4·5			
6	2 500	− − −	4·5			
	(still acceptable)					
			Poor cream			
1	20 000	− − −	>4·5	24 000	+ + +	1
2	20 000	+ − −	4·5	>1 600 000	+ + +	0
3	25 000	+ + −	4·5 (unclean odour and taste)			
4	67 000	+ + +	4·5			
6	560 000	+ + +	0			
	(sour)					

Coliforms: the three results indicate presence or absence in 1, 0·1 and 0·01 g.
MB = methylene blue test (Public Health Laboratory Service, 1971).

shelf-life is entirely a question of temperature. Thus, the rate of growth of all micro-organisms is controlled mainly by temperature, and each organism has a range of growth and an optimum. As far as bacteria are concerned, the psychrophiles can grow from 0–27 °C, the psychrotrophs from 0–45 °C, the mesophiles from 10–45 °C and the obligate thermophiles from 45–63 °C. These are not rigidly defined limits but serve as a guide, and there are also a large number of organisms which cannot be so classified. With cream and other refrigerated perishable foods, the problem is to minimise the rate of growth of the psychrophiles and psychrotrophs. There are two key temperatures in this respect. Below 6 °C growth is very slow, and above 13 °C growth is rapid, with the generation times varying from several hours up to the shortest known which are about 15 min. For the low-temperature bacteria there are two ranges of importance—from 0–4 °C and from 5–10 °C, and while the rate of growth is very slow up to 4 °C, it increases progressively from 5–10 °C. This observation is important in that reported temperatures in the dairy industry are not always accurate, so that although refrigeration temperatures in the night may not exceed 4 °C, during working hours they can easily reach 7 °C or even higher (Airey, 1978; Jackson, 1978; Muir *et al.*, 1978).

It is also relevant that although modern instrumentation is very efficient, temperatures should *always* be checked with a mercury in glass thermometer which has been checked against a standard thermometer kept in the laboratory.

Microbiological Problems in the Distribution of Cream

Cream presents more problems than milk because of the methods of distribution, and the requirements for a longer keeping quality. Sales may be erratic depending on the weather, holiday seasons, local activities, and many other factors. Broadly speaking, cream sales peak at Christmas when not only is cream bought for home consumption, but cream cakes, etc., also enjoy an increased sale, and there is a lesser peak at Easter, for the same reason. Any increase in consumption of foods traditionally associated with cream will increase the demand, and so there is a period of increased sales during the soft fruit season, in the the UK usually during June and July. Strawberries and cream is one of the most popular sweets, and this accounts for a considerable consumption of cream. Unfortunately this increased demand often results in problems in methods of distribution, reduced control, and longer storage (to cover erratic consumption), and these adverse factors may occur during periods of high atmospheric temperatures, so that microbiological problems are enhanced and keeping quality may be seriously affected. There seems to have been a belief that cream keeps better than milk, other things being equal, but this is very doubtful. Bacteria may be more static in cream because of the greater viscosity, which incidentally results in larger clumps and so smaller colony counts for a given total number of bacterial cells, but this effect would not restrain metabolic activities, such as souring, or affect the results of the methylene blue test.

Once cream has been despatched by the manufacturer, he has no further control over the way it is handled. Cream may be passed to a wholesaler who passes it to a retail dairy or a shop where there is no temperature control, and the cream may be some days old when it finally reaches the customer. This aspect has been considerably improved in recent years, but outside the dairy industry, food stores may not appreciate how fast bacteria can grow unless a low temperature is maintained *at all times* during distribution. The crucial temperatures are 6 and 13 °C. Below 6 °C growth is so slow that an adequate shelf-life is usually assured, other factors being satisfactory. The nearer the temperature is to 0 °C the better, but such refrigeration is costly. Between 6 and 13 °C the bacteria grow at an increasing rate, and deterioration of quality accelerates. Above 13 °C bacterial growth rapidly produces souring or other taints, or at least a failure in the total count, coliform or methylene blue tests. Refrigeration is the only legally permitted method for controlling the shelf-life of cartoned cream during distribution and storage.

The above considerations do not apply to sterilised or canned cream, or UHT-treated cream which has been aseptically packed. However, if the former is contaminated through a leaking seam, temperature will affect it, and enzymes which survive UHT treatment will cause changes which proceed faster at higher temperatures.

Micro-Organisms Causing Defects in Cream

All milk and products made from it, such as cream, become contaminated by micro-organisms from the udder, or from the cow, or during the milking process. The 'original' flora in this sense consists mainly of lactic and other streptococci, micrococci, corynebacteria, and aerobic and anaerobic spore-forming bacteria (Crossley, 1948; Davis, 1971; Tekinsen and Rothwell, 1974). Thus, if cream is sold raw, all these organisms will generally be present as well as cow-derived pathogens. From the time the milk is put through the various stages necessary for the production of cream, a variety of organisms accumulate until the final heat-treatment destroys most of them. Gram-negative rods from watery environments (many of them psychrotrophic), staphylococci and lactobacilli are usually prominent, but the proportions depend not only on the level of hygiene but on the temperature of the cream. The last two types and *B. cereus* are favoured by failure to cool the cream rapidly to 5 °C or under.

Types of Organism Found in Cream

In their investigation of changes in the microflora in retail creams held at 5 °C for 5 days, Tekinsen and Rothwell (1974) found that initially counts were low, and the proportion of psychrotrophs small. After storage, the psychrotrophic count (at 5 °C) varied from about 100 to over 10^7 cfu ml^{-1}, and the mesophilic count (at 30 °C) from about 1000 to 10^8 cfu ml^{-1}. In fresh cream, the predominating organisms at 5 °C were *Pseudomonas*, *Alcaligenes*, *Acinetobacter*, *Aeromonas* and *Achromobacter*, and at 30 °C, *Corynebacterium*, *Bacillus*, *Micrococcus*, *Lactobacillus* and *Staphylococcus*. The distribution of types varied greatly with the source of the cream.

After holding for 5 days at 5 °C, non-fluorescent *Pseudomonas* tended to become the predominant type and *Corynebacterium* and *Micrococcus* were reduced in number, although there were still differences between samples. Thus it appears that, assuming an efficient heat treatment, the microflora of a cream depends on the types of post-pasteurisation contamination, and the length of time of storage at 5 °C. The mean shelf-life of the samples varied from 6·5–23 days (a considerable difference). At the end of shelf-life, *Pseudomonas* spp., and especially the non-fluorescent types, were very definitely the predominant flora.

In their study of the bacteriological condition of retail cream in Worcestershire, Colenso *et al.* (1966) found that 223 out of 540 samples failed the methylene blue test, and only 181 were satisfactory. Counts were fantastically high, 70 samples having counts of $200–5000 \times 10^6 \, g^{-1}$, and 137 with counts of over $10^6 \, g^{-1}$; large numbers of coliforms and *Bacillus* spp. were recorded. Barrow *et al.* (1966) reported that cream could have a count of $50 \times 10^6 \, g^{-1}$ without the flavour being affected.

Identification of Bacteria Causing Taints in Cream

It is not necessary to identify organisms causing taints in cream with academic precision. A broad typing sufficient to identify the genus and probable species is quite adequate to indicate the source, the reasons for development of the taint (e.g. dirty equipment, storage at too high a temperature) and the method which should be adopted to eliminate the taint.

The following publications describe a number of systems of varying complexity and degree of detail: Davis (1955); Jayne-Williams and Sherman (1966); Thornley (1968); Buchanan and Gibbons (1974); Harrigan and McCance (1976).

Psychrophilic and Psychrotrophic Organisms

All perishable foods have to be held at low temperatures, usually not above 5 °C, and this treatment constitutes a form of selective enrichment. Psychrotrophic organisms are defined as those *capable* of growing at low temperatures, although they may have a high maximum temperature, e.g. 50 °C and an optimum in the range 30–40 °C. Psychrophilic organisms are those growing best at low temperatures, e.g. 15–20 °C, and having a maximum of about 25 °C.

While a temperature of about 30 °C is generally suitable for making total counts on dairy products, an examination of long shelf-life products, e.g. keeping for 7–14 days at 5 °C, should include counts made at 20–22 °C as well as at 30 °C in order to check the numbers of organisms capable of growing at *c*. 5 °C. The correlation between counts at 5–10 °C and those at 22 °C is usually high.

There is no internationally agreed definition of a psychrophile, but any organism which grows readily at 7 °C but not at 0 °C may be considered to be a psychrotroph, and an organism which grows at 0 °C may be accepted as a psychrophile. The minimum, optimum and maximum growth temperatures of true psychrophiles are all lower than the corresponding temperatures of other psychrotrophs. Psychrophiles and psychrotrophs may conveniently be differentiated by a count at 7 °C for 14 days (= psychrotrophs + psychrophiles) and a count at 0 °C for 14 days (= psychrophiles).

'Pseudomonads' is a general term used to embrace Gram-negative, non-spore forming, oxidase-positive, catalase-positive rods which commonly contaminate dairy products from dirty water. They often produce pigments, and usually attack fat and protein but not lactose. They do not sour cream, but produce a variety of taints when the count approaches 10^8 ml^{-1}.

A count of oxidase-positive bacteria can be made by flooding the plate with a reagent which develops a colour under oxidase reaction, such as freshly prepared 1 % tetramethyl-*p*-phenylenediamine hydrochloride, which turns successively pink, purple and finally black.

A parallel flooding of another plate with 1 % hydrogen peroxide solution will identify catalase-positive colonies by a *continuous* evolution of tiny bubbles of oxygen.

The pseudomonads may be broadly distinguished from the enterobacteria by being oxidase- and catalase-positive, and by not fermenting lactose. The enterobacteria ferment lactose and are catalase-positive but oxidase-negative.

Taints in Cream

Taints in cream may be chemically or non-biologically produced, or they may develop as the result of the growth of micro-organisms.

Being an emulsion of fat, cream has an enormous surface area in relation to volume, and readily absorbs odours from the atmosphere. It is therefore of the greatest importance that cream should not be stored where any odoriferous material, such as disinfectants, paint, varnish, scents or strong 'smelling' foods are stored, as cream can rapidly become inedible. The taint is usually so characteristic that it is unmistakable, but an absorbed odour is readily distinguished from a microbiological taint because the former passes off on standing open to the atmosphere, whereas the latter steadily increases in intensity.

Another type of chemical taint can be caused by cows eating certain plants such as garlic, decayed fruits, etc. Occasionally an absorbed type of taint can be imitated by the growth of certain organisms. Thus yeasts can produce fruity flavours (sometimes quite specific), and some bacteria can produce taints resembling apples and other fruits.

Oxidised taints can occur in cream of very good microbiological quality held at low temperatures. These are catalysed by traces of copper in the ionic state, and also by light. Minute concentrations of copper and sunlight can render cream so oily as to be inedible in a few hours. The higher the bacterial count, the slower the development of oxidised taint, because bacteria consume oxygen and so lower the E_h of the cream.

The origin of a taint in cream may be difficult to elucidate, but some fundamental causes of taints are:

(1) abnormalities in the milk inside the udder due to mastitis, late lactation, method of feeding, weeds in pasture, etc.;
(2) failure to cool the milk immediately after milking, so permitting lipase and other enzymes to act on the milk; aeration and agitation may accelerate the chemical changes involved;
(3) high count milk and/or cream (lack of hygiene in production);
(4) use of stale milk;
(5) dirty separators and other equipment (this gives a characteristic 'dirty taint');
(6) failure to cool the cream;
(7) high temperature of holding during distribution and sale; and
(8) cream stale when sold.

Asking the following simple questions may be helpful.

1. Was the taint in the original milk?
2. How old was the milk when separated?
3. Was the taint in the cream after separation?
4. What time elapsed between pasteurisation and sale?
5. At what temperature was the cream held?
6. What was the bacterial count at the time of sale?

A critical survey of the answers to these questions will usually afford some clue as to the nature of the problem. If the taint has developed only after pasteurisation, it may be due to dirty equipment, thermoduric bacteria and/or holding at too high a temperature. If present in the cream immediately after separating, it may be due to enzyme action in the raw milk, or it may be inherent in the milk itself.

A more sophisticated approach is to test the milk, raw cream, and pasteurised cream at the point of sale for:

(1) thermoduric bacteria;
(2) lipolytic bacteria using tributyrin agar (5 days at 22 °C); and
(3) caseinolytic bacteria using caseinate agar (5 days at 22 °C).

Usually (2) and (3) give similar results. If large numbers are found, the taint may be of bacterial origin, but if not it may be due to lipase–oxidase action.

The coliform and methylene blue tests are usually of little use for this purpose, although a high coliform count suggests dirty equipment. Bitterness and other fat taints in cream may be caused by contaminated water infecting the milk or cream with *Proteus, Pseudomonas, Achromobacter*, or other proteolytic and lipolytic organisms able to grow at low temperatures.

In general, all bacteriological tests for faults in cream should be made at 22 °C, and never at above 30 °C. In all quality control work for milk and cream, odour and taste tests should always be made immediately after processing, and at a time corresponding to sale. The product should be adjusted to 20 °C for this purpose.

Specific Organisms and the Taints They Produce
Bitterness
A bitter taste can be developed in cream by a number of microorganisms. It usually results from biochemical attack on proteins to produce peptones and polypeptides. Specific substances having a bitter taste have also been isolated, and partial glycerides may be responsible. In addition to *Proteus* and other Gram-negative rods, some yeasts and moulds can produce bitterness, although associated growth may be necessary. For example souring by a lactic organism, such as *Streptococcus lactis*, may be necessary to allow *Rhodotorula mucilaginosa* to produce a bitter flavour.

Pseudomonas
This genus consists of Gram-negative non-spore-forming rods which attack proteins and fats strongly, but have little or no effect on sugars. They require air for growth and often produce greenish pigments and various taints. They are associated with watery environments and can grow at low temperatures, but are easily killed by pasteurisation. The most common in dairy products is *Ps. fluorescens* which attacks fat and produces rancidity.

Ps. fragi develops an apple-like ester taint before rancidity is detected by taste, and *Ps. putrefaciens* produces a putrid odour. *Pseudomonas nigrifaciens* can give a blackish discolouration on the surface of dairy products.

In general the genus is harmless, but *Ps. aeruginosa* can grow at 42 °C, is very resistant to antibacterial chemicals (antibiotics and quaternary ammonium compounds), and is now recognised as an opportunist pathogen.

Yeasts

Yeasts are not commonly the cause of defects in dairy products, because (with a few exceptions) they do not ferment lactose and grow comparatively slowly. If, however, organisms capable of hydrolysing lactose are present, or if sugar is added (as for whipped cream), then yeasts can grow rapidly and produce a characteristic yeasty or fruity flavour and obvious gas. *Torula cremoris* or *Candida pseudotropicalis* and *Torulopsis sphaerica* have been responsible for outbreaks of this defect.

The Importance of Aerobic Spores

The presence of bacterial spores in milk is of considerable importance for certain dairy products, particularly those which are sterilised such as sterilised milk, evaporated milk, sterilised cream, and also clotted cream. There are two reasons for this. The sterilisation process kills all vegetative cells and so leaves a clear field for the growth of the spores, and the heating gives a shock to the spores which stimulates germination (Davies, 1975).

Anaerobic spores are rarely of importance in cream, but aerobic spores can be a serious source of trouble. The latter can be classified in groups according to their heat resistance.

The spores of *B. cereus* are only moderately heat resistant, but can easily survive pasteurisation and somewhat higher temperatures. This and other *Bacillus* species can grow at low temperatures (Cox, 1975), although a temperature of 5 °C will retard their growth for some days (Davis, 1969b, 1971). *Bacillus cereus* is of particular importance for cream, because it can not only induce sweet curdling (or 'bitty cream' in milk), but can also reduce methylene blue and so lead to failure in the official PHLS test. Other species can be prominent in cream such as *B. licheniformis* and *B. coagulans* (Cox, 1975). *Bacillus subtilis* and its variants (Buchanan and Gibbons, 1974) are the most important species in terms of producing spores that are markedly heat resistant, and may be responsible for bitterness and thinning in sterilised cream.

Microbiological Associations

Organisms seldom occur in pure culture in nature, and the importance of inter-connecting activities or associated action is not always appreciated. Different types of organism may exert a powerful influence on the resultant biochemical activities by releasing a source of energy (hydrolysing lactose) for another organism to utilise, or by creating favourable or unfavourable conditions (acid reaction or anaerobiosis). For example, souring of milk by *Str. lactis* may inhibit growth of *Bacillus* spp. and other types sensitive to acid (Harman and Nelson, 1955).

Food Poisoning from Cream

Cream may differ from milk in this respect for two reasons: (1) cream is normally more severely heat treated than retail milk; and (2) cream is expected to have a longer shelf-life than retail milk. Thus, in practice there is less probability of food poisoning organisms surviving in cream, but any which do have more chance of proliferating and so causing trouble. Any contamination *after* heat treatment may be more serious for this reason, although if the cream is held at 5 °C, the chance of food poisoning organisms growing is slight.

The most common cause of deterioration of cream is water-borne contamination, the most common source of *Pseudomonas*, and if the water is contaminated with sewage, *Salmonella* and other faecal types may be present. In the past this was the usual cause of food poisoning from cream, but such incidents are now rare.

DAIRY DESSERTS

There is no legal definition of dairy desserts in the UK, but we may conveniently define, or describe them, as sweets in which milk ingredients constitute at least 40 % of the dry matter in the food. Skim-milk and cream figure prominently in the list of ingredients when given in descending order of concentration. The ingredients of some typical dairy desserts are given in Table IV.

Most dairy desserts are sweet, but there are a few savoury types containing cheese. Typical dairy desserts are junket, custards, blancmanges, trifles, soufflés, often with fruit, nuts, cereal foods, starch, gelatine, etc., and all based on milk, skim-milk and/or cream. Useful market surveys of dairy desserts and other fresh chilled dairy products have been published (Anon., 1979, 1980).

TABLE IV

INGREDIENTS OF SOME DAIRY DESSERTS

Dessert	Ingredients
Fresh cream trifle	Skim-milk, fruit cocktail (cherry, grape, peach, pear, pineapple), cream, sugar, sponge cake, raspberry jam, food starch, roast almond, gelatine, flavouring, citric acid, stabiliser, colour.
Lemon soufflé	Cream, sugar, dextrose, lemon, gelatine, dried egg white, flavouring, colour.
Chocolate soufflé	Cream, skim-milk, chocolate, sugar, egg yolk, gelatine, dried egg white.
Fresh cream syllabub	Cream, sugar, ginger biscuit, sherry, lemon juice, brandy, carrageen, pectin.
Strawberry delight dessert	Concentrated skim-milk, sugar, cream, strawberry, starch, stabiliser, flavouring, citric acid, colour.
Blackcurrant cheesecake dessert	Low fat soft cheese, sugar, cream, biscuit, blackcurrant, full cream milk powder, food starch, citric acid, gelatine, flavouring.
Chocolate delight dessert	Concentrated skim-milk, sugar, cream, chocolate, food starch, stabiliser, cocoa.
Fruit jelly and custard	Skim-milk, fruits (cherry, grape, peach, pear, pineapple), sugar, cream, food starch, gelatine, citric acid, flavouring, colour.

The general description above would include ice creams (Chapter 1) and fruit yoghurts (Chapter 6), but these are usually described as such and may be defined legally in some countries.

Dairy desserts, ice creams and flavoured milks contain basically the same ingredients, and if they are held at low temperatures their microbiology should be fairly similar, allowing for the fact that ice cream is normally held at the lowest temperature. Fruit yoghurts also have the same ingredients, but as cultured products (pH 4), their microbiology is quite different.

Although there may be no specific regulations for dairy desserts in the UK, they have to comply with the Food and Drugs Act (1955), the Labelling of Food Regulations (1970), Etc. and others such as the Preservatives in Food Regulations (1979). The label must give an appropriate designation, and give the ingredients in absolute terms or in descending order of concentration.

In the USA, there are definitions and standards for frozen desserts under the Federal Food, Drug and Cosmetic Act, Part 20, Title 21, which deals with ingredients, and a maximum bacterial count is mentioned for some. Regulations for food additives are given in Part 121, Chapter 1, Title 21. Standards of identity for frozen desserts given in USA (1978) include milk-derived ingredients. The EEC has issued draft directives for some foods which are ingredients of dairy desserts. Standards in force in Czechoslovakia (1976) cover desserts made from cream. In Germany, flavoured and fruit milks, 'sweet' cheese puddings and special desserts may be included in the group 'sweet dairy desserts' (Otte *et al.*, 1979*a*).

Microbiology

It is obvious that products of this nature which are high in moisture, approximately neutral in reaction, contain good quality protein and have a high water activity, will be highly perishable. The product itself cannot normally be heat treated either just before or just after packaging, and although the individual ingredients are nearly always sterilised or pasteurised, or heated in the course of preparation, any organisms which gain access to the product afterwards will have ample opportunity for proliferation unless certain precautions are taken. There are four main methods for preventing, or at least minimising, this:

(1)   purchase of ingredients on microbiological specification;
(2)   efficient heat treatment of ingredients as may be appropriate;
(3)   extreme care in hygienic control during processing and packaging; and
(4)   efficient refrigeration during storage, distribution and sale.

With some desserts, hot filling may be possible, but with rare exceptions, preservatives cannot be used in dairy products. However, certain preservatives can be used for foods which may be incorporated in dairy desserts, for example flavourings, flour confectionery, fruit, gelatine, jam, pectin, starch and sugars. No objection is usually made to the presence of permitted preservatives in proportion to the amount of these foods present. Thus, the 1979 Preservatives in Food Regulations in Britain permit 100 ppm sulphur dioxide or 300 ppm sorbic acid for fruit-based milk and cream desserts.

In practice, the temperature of holding is crucial, and distributors usually maintain desserts at 5 °C or lower and give a limit for time of use after sale. The purchaser may be told to keep the food refrigerated and to consume it within 2 days. Provided the precautions mentioned above have been observed by the manufacturer, this should guarantee a safe product of adequate shelf-life.

Microbiological Examination

This may be considered under two headings—safety and shelf-life.

Safety

Assuming adequate heat treatment of all ingredients, the only danger will arise from inadequate sanitation of equipment, and human or animal contamination in some form.

Bearing in mind the nature of the product and the method of distribution and consumption, the most useful tests are the total colony count, and the enumeration of coliforms or *E. coli* or faecal coli and of staphylococci. The total count and coliform tests should be regarded as *indicator tests*. Low counts indicate a probably satisfactory condition, whereas high counts can be taken as a warning that something is unsatisfactory. It is unlikely that a product giving good results on these two tests would be a cause of food poisoning. More specific information is given by the examination for faecal coli and staphylococci. Faecal coli are generally used as a specific indicator for the presence of *Salmonella* spp. Staphylococci represent a different source of contamination (from skin, nose and throat).

The enumeration of total viable organisms, faecal coli and staphylococci, would thus afford an adequate control system for safety. The risk of food poisoning from other types, such as *Clostridium* and *Bacillus*, is so remote that it can be ignored. The presence of faecal coli in 1 g indicates the possibility of the presence of *Salmonella*, and the food should be rejected. Other organisms sometimes used as indicators of possible faecal contamination are faecal streptococci (enterococci or *Str. faecalis*, etc.) and *Clostridium perfringens* (*welchii*). These tests are commonly used in water control work, and so naturally find application in those branches of the food industry where water is used for the reconstitution of solid foods, such as milk powder. If solid ingredients are reconstituted and the product is not afterwards subjected to heat treatment, the water should be chlorinated to give a *residual* of 3 ppm available chlorine. At this level the chlorine will not give any taint.

In addition to the well recognised pathogens, there are some organisms which can be the cause of food poisoning, especially for the very young and the very old, and these are called *opportunist pathogens*. They include types formerly regarded only as indicator or undesirable organisms, such as enteropathogenic *E. coli*, *Ps. aeruginosa* (*pyocyanea*), which have been shown during the last 30 years to be capable of causing food poisoning or enteric infections, especially in babies, and sometimes with fatal consequences. Other organisms such as *Serratia* have occasionally been known to cause food poisoning or infections. When this occurs the relevant organism often constitutes a large proportion of the bacteria present, and this phenomenon should be regarded as a danger signal.

Shelf-Life or Keeping Quality

In selecting microbiological tests from this point of view, one has to consider the ingredients of a food and the temperature of holding. The essential difference between dairy desserts and plain dairy products is the presence of sugars (sucrose, fructose and dextrose) and fruit. This makes a dairy dessert, like fruit yoghurt, an ideal medium for the growth of yeasts and moulds. The most useful tests for keeping quality are therefore total count, coliforms, yeasts and moulds.

With a food held at 5 °C, only psychrophiles and psychrotrophs will grow in 2 or 3 days, and then only to a limited extent. The simplest check is therefore to make a total count at 20–22 °C as an index of the probable psychrotroph count. As an incubation time of 4–5 days is necessary for this (ideally), a parallel count can be made at 27–30 °C with an incubation time of 48 h, and counting colonies with a lens at least × 4. In this way an abnormally high count can often be detected quickly, and appropriate action taken. It is not uncommon to obtain a count 10 times higher at 20 or 30 °C than at 37 °C, and on occasions we have found a ratio of 100. A low count at 37 °C can, therefore, be misleading.

A possible quick method is to disperse the solid ingredients, such as fruit, milk powder, stabilisers, etc., in a 0·1 % glucose/0·1 % peptone solution and make a resazurin test at 30 °C. An empirical test such as this should only be regarded as a simple method for the rapid detection, and possible rejection, of really bad samples.

When considering microbiological standards for ingredients of mixed foods such as dairy desserts, two important aspects must be borne in mind. (1) What will be the likely growth rate of the organisms when the ingredients are mixed? (2) How do the original counts of the ingredients compare?

For example, a low count of a dangerous organism which cannot grow in one ingredient, may become a matter of concern when the ingredient is mixed with another. Dryness and acidity may be important factors operating in this connection, because moisture and a reduction in acidity may permit growth when dry and moist ingredients are mixed. A fairly high count of spores, or of an especially resistant organism, in one ingredient may be of little consequence if another ingredient usually has much higher counts.

In the absence of any legal or accepted standards, each manufacturer should work out his own microbiological standards in the light of the shelf-life required for each product at a particular temperature.

Microbiological Examination of Dairy Desserts
Techniques

When a product consists of two or more phases, such as cream on top of chocolate mousse, or custard on top of a fruit jelly, it may be useful to examine each phase separately in order to be able to identify the ingredient responsible for a high count, or similar fault. Specific selective media or techniques may then be used for each ingredient, for example testing for yeasts and moulds in fruit products, spores in chocolate, staphylococci and coli in cream and custard, etc. Although heat treatment may kill all organisms, except possibly resistant spores, an important aspect is the rate of growth of various organisms in the ingredients used. Thus a small contamination during filling may be very serious under some circumstances and of little consequence under others.

The ingredients or phases of a dairy dessert are usually easily dispersible in the Ringer diluent, but a few, such as nuts and whole cereal grains, have to be macerated, and a mixer or stomacher should be used. As only small numbers of organisms are usually involved, a membrane filtration method may be suitable, or alternatively the MPN method may be used, in which three aliquots of 10, 1 and 0·1 g are inoculated into glucose tryptone broth for a total count, a bile salt broth for coliforms, a malt extract or unhopped beerwort at pH 4 for yeasts and moulds, and a 6·5% sodium chloride mannitol broth for staphylococci; with the 10 and 1 g quantities in the coliform test, checks should be made for false positives. Usually the characteristic appearance of the growth in the tubes is adequate for quality control, but if desired, confirmation may be obtained by streaking the positive tubes on the appropriate solid media.

TABLE V
FOUR GRADES OF STANDARD SUGGESTED FOR DAIRY DESSERTS
(COUNTS PER GRAM)

	Target	Acceptable	Doubtful	Reject
Viable count	<1 000	1 000–5 000	5 000–20 000	>20 000
Coliforms	<10	10	10–100	>100
Yeasts	<10	10–50	50–100	>100
Moulds	<10	10–50	50–100	>100

Peterkin and Sharpe (1980) have shown that incubation of frozen dairy products with a protease or Tween 80 or both, permitted the membrane filtration method to be used with quantities up to 5 g without affecting the viability of common bacteria.

One firm, renowned for its high standards of hygiene in foods, operates on the basis of four grades as shown in Table V.

Microbiological Standards for Ingredients

Skim-milk powder and cream are commonly used in the preparation of dairy desserts, so their quality is of great importance for these products. The standards shown in Table VI have been suggested for milk powders.

TABLE VI
SOME SUGGESTED MICROBIOLOGICAL STANDARDS FOR MILK POWDERS (PER GRAM)

	Satisfactory	*Doubtful*	*Unsatisfactory*
ADMI Viable Count	From 30 000–100 000 according to type		
Spray-dried powders			
Viable Count	< 10 000	10 000–100 000	> 100 000
Coliforms	< 10	10–100	> 100
Yeasts	< 10	10–100	> 100
Moulds	< 10	10–100	> 100
Staphylococci (coagulase-positive)	< 10	10–100	> 100
Faecal streptococci	< 10	10–100	> 100
Direct microscopic count	$< 10 \times 10^6$	$10–100 \times 10^6$	$> 100 \times 10^6$
Roller-dried powders			
Viable count	1 000	1 000–10 000	> 10 000
Other tests	As for spray-dried powders		

After: Davis (1968); ADMI (1971).

Laboratory Control Methods

There is no point in making any laboratory test unless the result gives useful information. For foods, such information may be categorised under four headings as follows:

(1) to confirm that the product comes within the normal range of properties;
(2) to ensure that it will not harm the consumer;
(3) to ensure that it meets all legal requirements; and
(4) to give advance warning of any likely deterioration in properties.

No one schedule of tests will be applicable to all products, not even to similar ones. Tests and techniques may have to be modified according to the conditions of a product, for example, source, age, method of packaging and purpose for which the product is to be used. All testing should be aimed at improving or maintaining quality.

The Limitations of Microbiological Tests

There is often misunderstanding about the true nature and significance of microbiological tests, even among chemists. Whereas different methods for measuring, say, fat or moisture should all give closely similar results, microbiological tests cannot be expected to do this because they often

TABLE VII
PROPERTIES MEASURED BY VARIOUS MICROBIOLOGICAL TESTS

Test	*Property measured*
Plate or colony count	Cells or clumps forming colonies under the conditions of the test (medium, temperature, etc.).
Direct microscopic count	Cells (or clumps) taking the stain, including living and dead cells.
Dye reduction tests (methylene blue and resazurin)	Metabolic activity (reducing power) of active cells.
Increase in acidity or other metabolic product	Number of bacteria able to produce the substance measured, e.g. lactic acid.
Modern refined instrumental methods for measuring metabolites, increasing conductivity, producing heat, etc.	Production of metabolites or changes in physical properties.

measure different things. This is illustrated in Table VII. One test cannot be equated in terms of others. In addition, the error, as measured by the repeatability and reproducibility of microbiological tests, is usually much greater than that of physical and chemical tests. It is important that this fact should be kept in mind when operating quality control schemes and setting standards. The simplest and safest way to allow for this aspect is to think logarithmically. For example, the result of a single count should only be accepted as significantly different from another if it is at least 10 times greater or smaller.

All laboratories have their methods of choice for the quality control of foods, and some methods selected for the microbiological examination of cream have been described by Davis (1969c, 1971), Cox (1970, 1975), Kleeberger (1978) and Otterholm (1978).

Total Counts: Enumerative Media

After about 60 years experience in food control work, the total colony count is still the most popular test for microbiological quality and hygiene in production, although it gives (usually) no information about the safety of the food. Any count is entirely dependent on the medium used and the temperature of incubation. For dairy products, milk agar with incubation at 30 °C is commonly used, as this temperature can give counts up to 10 times higher than those obtained at 37 °C, and on rare occasions, up to 100 times higher.

Simple glucose tryptone agar, or a similar medium, is now firmly established for total counts in the food industry. More information about the types of organism in a food can be obtained by using a differential (but not selective) medium, in which certain reactions give a clue to the nature of the organism. Suggested media for the examination of cream and other dairy products are given by Donovan and Vincent (1955) and Davis (1955, 1959a). In general it is best to use a medium associated with the product under examination, for example milk agar or a casein digest medium for dairy products, peptone based media for animal products, and malt or unhopped beerwort agar for plant products.

In quality control work, much useful information can be gained by using two media, or one medium at two temperatures, or combining the two ideas. For example, gelatine or tributyrin agar at 20 °C and litmus lactose agar at 30 or 37 °C together give counts of the two most important types of bacteria in dairy products. The biggest cost item in microbiological laboratory work is time or labour. The extra cost of using two media or two conditions for one medium is small in relation to the whole, and is therefore well worth while for preliminary identification work (Table VIII).

Milk and other liquids can easily be measured volumetrically for bacteriological tests, but cream is better measured by weight (Brazis *et al.*, 1972).

Surface Culture

Most organisms of importance in cream and dairy desserts are aerobes or facultative anaerobes, and grow faster and form larger colonies on the surface of media than in the depth. Moreover, surface colonies are of greater usefulness in identification, and counts are usually higher. For some organisms, surface culture is much preferred; in fact, some organisms may

TABLE VIII
SOME RECOMMENDED MEDIA FOR COUNTS OF MICRO-ORGANISMS

Organism	Medium	Temperature (°C)	Reference
Total bacterial count	Milk agar Plate count agar	30	Davis (1959a)
Presumptive coliforms	MacConkey agar or broth	30	
E. coli	Violet red bile agar	37	
Faecal coli	Brilliant green bile broth	44	
Faecal streptococci (enterococci)	Kanamycin aesculin azide agar Slanetz and Bartley medium	37 44	Mossel et al. (1976) Slanetz and Bartley (1957)
Pseudomonas spp.	King's medium (modified)	37	Goto and Enomoto (1970)
Pseudomonas aeruginosa	Asparagine agar Cetrimide agar	42 42	Hedberg (1969) Lowbury and Collins (1955)
Psychrotrophs	Milk agar Nutrient agar	5 (14 days) 7 (10 days)	
Yeasts and Moulds	Uphopped beerwort or malt agar pH 4 Rose Bengal agar	27 22	Davis (1931) Mossel et al. (1975)
Aerobic spores	Milk agar Plate count agar	30	
Staphylococcus	Chapman medium Baird-Parker medium	37 37	Chapman (1945) Baird-Parker (1962)
Clostridium and other anaerobes	Cooked meat broth Reinforced clostridial agar	37 37	

Nearly all these media are described in: Ministry of Agriculture (1968); Difco (1971); FDA (1978) and Oxoid (1979).

be missed entirely in pour plates with incubation times of only 1 or 2 days. This applies particularly to staphylococci, yeasts and moulds. 'Spread plates' may be made by smearing up to 0·5 ml of a dilution over a well-dried plate or, if a count higher than 1000 is expected, the drop method may be used (Miles and Misra, 1938; Davis and Bell, 1958).

Control of 'Spreaders'

Aerobic spore-formers (*Bacillus*), if present in numbers in a dairy product, may form large, spreading colonies which inhibit the growth of other organisms and so invalidate the plate. It may be necessary to use a higher dilution for calculation of the count, which is then less accurate. Various devices may be used to minimise the spread of such colonies, such as a modified medium, 3 % agar, the use of small volumes of medium, pour-plating instead of surface culture, thorough drying of the plates immediately after inoculating, and pouring a thin layer of medium over the solidified inoculated medium. Lower temperatures of incubation may also be advantageous. *Proteus* colonies may also spread, although not so fast as *Bacillus*. The above methods will help to reduce this spreading.

It has long been recognised that some organisms can inhibit the growth of others, particularly some moulds, streptomycetes and *Bacillus*. Even similar types of bacteria can show this effect. Vanderzant and Custer (1968) found that some species of *Pseudomonas*, and also an *Achromobacter*, could inhibit other species; all microbiological tests are subject to this type of error. It is also important to remember that some organisms may not grow under the conditions of the test, but grow later in the product.

Although the colony count may be the most useful single test for cream and other dairy products, it has the disadvantage that it requires at least 2 days. A membrane filtration epifluorescent microscopy method, which gives a result correlating well with the colony count and requires only 25 min, has been described by Cousins *et al.* (1979), Pettipher *et al.* (1980) and Pettipher and Rodrigues (1981). It is suitable for counts of 5000 cells ml^{-1}, or more.

Diagnostic and Selective Media

Numerous media have been suggested for the selective cultivation of specific types of organisms (Davis, 1955, 1959a; Thatcher and Clark, 1968; Norris and Ribbons, 1969; Difco, 1971; Harrigan and McCance, 1976; FDA, 1978; Oxoid, 1979), and new media are constantly being described in the literature. These may be marginally superior to well established media of long standing, but there may be complications in preparation and some

J. G. Davis

ingredients may be unstable and/or expensive. Selective media may be slightly inhibitory for the organism concerned, in that sensitive strains or stressed cells may not grow. This point is easily checked by a parallel plating in a suitable but non-selective medium.

Microbiologists usually have their favourite media, the choice being largely influenced by their training and experience in a particular branch of the food industry. For quality control work it is better to keep to well-tried media rather than to be constantly changing to new media as they appear. '*Know your media*' is an excellent motto for all practising microbiologists. This is particularly true for the isolation and identification of pathogens. The media most commonly used in dairy bacteriology are listed in Table VIII, and of all the specific tests made for quality control in the dairy industry, the most common is that for presumptive coliforms. There are two reasons for this. Coliforms are abundant in milk production areas so that their numbers can be used as a measure of hygiene in production. Secondly, coliforms are killed by pasteurisation (with rare exceptions), so that their presence in a heat treated product indicates post-pasteurisation contamination.

Enumeration of Small Numbers
In the routine testing of pasteurised milk and cream and similar foods at the point of sale, the ordinary plate count and similar tests are adequate, but when testing products claiming to be sterile or to have very low counts, such as cream designed to last for 14 days at 5 °C, the ordinary count is inadequate, and it is necessary to employ a 'most probable number' (MPN) technique. In this method, three or five aliquots of varying size (by dilution) are tested, and the MPN value calculated statistically from the number of positive results obtained with each size of aliquot; for fairly low numbers, aliquots of 1, 0·1, 0·01 ml, etc., might be used. The MPN is then obtained by consulting the McCrady tables as given in Thatcher and Clark (1968), Davis (1956) or similar manuals. For the enumeration of very low numbers, aliquots of 100, 10 and 1 ml might be used, so that for a similar series of results, the MPN count would then be 100 times smaller (see Table IX).

The simplest way of identifying the positives is to incubate the aliquots at 27 °C for 24 h, and then streak three loopfuls on a plate and incubate. An alternative or additional method is to add sterile resazurin solution to the aliquots and incubate at 27 °C, when the positives will usually be detected by a change in colour from blue to purple or pink. This MPN method is especially applicable for in-line testing because high-temperature pasteurisation gives very low counts in pipe-line samples. If a count of a specific type of organism is required, the aliquot, after incubation, may be streaked on a diagnostic or selective medium for yeasts and moulds, coliforms, *Pseudomonas* spp., *Staphylococcus* spp., etc.

TABLE IX

ENUMERATION OF SMALL NUMBERS USING THE MPN TECHNIQUE

	Volumes in ml of three aliquots					
	1	*0·1*	*0·01*	*100*	*10*	*1*
No. of positives	2	1	0	2	1	0
Calculated count per 100 ml		150			1·5	

Aerobic spore-formers may be a problem in cream testing because they often form spreading colonies which may inhibit the growth of other organisms. Kleeberger (1976) recommended the incorporation of 33% violet red bile agar in a total count medium to prevent this.

Presumptive Coliforms
This test, together with the plate or colony count, has been used for hygienic control of foods and water for over 60 years, and in spite of numerous alternative tests which have been proposed, these two are still the favourites in the dairy industry. They are generally employed together, and the results are meaningful to an experienced dairy bacteriologist whether the product concerned is raw or untreated, or has been pasteurised or more severely heat treated.

The simplest form of the test is to add 1 ml of each dilution to a tube of MacConkey broth and incubate at 30°C for 2 or 3 days (Davis, 1959*a*; Ministry of Agriculture, 1968). Some coliforms from dairy products may not produce gas at 37°C (Hiscox, 1934). Although statistically of limited value, this very simple test permits classification into three grades (satisfactory, doubtful, unsatisfactory). When used with the plate count, the two tests together give considerable information about the quality of the cream (Table X). The accuracy can be increased by inoculating three or five tubes at each dilution and calculating the MPN of coliform bacteria.

False Positives in the Presumptive Coliform Test
False positive results may occur in a test which depends on the production of acid and gas in a bile salt lactose broth. There are three possible causes.

TABLE X

DIAGNOSIS OF REASONS FOR POOR QUALITY CREAM

Test results (comparative)	Interpretation
High count and high coliforms	Inadequate heat treatment and/or unhygienic manufacture and/or storage at high temperatures.
High count but low coliforms	Good hygiene but storage at high temperatures.
Low count but high coliforms	Poor hygiene in manufacture but storage at low temperatures ($<5\,°C$).
Low count and coliforms but high moulds	Good hygiene except aerial contamination in dairy.
Low count and coliforms but high yeasts	Good hygiene except contamination from fruit, etc., directly or indirectly.
Low count and coliforms but high aerobic spores	Cream made from milk having a high spore count.

1.   If the product contains sugars other than lactose, as dairy desserts usually do, these may be fermented by non-coliforms to produce acid and gas.

2.   Some anaerobes, such as *Clostridium*, may produce acid and gas from lactose, especially if the product is rich in protein. Neutral red is useful as an indicator here as anaerobes decompose it with the production of characteristic orange–green pigments.

3.   Other aerobes may ferment lactose.

In all three cases, a false positive is most likely to occur with 1 g of a high solids product, such as cream or a dairy dessert, in 10 ml of medium. A simple check test can be made by transferring three drops of the 'positive' tube to a fresh tube of medium and incubating as usual.

Choice of Media for the Coliform and Faecal Coli Tests

If a solid medium is used the colonies can be counted, but confirmatory tests are necessary for definite indentification; this involves picking off colonies, sub-culturing and further testing. A direct colony count can be made using an agar medium such as MacConkey or eosin methylene blue, and it is usually possible to differentiate *Escherichia*, *Klebsiella* and *Aerobacter*, and non-lactose fermenters, which may give a clue as to the source of the contamination. If a liquid medium is used (with aliquots of each dilution) the MPN may be calculated, but the 95 % confidence limit is quite broad. However, this method permits easy follow-up tests in lactose broth and tryptone broth for the standard confirmatory tests for faecal coli (*E. coli* type 1) based on the production of acid plus gas, and also indole, at 44 °C.

Faecal Coli

The use of faecal coli or *E. coli* type 1 as indicators of faecal pollution, and so the possible presence of *Salmonella*, is the most important and commonly used test of its kind in many branches of bacteriology. There are, however, differences in methods and interpretation. In the UK, the term 'faecal coli' is interpreted as meaning strains of *E. coli* type 1 which produce acid and gas from lactose *and also* indole from tryptone or peptone at 44 °C, but in some countries, the presence of *E. coli* producing acid and gas from lactose at 44 °C is accepted as the single criterion of faecal pollution.

In the UK, therefore, the double test is necessary. A shortened form may be used in which the cream (with aliquots) is incubated in MacConkey broth overnight (16 h), and then the positive tubes are sub-cultured into brilliant green lactose broth and tryptone water with incubation at 44 °C for 8 h. Only if both tubes give positive reactions (AG + and indole +) is the result regarded as positive for faecal coli.

Other Indicators of Faecal Contamination

Faecal streptococci and *Clostridium perfringens* (*welchii*) are sometimes used as indicator organisms for possible faecal contamination, which usually occurs through sewage contaminated water supplies. Clostridia and enterococci are not normally of any importance in cream. Faecal streptococci (enterococci) can be enumerated using a thallous acetate (Barnes, 1959) or azide (Slanetz and Bartley, 1957) medium. If faecal contamination is suspected, the faecal coli test is usually adequate for routine control.

Psychrotrophic and Lipolytic Counts

The water-borne Gram-negative bacteria such as *Pseudomonas*, *Achromobacter*, *Flavobacterium*, etc., are among the most active fat-splitting types occurring in dairy products. Most are capable of growing at low temperatures and may be fairly vigorous in the range 5–10 °C. The improved efficiency in refrigeration of milk in the dairy industry since 1950 has almost eliminated souring, but the improved keeping quality of raw milk has led to lengthened storage in tanks and silos, sometimes up to 2 days or longer at 5 °C, so that psychrotrophic bacteria are able to grow steadily, even if slowly. With its increased fat content, cream is naturally sensitive to lipolytic bacteria and lipases from any source.

The simplest medium for detecting lipolytic bacteria is well-emulsified tributyrin agar, but as tributyrin is much more easily hydrolysed by lipases than milk fat, the results may be regarded as giving a false picture of the potential hazard of the lipolytic bacteria in the cream. Thus a more accurate picture is obtained by using a medium incorporating cream, olive oil or butter fat, and a suitable indicator such as Victoria blue, but the slow lipolysis (and consequently narrow zones) and the inhibitory effect of the medium on some organisms have resulted in some authorities questioning the validity of the results obtained.

Psychrotrophic counts may be made by incubating plates at 10 °C for 7 days or at 5 °C for 14 days. While it is universally admitted that psychrotrophs are of the greatest importance in all perishable foods, the long incubation time required for their enumeration by the ordinary test is a major disadvantage in quality control work. Hankin and Dillman (1968) showed that a good correlation existed between psychrophiles and oxidase counts in dairy products. They suggested that the latter could be used as a rapid measure of potentially harmful psychrophiles. Hankin and Ullman (1969) used the total count, coliform and oxidase counts for assessing the possible sources of contamination in refrigerated delicatessen foods, and emphasised the advantages of the oxidase count as a rapid test for enumerating psychrotrophic pseudomonads.

Yeasts and Moulds
Yeasts and moulds are rarely the cause of taints in retail cream, although they are very important for butter cream. Selective media are usually based on a low pH value (Davis, 1931) or incorporation of antibiotics (Mossel *et al.*, 1975).

Trouble is sometimes experienced when a product is heavily contaminated with *Mucor* and *Rhizopus* which form spreading colonies. King *et al.* (1979) recommended the inclusion of 2 ppm dichloran and 25 ppm rose Bengal in a standard mould medium to restrict the colony size of these moulds. However, as few yeasts ferment lactose, and moulds usually require air for growth, deterioration of retail cream by either indicates very poor hygiene.

Yeasts and moulds are a more serious hazard for dairy desserts, and all products containing sucrose and fruit.

Thermoduric Organisms
Bacteria which do not form spores but which can survive pasteurisation were for many years a cause of taints and other defects in pasteurised milk

and cream. The most common are the thermoduric cocci, including micrococci and sarcinae, faecal streptococci and certain rod-forms, such as *Microbacterium* and *Corynebacterium*. They are usually not very active biochemically, and may grow only slowly, so that they are not always detected by dye reduction tests and colony count tests, unless the latter have a full incubation period, e.g. 3 days at 30 °C.

The factor mainly responsible for high thermoduric counts is failure to clean equipment properly on the farm and in the dairy. A common sign of this condition is the build up of a yellowish slime which often contains astronomical numbers of *Sarcina lutea*. Dirty farm and dairy equipment may be detected by being stained amber with iodophors. Thermoduric organisms are not only resistant to heat but are also usually resistant to disinfectants. Cleaning (removal of all soil) is, therefore, a most important part of the sanitising process.

Spores

Spores constitute a highly specialised problem in all branches of microbiology for two reasons. They all survive pasteurisation, and germination, and so detection may be uncertain and controlled by conditions in the processing (see later) and treatment in the laboratory. For the same reason, faults produced in foods by spores are an uncertain phenomenon. Outbreaks may be spasmodic, but on rare occasions can assume epidemic proportions.

In cream two species of aerobic spore-formers are of particular interest. *Bacillus cereus* is often found in milk, the source being soil, cereals and unsterile equipment, and as it usually ferments lactose and can grow well at 15 °C, it can increase in numbers quite rapidly unless the milk, and subsequently cream, are kept at 5 °C or lower. Although *B. cereus* can be a cause of food poisoning in meat and other foods, it does not appear to be a cause in dairy products. However, because it grows well in cream, survives heat treatment, ferments lactose, and reduces methylene blue, it can give a failure in the PHLS test (Davis, 1969*c*, 1971; Cox, 1970, 1975), although the result has no public health significance.

In cream, *B. cereus* is the most important aerobic spore-former and is often present in milk and cream. In Denmark, Søgaard and Peterson (1977) found it in 8 % of pasteurised samples, and after 24 h at 17 °C, positive samples and counts were much higher. The proportion of positive samples rose to 40 % in July–September. Although the spores survive ordinary pasteurisation, they are not very heat-resistant, so that the classical method for enumerating spores in cream, i.e. by heating it to 80 °C for 15 min prior to plating, may kill a large proportion. For the enumeration of *B. cereus* spores, therefore, the cream should be held at 80 °C for only 1 min, and then the cream plated as usual using surface culture or spread plates. Although some resistant cocci (such as *Sarcina lutea*) and resistant rods (such as *Microbacterium* or *Corynebacterium*) may survive and form colonies, there should be no difficulty in distinguishing colonies of *B. cereus*. The colonies have a characteristic appearance and produce rennin-type and lecithinase enzymes. An egg yolk medium may be used for the detection of lecithinase production (Stone and Rowlands, 1952). Cox (1970) used 2·5 % concentrated egg yolk emulsion (Oxoid) in milk agar (28 h at 30 °C) to identify *B. cereus* by the clear zones round the colonies; other types of bacteria producing the lecithinase reaction could be distinguished by colony appearance. Vegetative cells of all spore-formers are killed by the severe heat treatment usually given to cream, so that only the spores are of interest.

A point often not realised is that some aerobic spore-formers (*Bacillus*) are psychrotrophic. In Australia, Coghill and Juffs (1979*b*) found psychrotrophic spore-formers in 31 % of pasteurised milks and creams, mostly *B. cereus* but also *B. coagulans, B. licheniformis, B. megaterium* and *B. firmus*. An important observation was that spores could germinate and the organisms grow at temperatures in the range 1–7 °C; most could grow at 1–4 °C, and some could reduce methylene blue.

Bacillus subtilis, and other members of the *Bacillus* genus which form very heat-resistant spores, can be the cause of bitterness in sterilised cream. The spores may survive the autoclaving and germinate to attack the fat and protein, producing peptides, etc., and free fatty acids. In addition to producing a taint, the bacteria can bring about a thinning of the cream by breaking down the fat and the casein; other defects are clots and gas production. Nichols (1939) gives detailed information on the types of *Bacillus* responsible. Destruction of these spores by heat may be impracticable without affecting the quality of the cream, as some can survive 120 °C for 40 min (Nichols, 1938, 1939). The remedy is to use only milk of low spore content (less than 10 per ml), to practise the greatest care in hygiene in separation, standardisation and processing, and, in particular, avoid holding the cream or milk warm for more than a few minutes, and to ensure efficient sterilisation by checking thermometers and time–temperature conditions *in* the cans. It is not sufficient merely to record autoclave temperatures.

Anaerobic spores (*Clostridium*) are not normally of importance in cream and desserts.

Pathogens

With efficient hygiene and modern technology, pathogens are rarely if ever found in heat treated cream and dairy desserts, and so food poisoning from these is now virtually unknown. However, cream was formerly sometimes a cause of problems, and salmonellae or toxigenic staphylococci were usually responsible. Raw cream should always be regarded with suspicion.

Cream may be used in the manufacture of other foods, and heat treatment of the cream may then not be so efficient. The survival of pathogens is of the greatest importance in food manufacture, and high fat in foods may give an increased resistance to otherwise heat-sensitive bacteria.

Salmonellae

Konečný (1978) found that it was necessary to heat 12 and 33% fat creams to 80 °C for 30 min to kill salmonellae; none survived 10 min at 85 °C. For the detection of salmonellae, selenite and/or tetrathionate broth are generally used for enrichment, followed by streaking on desoxycholate, bismuth sulphite, XLD or Hektoen agar (Difco, 1971; Oxoid, 1979). Presumptive colonies are checked for biochemical reactions and finally tested by slide agglutination against appropriate antisera (FDA, 1978).

Rappold and Bolderdijk (1979) found that lysine iron agar gave better isolation results than the other, more usual, media tested. Their modified medium was: lysine iron agar (Oxoid), 34 g; bile salts, 1·5 g; lactose, 10 g; sucrose, 10 g; sodium novobiocin, 15 mg; and de-ionised water, 1 litre.

Staphylococci

After the salmonellae, staphylococci are usually the most important cause of food poisoning from dairy products. Like *Salmonella*, coliforms and *Pseudomonas*, staphylococci are killed by pasteurisation so they should never present a hazard in heat treated cream. This type of food poisoning arises from a toxin produced by the growth of some staphylococci in an unrefrigerated, non-acid food. As toxigenic staphylococci are a common cause of bovine mastitis and are carried by many humans on skin, nose and throat, raw cream can easily become infected, and if not promptly refrigerated, the cream can reach a count of 10^6 ml^{-1} in some hours, at which level the amount of toxin produced can cause food poisoning. The term '*Staphylococcus*' may be limited to toxin-producing types, and in practice, usually refers to coagulase-positive (CP) *Staph. aureus*. The coagulase test is highly correlated with toxin production, and is routinely used for identifying food-poisoning types. In our experience, about 70% of

Staph. aureus strains isolated from dairy products are coagulase-positive, but only about 10% of the *citreus* type, and none of the *albus* type give a positive reaction.

The simplest medium for routine enumeration is sodium chloride (6·5%), mannitol (1%), yeast autolysate agar, which is basically similar to Chapman's medium (Oxoid, 1979). Surface culture is essential. Incubation should be 1–2 days at 37°C followed by 1–2 days at room temperature. *Staphylococcus aureus* can be recognised by the characteristic dome-shaped, golden-orange colonies. The definite identification of toxigenic strains is expensive. The Baird-Parker (1962) medium reaction is claimed to be nearly 100% correlated with coagulase production, so that this is the most accurate practicable method for detection of toxigenic strains. Lachica (1980) has described a procedure for the enumeration of healthy and stressed cells of *Staph. aureus* in 29 h.

Staphylococcus aureus grows better and produces more Enterotoxin A in skim-milk and milk than in cream, and it appears that milk fat depresses growth and toxin production (Ikram and Luedecke, 1977).

Other Pathogens

Roberts and Gilbert (1979) studied the growth of non-cholera vibrios in cream and other foods. At suitable temperatures, a count of 10^8-10^9 g^{-1} could be obtained in 6–12 h, counts probably high enough to cause food poisoning. All were easily killed at 55°C.

In Australia, Hughes (1979) isolated 48 strains of lactose-fermenting *Yersinia enterocolitica* from pasteurised cream and other dairy products. Although apparently not pathogenic for man, their presence in pasteurised products and growth at 3°C may be a matter for concern.

Mycobacterium tuberculosis and *Brucella abortus* were formerly important pathogens for cream, but the eradication of the former in the UK and the almost universal heat treatment of cream and milk have removed these hazards.

Otte *et al.* (1979*b*) have described miniaturised test systems for the culture and identification of pathogens in milk products.

Enzymes

Phosphatase

This enzyme is present in all living tissues, and as the alkaline phosphatase is destroyed in milk by pasteurisation, this destruction has formed the basis of a statutory test in nearly all countries (HMSO, 1963). However, some precautions are necessary in applying the test to cream because of possible reactivation (Wright and Tramer, 1953, 1954, 1956). The AOAC (Association of Official Agricultural Chemists) method for differentiating residual and reactivated phosphatase can be used for milk and cream prepared from it (Serebrennikova *et al.*, 1978). These authors have described a modification which can demonstrate the absence of residual phosphatase in cream from milk pasteurised at 95 °C for 15 s. The relative increase in activity of the enzyme in the presence of Mg^{2+} can be used to distinguish reactivated and residual phosphatase (Kleyn and Ho, 1977).

Lipases and Proteinases

Raw milk contains a very variable amount of natural lipase and other enzymes. In addition, some organisms produce lipase, and psychrotrophic bacteria can grow slowly in milk or cream held at 5–10 °C and produce enzymes which may survive pasteurisation. Proteinases may also be produced, and survive. In addition, oxidising enzymes or systems may be present and work in association with the lipases. It is often difficult to identify the specific responsibility of these enzymes for chemical changes in cream, but the overall effect may be a peculiar taint which is difficult to describe, but which effectively renders the cream inedible. Some cows may produce milk rich in lipase under certain conditions, such as certain stages of lactation, and slow cooling intensifies the effect of any enzymes. Oxidative enzymes may also be present, and soluble copper can readily induce an oily taint, especially in the presence of oxygen, and these reactions are intensified by direct sunlight.

The best solution to this problem is to perfect the hygiene in all stages of processing, keep the milk and cream as cold as possible, and carry out the separation, homogenisation and heat treatment as quickly as possible in rapid sequence.

The Survival of Enzymes After Heat Treatment

With raw milk and cream, and with heat treated products before the introduction of the UHT method, it was assumed that all failures in keeping quality (other than obvious chemical taints) were caused by the growth of micro-organisms, and the action of enzymes as such was ignored. However, the development of bulk collection, and the UHT process which kills all bacteria, have served to draw attention to the importance of certain enzymes which can survive heat treatment. Thus in UHT-treated cream, peculiar flavours may slowly develop, and the type and intensity of such flavours will depend upon the initial content of enzymes in the milk.

Tests for the Measurement of Keeping Quality

The problem of measuring (or predicting) keeping quality of milk products has been studied since the beginning of dairy bacteriology. One very obvious point was missed for many years, and that is that the tests should be based on the method of storage and distribution of the product. Plate counts and methylene blue tests at 37 °C have little application for milk and cream held at 5 °C. Resazurin and methylene blue tests at 18 °C take three times as long, but give much better correlation with keeping quality (Davis, 1959a).

Moseley (1975) stored the product at 7 °C for 7 days and then made a bacterial count, the total time required being 10 days. Felmingham and Juffs (1977) found that this test was superior to the MB test. Blankenagel (1976) and Freeman et al. (1964) described 'inhibitory tests' in which the sample was incubated with sodium desoxycholate and then tested. Coghill and Juffs (1979a,b) compared a number of such tests, and found good correlations on the whole, and they recommended a modified Moseley test (4 days incubation) with a standard of 500 000 cfu ml⁻¹ or the Freeman test with a standard of 100 000 cfu ml⁻¹.

Dye Reduction Tests

It has long been recognised that dye reduction tests do not detect 'dirty milks' if they are refrigerated immediately after production, and they are over-sensitive if refrigeration is inadequate and atmospheric temperatures are high. The same result is found with cream. Bacteriologists have attempted to overcome this difficulty by various forms of 'temperature compensation' devices, such as the MB test for cream. None of these methods has been found entirely acceptable; indeed the very basis of these tests is unsound. If milk or cream is efficiently refrigerated (5 °C), a satisfactory KQ or shelf-life is usually obtained and the sample passes the dye tests. Failure to ensure adequate refrigeration generally results in a short KQ and failure to pass laboratory tests. Many workers in this field have drawn attention to discrepancies between various types of test, especially if samples are incubated at an elevated temperature. It is not always realised that different bacteriological tests measure different properties. The only sound approach is to use a specific test for measuring a specific property. Thus, the colony count and coliform tests are reliable measures of cleanliness in production and processing. A dye reduction test at a particular temperature can usually give an accurate measure of KQ *at that temperature* (Hiscox et al., 1932; Davis 1959a, p. 151), but a dye test after incubation at one temperature is of little use for assessing either the initial type of contamination or the likely KQ at a different temperature.

The PHLS Methylene Blue Test

This test is an adaptation of the statutory test for designated milks (HMSO, 1963), but has no legal standing for cream. It was adopted by the Public Health Laboratory Service (PHLS, 1971) as a screening test for creams of unsatisfactory bacteriological quality. As such it can serve a useful purpose, but unfortunately its significance is often misunderstood. Some think it is a legal test and, because it is used by the PHLS, has a public health significance. This is not true. The test simply measures the number of bacterial cells which can reduce methylene blue, and thus gives a weighting to those which ferment lactose such as lactic acid bacteria, coliforms, staphylococci and certain thermoduric and spore-forming organisms. The PHLS Working Party tested 4385 heat treated, 282 clotted and 517 untreated cream samples (HMSO, 1970). *Staphylococcus aureus* was isolated from 54 samples of untreated and five of heat treated cream, and of other pathogens, only one each of *S. typhimurium*, *Br. abortus*, *E. coli* and *Cl. perfringens* were isolated. From the public health point of view, cream in Britain can be considered satisfactory.

The samples were also examined by the methylene blue, colony count, coliform and *E. coli* type 1 tests. The PHLS chose the methylene blue test as a screening test because it was cheap and easy to carry out. Samples were held at 20 °C overnight (from 5 p.m. to 10 a.m. next day, i.e. 17 h) and then tested by the MB test at 37 °C. The following gradings were proposed

Not decolourised in 4 h — satisfactory
Decolourised in 0·5–4 h — fairly satisfactory
Decolourised immediately — unsatisfactory

The PHLS found that, although in general the agreement was good, there were anomalies when the MB results were compared with the count and coliform tests and, for this reason, emphasised that samples giving an unsatisfactory result should be further examined. The general findings of earlier workers were confirmed (Colenso *et al.*, 1966; Barrow and Miller, 1967; Jenkins and Henderson, 1969).

Few samples graded as satisfactory had counts above 100 000 cfu ml^{-1} or coli in 0·1 ml. Untreated creams were clearly worse bacteriologically than heat treated or clotted creams. As might be expected, counts at 4, 20 and 30 °C were often much higher than those at 36 °C, especially when *Pseudomonas* predominated. Large dairies, which had mechanical fillers, usually had better results than small dairies which usually filled cartons by hand.

When cream samples gave a low count at 37 °C but failed the MB test, incubation of the plates at 4 or 20° often gave a high count. An explanation for the converse was not found, but almost certainly the explanation would be that biochemically inactive organisms predominated in the microflora.

Rothwell (1969) observed that creams which failed the MB test were nearly always 'off' in both odour and taste after 7 days at 5 °C, and creams which did not reduce the dye in 4 h were nearly always in good condition.

Cox (1970) used a 'full cream' modification of the MB test in his study of bacterial factors affecting the result of this test. Like other workers, he found discrepancies between the MB results and plate count and coli figures, poor MB results with creams of good bacterial quality being most frequent in July–September. In his pure culture studies he found that *B. cereus* was most active, a count of $100\,000\,ml^{-1}$ being correlated with an MB time of 1 h. Of other species from cream, *Aerobacter*, *Achromobacter* (or *Acinetobacter*), *Pseudomonas*, *Micrococcus* and *Corynebacterium* were appreciably less active, counts of 10^7–$10^8\,ml^{-1}$ being necessary to give an MB time of 1 h.

Cox studied the frequency distribution of types of bacteria in cream on receipt, and after 17 h incubation at 20 °C. He found that 40 % contained coliforms and 69 % oxidase-positive bacteria, and although *Corynebacterium* tended to dominate the flora in fresh samples, the genus was quickly overgrown by other types. Conversely, *B. cereus* did not predominate in fresh samples, but did in 36 % after incubation, but strains appeared to vary in their power to reduce MB. Rather surprisingly coliforms did not dominate the flora of any creams after incubation, and *E. coli* did not appear to influence the MB test. Although not active biochemically, *Acinetobacter* (Thornley, 1968) had a marked effect on the MB test; an effect attributed to the high initial counts.

It is surely significant that although the MB test has been known and extensively used in the dairy industry for 80 years, the most reputable companies still prefer the colony count and coliform tests for the assessment of hygiene, expected keeping quality, and quality control in general for milk and cream.

MICROBIOLOGICAL STANDARDS

The Dynamic Nature of the Microbiological Population of a Food
In the microbiological control of foods, it is important to realise at all times that whereas the chemical composition of a food is (for all practical purposes, apart from loss of moisture) a *static* property, the microbiological composition is a *dynamic* property. In other words the microbial population is always changing, and the various types of organisms present will change in numbers in various ways. Foods may be placed in three groups from this point of view.

1. Foods in which bacterial numbers increase at room temperatures (15–25 °C). Those foods in which the numbers increase rapidly are commonly called 'perishable foods', and cream and dairy desserts come into this group.
2. Foods in which bacterial numbers are approximately static. Some solid foods, and foods high in sugar, acid, salt, etc., come into this category; sweetened condensed milk is an example.
3. Foods in which micro-organisms steadily die out. Foods very low in moisture or very high in sugar, salt, acid, etc., come into this category, and milk powder is a typical example. Resistant organisms such as some cocci may persist for a long time, and spores survive almost indefinitely.

As might be expected for perishable foods, counts reported in the literature vary widely. In German whipping cream, Otte (1979*b*) found total counts up to 719×10^6 ml^{-1}, psychrotrophs up to 100×10^6 ml^{-1}, pseudomonads up to $6\cdot3 \times 10^6$ ml^{-1}, coliforms up to 790 ml^{-1}, fungi up to 50 ml^{-1} and virtually no enterococci. They also gave microbiological findings for milk drinks, puddings, desserts, fresh cheese 'sweets' and sweet yoghurt preparations.

Microbiological Standards
The standards shown in Table XI have been suggested for cream at the point of retail sale by various workers or organisations.

Bacteriological standards for cream in other countries have been laid down as follows:

Northern Ireland—Untreated: count less than $50\,000$ g^{-1}. Pasteurised: no coliforms in 1 g.

TABLE XI
SOME SUGGESTED MICROBIOLOGICAL STANDARDS FOR CREAM AT THE POINT OF
SALE TO THE CONSUMER

	Colony counts per gram		
	Satisfactory	Doubtful	Unsatisfactory
Total colony count			
SDT (1975)	<10 000	10 000–100 000	>100 000
Davis (1968)	<10 000	10 000–100 000	>100 000
Jackson (1978)	<50 000	50 000–250 000	>250 000
Jackson (1978)(MB test)	2·5–4 h	–2 h	0 h
Coliform bacteria			
SDT (1975)	<10	10–100	>100
Davis (1968)	<10	10–100	>100
Jackson (1978)	Absent in 0·1 g	Present in 0·1 g	Present in 0·01 g

Canada   —Count less than 50 000 g^{-1}; no coliforms in 1 g;
             phosphatase negative.

Sweden   —Count less than 100 000 g^{-1}; coliforms less than
             10 g^{-1}; aerobic spores less than 100 g^{-1}.

Maximum counts for cream and other dairy products have been laid
down in Italy (1979). The tests include total count, coliforms, salmonellae
and *Staph. aureus*.

Bakers' Cream
When a dairy manufactures cream for bakers, catering or food manufacturing establishments, the cream may be bought on both chemical
and microbiological specifications. For the latter, a total bacterial count
of 2000 to 5000 cfu g^{-1} with coli, yeasts and moulds less than 10 g^{-1}
are typical of good practice.

When cream is used for bakery products it is likely to be the cause
of failure in any official microbiological examination of the food. In
Germany, Büning-Pfaue (1978) found that over 80% of the bakery
goods sampled failed the standards set by the Swiss Food Manual;
the percentages failing on specific tests were: total count, 25%; coliforms,
74%; *E. coli*, 19%; yeasts, 35% and coagulase-positive *Staph. aureus*,
<1%. Nevertheless, microbiological standards can be very useful for
informal purposes, for example in a contract between buyer and seller.
However, any attempt to lay down *legal* microbiological standards

for perishable foods would lead to hopeless complexities in legislation and enormous confusion in practice (Davis 1963, 1969*a*). Innumerable technical problems would arise.

Organoleptic or Sensory Examination
Whatever may be the results of any laboratory examination, the ultimate effective criterion is acceptability by the consumer. Regular sensory testing by a panel of at least three experienced tasters should be an essential feature of quality control in any dairy. In practice any unsatisfactory condition in cream is soon noticed by customers, but it is better for the dairy to detect any taint before distribution.

An elaborate, statistically controlled system is not neccessary for daily routine control, but it is important that all tasters are checked for ability to smell odours and detect abnormal tastes. This is easily done by forming an experimental panel to examine a number of creams over a week. Each member should assess visual appearance, colour, smoothness, odour and taste independently, using a code drawn up by the head of the laboratory. It is usually found that most people agree broadly on the quality of a cream, but a minority may differ by being less able or even completely unable to detect odours and abnormal tastes. These persons should then be excluded from the working panel. Even highly skilled laboratory workers may come into this category.

No attempt should be made to place any specific quality in more than six grades, five marks indicating excellence and zero marks indicating extremely unpleasant or bad.

Legal Aspects
All sales of food are controlled by the Food and Drug Act (1955) and various regulations in the UK, and by official statutes in most developed countries. In addition to chemical standards, there may be microbiological standards for cream and dairy desserts, and also regulations in respect of the use of preservatives, anti-oxidants, emulsifiers, colours and other food additives. The only preservative permitted for cream in the UK is nisin (an antibiotic) for clotted cream.

REFERENCES AND BIBLIOGRAPHY

ADMI (1971) *Standards for Grades of Dry Milks.* American Dry Milk Institute, Chicago.

AIREY, F. K. (1978) *J. Soc. Dairy Technol.*, **31**, 148.
ALAIS, C., LORIENT, D. and HUMBERT, G. (1978) *Ann. Nutrit. Aliment.*, **32**, 511.
ANON. (1979) *Dairy Ind. Internat.*, **44**, 21.
ANON. (1980) *Dairy Ind. Internat.*, **45**, 5.
AOAC (1980) *Official Methods of Analysis*. Association of Official Agricultural Chemists, Washington, DC.
BAIRD-PARKER, A. C. (1962) *J. Appl. Bacteriol.*, **25**, 12.
BARNES, E. M. (1959) *J. Fd Sci. Agric.*, **10**, 656.
BARROW, G. I. and MILLER, D. C. (1967) *Monthly Bull. Min. Health and Public Health Lab. Service*, **26**, 254.
BARROW, G. I., MILLER, D. C. and JOHNSON, D. L. (1966) *Lancet*, **2**, 802.
BLANKENAGEL, G. (1976) *J. Milk Fd Technol.*, **39**, 301.
BOARD, P. W. and STEEL, R. J. (1978) *Fd Technol. Australia*, **30**, 169.
BRAZIS, A. R., MESSER, J. W. and PEELER, J. T. (1972) *J. Milk Fd Technol.*, **35**, 730.
BROWN, J. V., WILES, R. and PRENTICE, G. A. (1979) *J. Soc. Dairy Technol.*, **32**, 109.
BROWN, J. V., WILES, R. and PRENTICE, G. A. (1980) *J. Soc. Dairy Technol.*, **33**, 78.
BSI (1968) *Methods of Microbiological Examination for Dairy Purposes*, BS 4285. British Standards Institution, London.
BUCHANAN, R. E. and GIBBONS, N. E. (Eds.) (1974) *Bergey's Manual of Determinative Bacteriology*, 8th edn. Williams & Wilkins Co., Baltimore.
BÜNING-PFAUE. H. (1978) *Deutsche Lebensm.-Rund.*, **74**(2), 38.
CHAPMAN, G. H. (1945) *J. Bacteriol.*, **50**, 201.
COGHILL, D. C. and JUFFS, H. S. (1979*a*) *Australian J. Dairy Technol.*, **34**, 118.
COGHILL, D. C. and JUFFS, H. S. (1979*b*) *Australian J. Dairy Technol.*, **34**, 150.
COLENSO, R., COURT, G. and HENDERSON, R. J. (1966) *Monthly Bull. Min. Health and Public Health Lab. Service*, **25**, 153.
COUSINS, C. M., PETTIPHER, G. L., MCKINNON, C. H. and MANSELL, R. (1979) *Dairy Ind.*, April, **44**, 27.
COX, W. A. (1970) *J. Soc. Dairy Technol.*, **23**, 195.
COX, W. A. (1975) *J. Soc. Dairy Technol.*, **28**, 59.
CROSSLEY, E. L. (1948) *J. Dairy Res.*, **15**, 261.
CROSSLEY, E. L. (1955) In: *A Dictionary of Dairying*, 2nd edn. Leonard Hill, London, pp. 320, 322.
CZECHOSLOVAKIA (1976) *Czechoslovak Standard* CSN.57, 1143.
DAVIES, F. L. (1975) *J. Soc. Dairy Technol.*, **28**, 69.
DAVIS, J. G. (1931) *J. Dairy Res.*, **3**, 133.
DAVIS, J. G. (1935) *J. Dairy Res.*, **6**, 121.
DAVIS, J. G. (1940) *Zbl. Bakt. II*, **101**, 97.
DAVIS, J. G. (1955) *A Dictionary of Dairying*. Leonard Hill, London.
DAVIS, J. G. (1956) *Laboratory Control of Dairy Plant*. Dairy Ind., London.
DAVIS, J. G. (1959*a*) *Milk Testing*. Dairy Ind., London.
DAVIS, J. G. (1959*b*) *Proc. Soc. Water Treat. Exam.*, **8**, 31; *Dairy Ind.*, **25**, 828, 919.
DAVIS, J. G. (1963) *J. Soc. Dairy Technol.*, **16**, 150, 224.
DAVIS, J. G. (1965) *A Dictionary of Dairying. Supplementary volume*. Leonard Hill, London.
DAVIS, J. G. (1968) Dairy products. In: *Quality Control in the Food Industry* (Ed. Herschdoerfer, S. M.). Academic Press, London.

DAVIS, J. G. (1969a) *Lab. Practice*, **18**, 749, 839.
DAVIS, J. G. (1969b) *Dairy Ind.*, **34**, 555.
DAVIS, J. G. (1969c) *Med. Officer*, **122**, 115,
DAVIS, J. G. (1971) *Dairy Ind.*, **36**, 267.
DAVIS, J. G. (1972) *Dairy Ind.*, **37**, 212, 251.
DAVIS, J. G. and BELL, J. S. (1958) *Lab. Practice*, **8**, 58.
DHAMANGAONKAR, A. D. and BRAVE, O. (1978) *XXth Internat. Dairy Congr.*, E, 846.
DIFCO (1971) *Difco Manual*. Difco, Detroit.
DONOVAN, K. O. and VINCENT, J. M. (1955) *J. Dairy Res.*, **22**, 43.
DUKE, M. (1980a) *Lab. Practice*, **29**, 377.
DUKE, M. (1980b) *Med. Lab. World*, **4**(1), 33.
FDA (1978) *Bacteriological Analytical Manual*. Association of Official Agricultural Chemists, Washington, DC.
FELMINGHAM, D. M. and JUFFS, H. S. (1977) *Australian J. Dairy Technol.*, **32**, 158.
FLÜCKIGER, E. (1979) *L.T. (Switz.)*, **12**, No. 2.
FOSTER, E. M., NELSON, F. E., SPECK, M. L., DOETSCH, R. N. and OLSON, J. C. (1958) *Dairy Microbiology*. Prentice Hall, New Jersey.
FRANKLIN, J. G. (1969) *J. Soc. Dairy Technol.*, **22**, 100.
FREEMAN, T. R., NANAVATI, J. V. and GLENN, W. E. (1964), *J. Milk Fd Technol.*, **27**, 304.
GOTO, S. and ENOMOTO, S. (1970) *Jap. J. Microbiol.*, **14**, 65.
GRIFFITHS, M. W., PHILLIPS, J. D. and MUIR, D. D. (1980) *J. Soc. Dairy Technol.*, **33**, 8.
GRINSTEAD, E. and CLEGG, L. F. L. (1955) *J. Dairy Res.*, **22**, 178.
HAMMER, B. W. and BABEL, F. J. (1957) *Dairy Bacteriology*. Wiley, New York.
HANKIN, L. and DILLMAN, W. F. (1968) *J. Milk Fd Technol.*, **31**, 141.
HANKIN, L. and ULLMAN, W. W. (1969). *J. Milk Fd Technol.*, **32**, 122.
HARMON, L. G. and NELSON, F. E. (1955) *J. Dairy Sci.*, **38**, 1189.
HARRIGAN, W. F. and McCANCE, M. E. (1976) *Laboratory Methods in Food and Dairy Microbiology*. Academic Press, London.
HEDBERG, M. (1969) *Appl. Microbiol.*, **17**, 481.
HERSOM, A. C. and HULLAND, E. D. (1980) *Canned Foods*. Churchill Livingstone, Edinburgh.
HISCOX, E. R. (1934) *J. Dairy Res.*, **5**, 233.
HISCOX, E. R., HOY, W. A., LOMAX, K. L. and MATTICK, A. T. R. (1932) *J. Dairy Res.*, **4**, 105.
HMSO (1963) *The Milk (Special Designation) Regulations*. HMSO, London.
HMSO (1965) *The Milk (Special Designation) (Amendment) Regulations*. HMSO, London.
HMSO (1970) *The Cream Regulations*. HMSO, London.
HMSO (1979) *The Preservatives in Food Regulations*. HMSO, London.
HOBBS, B. C. and GILBERT, R. J. (1978) *Food Poisoning and Food Hygiene*. Arnold, London.
HUGHES, D. (1979) *J. Appl. Bacteriol.*, **46**, 125.
IKRAM, M. and LUEDECKE, L. O. (1977) *J. Fd Protect.*, **40**, 769.
ITALY (1979) *Gelatiere Italiano*, **15**(1/2), 81.
JACKSON, A. C. (1978) *J. Soc. Dairy Technol.*, **31**, 80.

JAYNE-WILLIAMS, D. J. and FRANKLIN, J. G. (1960) *Dairy Sci. Abstr.*, **22**, 215.
JAYNE-WILLIAMS, D. J. and SHERMAN, T. M. (1966) *J. Appl. Bacteriol.*, **29**, 72.
JENKINS, H. R. and HENDERSON, R. J. (1969) *J. Hyg. Camb.*, **67**, 401.
KIESEKER, F. G., ZADOW, J. G. and AITKEN, B. (1979) *Australian J. Dairy Technol.*, **34**, 21, 112.
KING, A. D., HOCKING, A. D. and PITT, J. I. (1979) *Appl. Envir. Microbiol.*, **37**, 959.
KLEEBERGER, A. (1976) *Molk. Z. Welt d. Milch.*, **50**, 1539.
KLEEBERGER, A. (1978) *Deutsche Milchwirtschaft*, **29**, 1380.
KLEYN, D. H. and HO, C. L. (1977) *J. Assoc. Offic. Agric. Chem.*, **60**, 1389.
KONEČNÝ, S. (1978) *Veterinářství*, **28**, 319.
KONIETZKO, M. and REUTER, H. (1980) *Milchwissenschaft*, **35**, 274.
LACHICA, R. V. (1980) *Appl. Envir. Microbiol.*, **39**, 17.
LANGEVELD, L. P. M., BOLLE, A. C. and CUPERUS, F. (1978). *Neth. Milk Dairy J.*, **32**, 69.
LOWBURY, E. J. L. and COLLINS, A. G. (1955) *J. Clin. Path.*, **8**, 47.
MILES, A. A. and MISRA, S. S. (1938) *J. Hyg. Camb.*, **38**, 732.
MINISTRY OF AGRICULTURE (1968) *Bacteriological Techniques for Dairy Purposes.* HMSO, London.
MOSELEY, W. K. (1975) *Dairy Ice Cream Field*, **158**, 44.
MOSSEL, D. A. A., VEGA, C. L. and PUT, H. M. C. (1975) *J. Appl. Bacteriol.*, **39**, 15.
MOSSEL, D. A. A., EELDERINK, DE VOR H. and KEIZER, E. D. (1976) *Lab. Practice*, **25**, 393.
MUIR, D. D., KELLY, M. E. and PHILLIPS, J. D. (1978) *J. Soc. Dairy Technol.*, **31**, 203.
NICHOLS, A. A. (1938) *Hannah Dairy Res. Inst. Rep. No. 75.*
NICHOLS, A. A. (1939) *J. Dairy Res.*, **10**, 231.
NORRIS, J. R. and RIBBONS, D. W. (Ed.) (1969) *Methods in Microbiology.* Academic Press, London.
OTTE, I., TOLLE, A. and HAHN, G. (1979a) *Milchwissenschaft*, **34**, 152, 213.
OTTE, I., SUHREN, G., HEESCHEN, W. and TOLLE, A. (1979b) *Milchwissenschaft*, **34**, 463.
OTTERHOLM, B. (1978) *Meieriposten*, **67**, 583, 664.
OXOID (1979) *The Oxoid Manual.* Oxoid Ltd, Basingstoke.
PETERKIN, P. I. and SHARPE, A. N. (1980) *Appl. Envir. Microbiol.*, **39**, 1138.
PETTIPHER, G. L. and RODRIGUES, U. M. (1981) *J. Appl. Bacteriol.*, **50**, 157.
PETTIPHER, G. L., MANSELL, R., MCKINNON, C. H. and COUSINS, C. M. (1980) *Appl. Environ. Microbiol.*, **39**, 423.
PIEN, J. (1977) *La Technique Laitière*, **904/905**, 7.
PUBLIC HEALTH LABORATORY SERVICE (1971) *J. Hyg. Camb.*, **69**, 155.
RAMUZ, F. J. (1979) *J. Soc. Dairy Technol.*, **32**, 187.
RAPPOLD, H. and BOLDERDIJK, R. F. (1979) *Appl. Envir. Microbiol.*, **38**, 162.
ROBERTS, D. and GILBERT, R. J. (1979) *J. Hyg. Camb.*, **82**, 123.
ROTHWELL, J. (1969) *J. Soc. Dairy Technol.*, **22**, 26.
SDT (1959) *In-place Cleaning of Dairy Equipment.* Society of Dairy Technology, London.
SDT (1966) *Pasteurising Plant Manual.* Society of Dairy Technology, London.
SDT (1975) *Cream Processing Manual.* Society of Dairy Technology, London.

SEREBRENNIKOVA, V. A., PATRATII, A. P., RASHKINA, N. A. and KRAVTSOVA, A. M. (1978) *Molochnaya/Promyshlennost No.* 10, p. 23.
SLANETZ, L. W. and BARTLEY, C. H. (1957) *J. Bacteriol.*, **74**, 591.
SØGAARD, H. and PETERSON, E. (1977) *Nordeuro. Mejeri Tids.*, **43**, 183.
STEWART, D. B. (1975) *J. Soc. Dairy Technol.*, **28**, 80.
STONE, M. J. and ROWLANDS, A. (1952) *J. Dairy Res.*, **19**, 51.
STORCK, W. (1961) *Whipping Cream—Sterilised Milk*. Hildesheim, Germany.
TEKINSEN, O. C. and ROTHWELL, J. (1974) *J. Soc. Dairy Technol.*, **27**, 57.
THATCHER, F. S. and CLARK, D. S. (Eds.) (1968) *Micro-Organisms in Food*. University of Toronto.
THORNLEY, M. J. (1968) In: *Identification Methods for Microbiologists*, Eds. Gibbs, P. A. and Shapton, D. A. Academic Press, London.
TRESSLER, D. K., VAN ARSDEL, W. B. and COPSLEY, M. J. (Eds.) (1968) *The Freezing Preservation of Foods*. AVI, Westport.
USA (1978) *Federal Register*, **43**, 4596.
VANDERZANT, C. and CUSTER, C. S. (1968) *J. Milk Fd Technol.*, **31**, 302.
WILCOX, G. (1971) *Milk, Cream and Butter Technology*. Noyes, New Jersey.
WRIGHT, R. C. and TRAMER, J. (1953) *J. Dairy Res.*, **20**, 177, 258.
WRIGHT, R. C. and TRAMER, J. (1954) *J. Dairy Res.*, **21**, 37.
WRIGHT, R. C. and TRAMER, J. (1956) *J. Dairy Res.*, **23**, 248.

3

Microbiology of Butter

M. F. MURPHY

Department of Dairy and Food Technology,
University College, Cork, Ireland

Butter has been made from milk from earliest times. Ancient manuscripts indicate that butter was used as a food in India around 2000 BC, and the product is mentioned in Genesis and other biblical writings. It seems probable that the practice of using butter for food was introduced into Western Europe via Scandinavia. Butter has long been an important article of commerce, and was being exported from Norway and Sweden as early as the 14th Century.

The keeping quality of butter compared to milk has been known since early times. In Ireland, it was stored in wicker baskets in bogs, and produce buried between 600 AD and 1800 AD is still found during peat harvesting. It is presumed that such burial kept the product cool and allowed desirable flavours to develop.

Originally, butter was produced directly from milk, and the method of churning usually involved a stationary churn with rotating agitators operated manually.

The development of the buttermaking industry dates from about 1850 when gravity settling of cream was practised, but the invention of the mechanical separator in 1877 heralded the real beginning of the butter factory. About 1890, pasteurisation of cream for buttermaking was introduced with consequent improvements in keeping quality.

The original churns were made of timber, and were fitted with rollers for 'working' the aggregated granules. In about 1935, stainless steel churns were developed, and this type gradually replaced the wooden churn, principally because of the ease of cleaning and sanitising.

The continuous buttermaking method developed in the 1940s, based on the Fritz principle, quickly gained wide acceptance in countries

91

manufacturing unsalted or lightly salted butter. However, it was not until the late 1960s that satisfactory salting methods were developed, and the continuous machine has now largely replaced the stainless steel churn in all major butter producing countries.

The production of butter on a world scale amounted to some 6·1 million tonnes in 1980, and manufacture is presently increasing at about 1% per annum. Europe manufactures about 50% of total production, while the nine EEC member countries account for one-third of world production. In the EEC, most of the annual increase in milk supplies is converted to butter. The three leading producers of butter on a world basis are the Soviet Union, France and the Federal Republic of Germany.

The world stocks of butter continue to rise, and butter is consequently being stored for increasing periods of time with additional demands on overall quality. Stocks of EEC butter, which presently comprise 50% of the world stocks, are very high, and amounted to 358 000 tonnes in 1980.

COMPOSITION

The almost universally accepted standard for water content in butter is 16% maximum. Some countries also specify for fat (usually 80% minimum, exceptionally 78% in the UK) and milk solids-not-fat (MSNF) (usually 2% maximum). The latter limit of 2% MSNF will not be exceeded by normal practice when manufacturing sweet cream butter, and the usual level is about 1·5%, depending on percentage SNF in the original milk. It is not common practice to legislate for the salt content in butter, though there are usually accepted consumer maxima in different markets.

Many countries allow sodium chloride and lactic acid cultures as the only non-milk additives in butter. Several countries, including the UK (though not Ireland), allow neutralisation of the cream and the addition of the colouring agents annatto, carotene and turmeric. It is normally not permitted to add preservatives, other than salt, or anti-oxidants to butter.

TYPES OF BUTTER

Traditionally, butter was manufactured from ripened cream produced by natural souring. Nowadays, ripened cream butter is made by the careful use of specific starter bacteria added to the cream, which produce lactic acid and various flavour compounds, notably diacetyl.

Many countries, particularly those of continental Europe, still favour ripened cream butter which is usually unsalted or, at most, salted to a level of 0·5%. In Ireland, UK, North America and Australasia, sweet cream butter is the preferred product, usually with 1·5–2% added salt. In exceptional areas (e.g. certain mining districts), higher salt levels are desired, sometimes as high as 3·5%.

In the recently developed Dutch method (the so-called NIZO technique), flavours are added to sweet cream butter after manufacture. In this method, the cream is not ripened, but is churned in the usual way and the sweet buttermilk drawn off. Subsequently, a mixture of the starters and culture concentrate is worked into the butter at the correct rate to yield butter with the characteristics of normal ripened cream butter made by the traditional process. Advantages claimed for the method include the production of sweet buttermilk, and lower copper levels in the butter (about 50% lower) due to the non-migration of copper from the serum to the fat; during normal ripening, extensive movement of copper into the fat occurs.

Milk and Cream Treatment
The main factors affecting butter quality have been reviewed by Murphy and Mulcahy (1965). The quality of the manufactured butter is obviously influenced by the nature and quality of the original milk. In the past, butter was often a means of disposal of milk of doubtful quality, but nowadays, with increased requirements for better quality and longer shelf-life, this is no longer true, and the improved methods and control of milk production and collection ensure a high quality raw milk for butter and other product manufacture.

Milk, after reception at the factory, is usually chilled and held at about 5°C before separation. In the case of continuous buttermaking, separation to give a fat content in the cream in the range 38–42% fat is usual, and it is important to select the value best suited to the fat quality and the equipment being used. Furthermore, the actual percentage of fat must be known (see section on Cream Cooling), and for butter manufacture in conventional stainless steel churns, the fat content of the cream is usually 45–50%.

The modern development of efficient, high-speed, automatic desludging and self-cleaning separators has significantly improved this operation. The milk temperature will critically affect the efficiency of separation, the ideal temperature being 46–49 °C. In some plants, the milk is first pasteurised and then cooled to separation temperature. In others, the milk is heated to separation temperature, and then the cream, and perhaps the separated milk, is pasteurised subsequently. In the case of the manufacture of ripened cream butter, the former procedure is considered undesirable since it may favour the migration of copper from the milk serum to the fat globules, and so increase the propensity of the subsequent butter to oxidation.

Cream after separation is normally pasteurised, even in those cases where the original milk was also pasteurised. The temperatures of cream pasteurisation vary, but in recent years, the trend has been towards high heat treatments for cream. In several countries, heat treatments between 88 and 93 °C are usual. The oxidative stability of cream has been shown to decrease progressively as temperatures increase in the range 60–95 °C. The pro-oxidant influence is associated with the migration of copper from the cream serum to the fat globule phase when cream is heated. However, as a minimum, cream for buttermaking should be pasteurised so that it is peroxidase-negative, and coliform bacteria are absent in 1 ml; Davis (1963) suggested the following standards for cream for buttermaking (Table I).

TABLE I
SUGGESTED STANDARDS (cfu ml^{-1}) FOR CREAM FOR BUTTERMAKING

	Satisfactory	Doubtful	Unsatisfactory
Yeasts	<1	1–10	>10
Moulds	<1	1–10	>10
Coliforms	<1	1–10	>10
Total colony count	<1000	1 000–10 000	>10 000

In addition to destroying bacteria and enzymes, pasteurisation of the cream can aid the flavour of sweet cream butter. High heat treatments may tend to accentuate the 'nutty' flavour, due, in part at least, to the formation of sulphydryls, which are often associated with high quality, sweet cream butter.

In certain countries, notably New Zealand, vacreation of the cream after pasteurisation is common. The technique is used, in particular, to rid cream of undesirable feed taints which are prevalent in some climates at certain times of the year. If such off-odours are absent, the use of deodorising systems may result in the removal of desirable flavours from the cream, and the indiscriminate use of vacreation is not recommended.

Cream Cooling
Sweet Cream Butter

The cream is cooled immediately after pasteurisation. This stage, properly carried out, is singly the most vital step in quality butter manufacture, and affects the physical, chemical and bacteriological properties of the final product. The cream is normally cooled quickly in the plate pasteuriser, and finally in the holding tank. While the rate of cooling can have certain effects, it is the final temperature and holding time which have the major influence on the subsequent manufacturing stages; cooling and ageing are equally important for both conventional and continuous buttermaking. Cream is normally aged at temperatures less than the churning temperature, and may be allowed to warm up to the latter temperature naturally, or may be warmed gently immediately prior to manufacture.

Cream is usually cooled and held at 3–5°C, and aged for an absolute minimum of 4 h (preferably longer) to allow an extensive fat crystal network to grow, and to enable stable crystal modifications to develop. This ageing stage has a considerable influence on the texture of the resulting butter.

For continuous buttermaking, the actual churning temperature is related to the hardness of the fat and the fat content of the cream. The appropriate temperature may be roughly gauged from the following relationships:

$$F + 2T = 56 \text{ in summer}$$

and

$$F + 2T = 58 \text{ in winter}$$

where $F = \%$ fat in the cream, and $T = $ cream temperature in °C (to the nearest 0·5°C).

These formulae are approximate, and can be no more than guides in determining the ideal conditions for individual machines.

In the case of conventional buttermaking, the churning temperature is not dependent on fat content of the cream, and the temperature is about 5–7°C (i.e. the ageing temperature). The actual temperature used will depend on such factors as the hardness of the fat (which is season-dependent), size of churn and pH of the cream. The temperature chosen is commonly that which results in the lowest fat losses into the buttermilk.

Ripened Cream Butter

After pasteurisation, the cream is cooled to 16–21 °C and inoculated with about 4 % of a mixed starter culture containing the acid-producers *Streptococcus lactis* and/or *Str. cremoris*, and the flavour-producers *Leuconostoc cremoris* and/or *L. dextranicum* and *Str. lactis* sub-sp. *diacetylactis*. As a general rule, the temperature of ripening will depend on the season, and may be 16–18 °C in summer or 19–21 °C in winter. Higher temperatures favour rapid ripening, while the lower levels result in easier cooling subsequently.

The ripening process is always carried out in two or three stages in order to facilitate the cooling of the highly viscous ripened cream. This is the so-called Alnarp process which, in addition to assisting cooling of the cream, can be used to modify the hardness characteristics of the final butter. The improvements on the latter are minor, but are worth achieving in ripened cream butter manufacture since a step-wise system is, in any case, essential to achieve cooling of the highly viscous cream. For sweet cream butter, the extra time and cost of step-wise cooling is usually not considered worthwhile. To produce a firm butter proceed as follows:

(1) cool the cream, after pasteurisation, to 19 °C and inoculate with the required amount of starter, and hold at 19 °C until the pH falls to 5·2;

(2) cool to 14–16 °C and hold for 2 h;

(3) cool to churning temperature.

To produce a soft butter proceed as follows:

(1) cool the cream to 6–8 °C after pasteurisation, inoculate with starter and hold for 2–3 h to form an intensive crystal network in the fat;

(2) carefully warm (using warm water at 25 °C) to 19 °C, this stage aims to melt the small crystals and create a network of predominantly large crystals, hold at 19 °C until the pH falls to about 4·9;

(3) cool to 15–16 °C;

(4) before manufacture, cool to churning temperature.

The actual churning process for ripened cream butter is similar to that for sweet cream butter, with the following notable exceptions:

(1) the churning time is considerably quicker because of the weakened fat globule membrane at the low pH;

(2) fat losses in buttermilk are lower due to more efficient de-emulsification and coalescence of the fat globules;

(3)   the churning temperature is usually slightly higher because of the high viscosity of the cream;

(4)   ripened cream butter is normally unsalted, and if salted, it is not usual to add in excess of 0.5%. If greater than this value, the butter will be particularly prone to oxidation, because of the combined effects of low pH and the catalytic action of the salt. If higher salting levels are desired, it is recommended that neutralisation to pH 6 (or above) be done.

General Principles Regarding Cooling/Ageing

With the advent of more efficient heat exchangers, plate cooling of cream is common. Rate of cooling will naturally affect the crystallisation process, but experience suggests that the holding time after cooling is much more significant than rate of cooling. Traditionally, cream was aged overnight, but, in modern practice, this is not always possible, and consequently shorter and shorter ageing times are sought. However, it should be considered a rigid rule that cream be aged for a minimum of 4 h. Ageing is best done at low temperatures (e.g. 5 °C), and consequently the cream must be heated-up prior to churning in the continuous system. The exception to this rule is the Alnarp process (see above), where the holding/ageing stage involves rather higher temperatures. In modern practice, cooling rates and holding temperatures are of little significance from a microbiological viewpoint, since bacterial counts are low and consist, mainly, of thermophilic organisms which present few problems, if any, in butter.

The rewarming of the cream prior to churning in the continuous machine may be done by warm water circulating in the cream tank jacket, or by means of an in-line heater *en route* to the continuous machine. In either case, care must be taken to accurately control the final temperature and the rate of heating. The temperature difference between the cream and heating water must be kept to a minimum in order to prevent changes in the crystalline state of the fat.

MANUFACTURE

Butter was traditionally manufactured by a batch method. Formerly, wooden churns were used, and these usually employed pairs of wooden rollers within the churn for working the product. Stainless steel churns were developed in about 1935, and these largely replaced the wooden

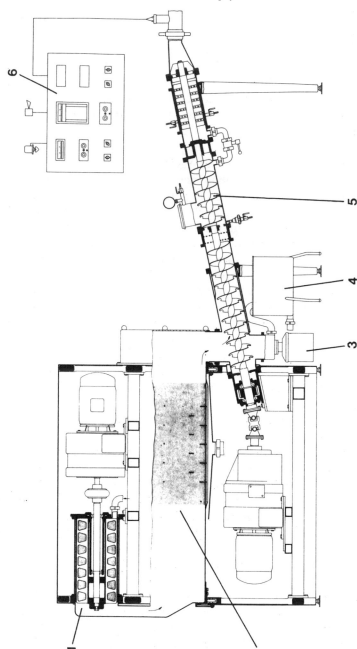

FIG. 1. Schematic diagram of the Westfalia continuous buttermaker: (1) primary churning cylinder; (2) secondary churning cylinder; (3) buttermilk clarifying device; (4) buttermilk vat and strainer; (5) texturising section; (6) moisture measurement and recording unit. Reproduced by courtesy of Westfalia Separator AG, Oelde, W. Germany.

churn, particularly for reasons of hygiene. Because of weight considerations, the metal churn did not employ rollers, but relied on the falling impact of the butter to achieve working. The size of churns varied considerably, and were designated by the number of boxes (formerly 56 lb, now 25 kg) capable of being made in one churning. The size ranged from about '10 box' to as high as '100 box' churns.

In general, the shape of the wooden churn was cylindrical—its shape not being a factor in working because of the rollers—while the stainless steel churn tends to be somewhat irregular in shape to facilitate working. Consequently various shapes, such as cubic and single or double conical types, were developed by different manufacturers.

Continuous buttermaking machines were developed during, and after, the Second World War, and three main types have evolved.

(1) Simple phase inversion: this is a concentration process whereby the cream (30–40% fat) is concentrated to between 80 and 82% fat depending on the desired composition of the final butter. The concentrated cream is then phase inverted from an oil in water (O/W) emulsion to a water in oil (W/O) type.

(2) The emulsification process also involves a concentration of the cream (30–40% fat), and during this stage the emulsion is broken and standardisation of the fat, water and salt contents is carried out. This is followed by re-emulsification, cooling and working.

(3) The 'accelerated churning' or Fritz process is the basis of a number of modern machines, including the Westfalia system. The system involves the use of high-speed beaters to form butter grains from the cream (40% fat). The buttermilk is then drained off, and the resulting grains, which may be washed, are worked (with or without salt addition) into butter.

The Fritz process has found general favour over the other continuous methods, since a product of suitable structure and consistency is produced under hygienic conditions. There is a definite trend in the major butter manufacturing countries, especially in the case of larger factories, to use the Fritz method in place of the conventional stainless steel churn, and most butter is now manufactured using the Fritz process.

Though the various Fritz-principle machines made by various manufacturers differ in several details from each other, they all consist of the following basic sections (see Fig. 1):

(1)   primary churning,
(2)   secondary churning,

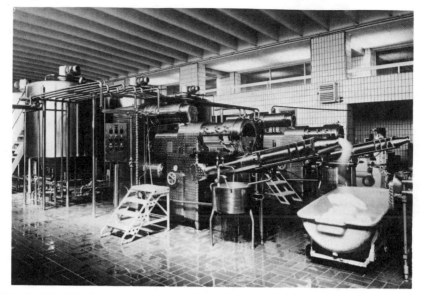

FIG. 2. The Westfalia continuous buttermaking machine is capable of producing 1·5 tonnes of butter per hour (minimum); the various stages of the process can be seen in Fig. 1. Reproduced by courtesy of Westfalia Separator AG, Oelde, W. Germany.

(3)   buttermilk drainage,
(4)   salt addition,
(5)   working.

The basic operation is as follows. The cream flows from the cream storage tank via a balance tank (to maintain a constant head), and is fed by means of a positive displacement pump of infinitely variable capacity to the rear of the primary churning section. The feed rate may be adjusted in the light of actual day-to-day manufacturing conditions (see Fig. 2). The primary churning section (see Fig. 1) consists of a horizontal cylinder with rapidly rotating beaters infinitely variable in the range 1000–3000 rpm. The beater speed is adjusted to give butter grains which form within the few seconds that it takes for the cream to pass through this section. The prechurned cream then falls by gravity into the secondary churning section, which also consists of a horizontal cylinder fitted with slowly rotating beaters (also variable) which complete the churning to give precisely the correct grain size. The grains and buttermilk pass into a continuation of this section which has a perforated base to allow drainage of the buttermilk, and the grains are gently tumbled to facilitate this separation. The grains may be washed by spraying with chilled water as they emerge from this section and drop into the working area. The working section is inclined slightly upwards which allows buttermilk drainage, and the grains are pushed gently forward by two counter-rotating augers. In some machines, the inclination of this section may be varied in order to vary the degree of drainage. The buttermilk drains off via a variable height 'goose-neck' or syphon, whose function is also to alter the degree of drainage if necessary. The gently-compacted grains now pass through a series of perforated plates which have rotating vanes between them, and virtually all the working is achieved by passage through these plates. Salt is added as a slurry after the first working plate. The end of the working section may be fitted with throttling plates which apply a back pressure on the butter, thereby controlling the degree of working.

Control of Final Moisture Content

The moisture content of continuously manufactured butter is largely controlled by the level of the 'goose-neck' and the speed of the augers in the working section. By changing auger speed, the drainage time for the buttermilk may be altered, while the degree of drainage which occurs will be affected particularly by butter grain size, which must be carefully controlled. In the case of unsalted butter, the final water content may be finely adjusted by water or buttermilk injection.

Salting

This stage is one of the most vital stages in continuous butter manufacture. In order to prevent salt getting into the buttermilk, the salt must be added as late as possible in the process. Experience shows that the butter reaching the salting stage will contain up to 14% moisture, which means that in order to salt to 2% and maintain the legal limit of 16% for water, the salt must be injected as a 50:50 salt:water slurry. Since salt is only 26% soluble in water, approximately half the salt added will be undissolved. In the subsequent butter, osmotic gradients between salt and buttermilk will tend to aggregate water droplets, so giving 'loose' moisture and mottling. However, by careful injection of the salt, by using a very finely ground salt (40 nm nominal particle size with none above 50 nm), and employing no more than 55% salt in the slurry, a satisfactory product may be produced. The salt/water mixture should be made up at least 30 min before churning begins, and should be violently agitated throughout; in this way, maximum

solution and distribution of the salt is assured. Nevertheless, 50%
of the salt added is in the form of a saturated solution dissolved
in approximately one-eighth of the overall butter moisture, the remaining
salt being added as undissolved particles. In the short period while
the butter is passing through the working plates, every endeavour is
made to distribute the salt as completely as possible, but it is to
be expected that some migration of salt and moisture will occur for
some time after manufacture. Though theoretically the moisture in
the butter represents a 12·5% salt solution (i.e. 2 in 16), which would
act as quite an efficient bacteriostat, the salt, in practice, will not
be perfectly distributed, and it is to be expected that little or no
salt will be present in some portions of the aqueous phase. Therefore,
the preservative effect of salt, while considerable, should be seen as
being less than would be expected from purely theoretical, gross quantity
considerations.

SNF Content in Continuous Buttermaking
Traditionally, the butter grains were washed using the conventional
churn, and some continuous machines are fitted with a means of spraying
chilled water on the butter grains as they fall into the working section
from the secondary churning section. Wash water is a potentially serious
source of contamination, and should be chlorinated effectively (see
section on Control of Water Supplies). With modern equipment, methods
and hygiene, it is usually deemed unnecessary to wash butter. The
effect of not washing is an improved yield and flavour, a reduction
in refrigeration and manufacturing costs, and the removal of perhaps
the most serious potential source of contamination. Since 2% milk
SNF is permitted in butter, and this limit is unattainable using normal
methods for sweet cream butter manufacture (the maximum is about
1·5% and depends on the SNF content of the original milk), additional
SNF is sometimes added in concentrated form. In addition, if the
solids in the buttermilk are for direct recovery, then especial care
must be taken to avoid possible contamination.

Air Content
Some continuous buttermaking machines provide a means whereby
the butter can be subjected to reduced pressure, thereby removing
air from the product. Reducing the air content has no effect on keeping
quality, and its only purpose is to produce a more dense, finely textured
butter.

After emergence from the continuous buttermaking machine, the butter is commonly transported on a slow-moving stainless steel conveyor from which it may be diverted for packaging. Careful consideration needs to be given to the hygiene of this section. In addition to microbial contamination from the air, contamination from ceilings needs to be considered as well as protection from light. It is recommended that the conveyor be fitted with a suitable cover, and should be cleaned regularly. Attention must also be given to cleaning and sanitising packaging equipment. In particular, difficulties often arise in cleaning high speed packaging lines, since the freshly cleaned machine may give rise to considerable start-up problems subsequently.

Cleaning

The general principles of cleaning and sanitising apply to all equipment and holding tanks for use in buttermaking, except in the case of the actual continuous buttermaking machine. The modern machine may be cleaned without dismantling, and after any remaining product has been removed by hot water or steam and collected, a detergent is circulated through the various sections.

The minimum of sticking of the butter to the churn, especially at start-up, is important for correct operation of the machine, and for this reason, all butter contact surfaces are sand-blasted which helps them to retain a fine film of water which reduces sticking. In addition, the use of an appropriate detergent will make a major contribution to reducing stickiness. Best results are obtained by circulating a cleanser/sanitiser containing silicate just prior to churning, and if this is followed by a cold water rinse, the non-stick effect of the silicate is not removed.

Ripened Cream Butter

The traditional method for the production of ripened cream butter was from cultured cream. This technique, while it produces butter of top class quality, suffers from certain disadvantages, e.g. problems with cooling the highly viscous ripened cream, and the production of a cultured buttermilk. Cultured buttermilk has limited uses, especially in those countries having no market for buttermilk, whereas fresh buttermilk may be used in a variety of products. To overcome these disadvantages, a new method (Veringa *et al.*, 1976) has been developed in which fresh cream is churned in the usual way, and a concentrated culture is added at the working stage. A cultured concentrate, containing 11–12% lactic acid and having a pH of 3, is added to the butter during working at a rate of about 0·6%, in addition to 2% of an aromatic starter culture. This addition lowers the pH of the butter serum to below 5·3. The aromatic starter used is added in sufficient quantity to give 1·5–2 mg kg^{-1} of diacetyl in the resultant butter.

This method of manufacture of ripened cream butter is gaining wide acceptance, and the manufacturers of continuous buttermaking machines are now making modifications to their equipment to accommodate the new technique.

PACKAGING

Butter may be packaged either in bulk, or in consumer-size containers. Traditionally, vegetable parchment was used to line bulk boxes, and also as a wrapper for consumer packs, and although it is still widely used, other more suitable materials are now available. These coverings include polyethylene film, aluminium foil, various plastics and compound laminates.

The replacement of parchment by polyethylene film for lining bulk boxes is now a well-established practice in several countries, and considerable savings in production costs can be achieved using this method. In addition, several other advantages accrue, including less contamination (e.g. it is virtually sterile and is copper-free) of the butter from the packaging materials. The recommended polyethylene is 0·001 gauge, food grade, low density, high impact film.

If parchment is used as a butter packaging material, care must be taken to minimise mould contamination of the product. Moulds will grow on parchment containing a high proportion of water-soluble organic matter, if humidity conditions are favourable, so that while it is important that proper handling and storage facilities be available for all wrapping materials, especial precautions are required for parchment.

Dry parchment or parchment treated in hot brine has often been used for bulk packaging. Calcium or sodium propionate may sometimes be utilised to treat the material, but any such treatment does not guarantee prevention of mould growth. It is normally recommended, therefore, that parchment be treated by immersion for 24 h in concentrated brine plus 0·5% sorbic acid, or failing this, that the butter containers should be lined with dry parchment.

In the case of high speed packaging of consumer size packs, there is no alternative to using dry wrapping material, and particular care in the handling and storage of the packaging material must be exercised. In particular, the store must be clean, and should be maintained at low humidity.

Aluminium foil and other laminates are gaining wide acceptance for consumer packs. Besides the attractive appearance of metal foil, it also provides a complete barrier to light, a fact that is important under the bright lighting conditions found in supermarket display cabinets.

From a microbiological and economic point of view, butter should be packed into the final consumer pack immediately on manufacture, without any intervening bulk packing and storage stage, for in this way, packaging costs and opportunities for contamination of the product are reduced. Perhaps even more significant in its possible effect on bacterial quality is the fact that repackaging after bulk storage almost invariably results in a redistribution and enlargement of water droplets in the butter, so increasing the danger of bacterial growth. All bulk butter should, after storage and immediately prior to repackaging, be vigorously remixed, for in this way, the semi-fluid state of freshly-made butter is restored, together with an even distribution of water. This reworking of 'set' butter improves plasticity, and has led to the practice of bulk packaging and holding before moulding into consumer packs.

In fresh butter with a varied microflora, deep cold storage will usually exert a selective killing effect, leaving only species tolerant of low temperatures, high salt concentrations and high acidity; in sweet cream butter the micrococci predominate, and in ripened cream butter, yeasts are dominant. Many of these contaminants are lipolytic, proteolytic and potentially taint-producing, but if reasonable care is taken (i.e. good moisture distribution, proper packaging and low contamination with micro-organisms), it is possible to manufacture butter which will withstand retailing without refrigeration.

CONTROL OF WATER SUPPLIES

The sanitary quality of water is often considered more important for food factories than for domestic use, because while pathogens cannot grow in clean water, spoilage organisms may do so. Contaminated water in a dairy can infect anything with which it makes contact, but of all dairy products, butter is perhaps the most vulnerable to spoilage by water bacteria, especially if the butter is washed. The most dangerous are those that are strongly lipolytic, proteolytic and grow at low temperatures. The following organisms in butter have often been traced to wash water: *Pseudomonas putrefaciens, Ps. fluorescens, Ps. mephitica* and *Ps. fragi.* These species, in addition to coliforms, are associated with defects in butter and can grow in water; and outbreaks of spoilage, especially surface taint, are often associated with contaminated water supplies.

The level of chlorination of water used for domestic purposes is usually inadequate for dairy use, since many spoilage organisms are more resistant to chlorine than are pathogens, and may survive the low chlorine treatment given to municipal water supplies. Dairy water can be chlorinated to high levels, even up to 20 ppm, with safety, and very high levels of chlorine (up to 200 ppm) in wash water have little or no effect on butter flavour.

Too little chlorine, in addition to not being effective in killing defect-producing organisms, can cause taints in butter through the formation of chlorophenols in the water. A medicinal flavour in butter may be evidence of such under-chlorination, and it is thus better to over-chlorinate than to under-chlorinate butter wash-water.

With improved methods of manufacture and hygiene, it is now usually considered unnecessary to wash the butter grains during manufacture, thereby eliminating this potential source of contamination. In continuous butter manufacture washing of the grains is not strictly possible. However, some machines are provided with a means of spraying the grains with chilled water as they leave the churning section, but this facility is not normally availed of. In good commercial practice, washing of butter grains would only be practised if (1) the cream had for some reason been inadequately cooled, or (2) cream having off-flavours was being churned. In the latter case, washing can only reduce the level of off-flavour if the causative agent is water soluble.

Davis (1956) suggested that a water supply for dairy purposes should be of the best bacteriological quality, and should have a hardness not exceeding 50 ppm calcium carbonate. He suggested the levels shown

TABLE II

| | Total colony count ml^{-1} | |
	$22°C$	$37°C$
Satisfactory	<100	<10
Doubtful	100–1 000	10–100
Unsatisfactory	>1 000	>100

in Table II as a guide to the bacteriological quality of water for dairy purposes.

AIR CONTAMINATION

Contamination of butter from the air can be a problem, especially in a factory using the batch process of manufacture. It is often considered that air contamination in a butter plant is more important than in any other dairy plant, because butter may be exposed rather extensively during packaging, and is frequently held for relatively long intervals under conditions conducive to the growth of organisms.

The contamination of the finished butter may be as harmful, or even more so, than that of either the cream or the butter during processing and, therefore, special care during handling and packaging is necessary. The air in the butter and packaging rooms normally carries mould spores originating from the outside air, or from its passage over contaminated surfaces as it passes indoors. The mould content of outdoor air is usually at a maximum at noon during the summer, and the dominant mould flora of the air is particularly harmful to butter. Contamination from the air is so important that the butter department should ideally be well ventilated by sterile filtered air, and be held under a slight positive pressure. The extent of mould growth on small packages depends partly on the degree of contamination of the air with mould spores, and partly on the cleanliness of the packing machine.

Air contamination can be especially serious with unsalted or lightly salted butter, since moulds develop more easily in such butters. Svedberg (1955) found that unsalted butter contaminated with spores of *Cladosporium herbarum* during manufacture developed a deep discolouration throughout the mass in 4–6 days, lightly salted butter in 6–10 days, and normally salted butter in approximately 3 weeks after removal from the cold store. If such discolourations occur in butter, then the manufacture of unsalted butter should be stopped until the source of the infection has been eliminated. It follows, therefore, that the site of the packing operations within the factory should be carefully chosen, and any likely source of mould contamination should be removed. In addition, an important factor in the extent of mould growth in small packages of butter may be water condensation on the surface of bulk butter exposed unduly to relatively moist air after removal from cold store.

PLANT LAYOUT AND HYGIENE

Careful consideration should be given to several factors regarding the layout and operation of butter manufacturing facilities. In processing areas where the risk of cross-contamination exists, separate rooms should be provided as far as possible under the following headings.

1. Individual room for actual butter manufacture.
2. Storage for working stocks—separate compartments—subdivided as follows:
    (i)   cleaning agents—detergents and sterilants;
    (ii)   packaging materials, e.g. fibreboard boxes, polythene liners, parchment and aluminium foil, etc.;
    (iii)   transport and storage requisites, e.g. slats, pallets, etc.;
    (iv)   additives, e.g. salt, colouring matter;
    (v)   utensils, e.g. buckets, brushes, routine cleaning equipment;
    (vi)   cold stores—no other product should be stored in a butter cold store in order to prevent possible flavour uptake by the butter;
    (vii)   toilets—these should be isolated from neighbouring areas by ventilated lobbies;
    (viii) dispatch area, loading docks and bays;
    (ix)   all services, e.g. boilers, refrigeration, maintenance, etc., should be in separate rooms.

The direction of prevailing winds should be taken into account in planning the inter-relationship of the various facilities, so as to avoid contamination. Compass orientation should be considered in planning the direction in which the butter manufacturing area should face, so as to avoid, as far as possible, excessive direct sunlight in the area, especially on the exposed product.

All floors in the manufacturing area should be constructed of impervious, dense-quality concrete. Falls, where needed, should be not less than 1:60 towards drainage outlets. Several floor finishes are suitable. Consideration should be given to the use of a 'sandwich' floor construction —a floor in two layers with a polythene isolating layer between—which is a suitable construction for tiling, and facilitates maintenance if a concrete finish is adopted.

Suitable finishes for floors include the following.

1. Compound screeds based on resins, and topped with a dense material resistant to hot water, acid and alkaline detergents.

2. Terrazzo.
3. Acid-resistant clinker tiles bedded on screeds of acid-resistant compounds, and pointed also with acid-resistant compounds.

The walls in the processing area are normally tiled to 4 ft 6 in above floor level, and topped with bull-nosed tiles.

Every possible effort should be made to avoid crevices and other areas, e.g. under equipment, which may harbour contamination.

Butter churns and other equipment are normally rinsed with hot water in order to melt off remaining butter, and washing is done using a phosphate-based detergent providing good wetting properties.

STORAGE OF BUTTER

Since butter is a predominantly fat product, which in turn is a good insulator, it is always difficult to cool butter quickly in cold storage. The thermal conductivity of butter is approximately $0.15\,\mathrm{BTU\,ft\,ft^{-2}\,h^{-1}}$ $°\mathrm{F^{-1}}$, and based on this value, a single box (25 kg) at 9 °C will take in excess of 3 days to reach $-8\,°\mathrm{C}$ at an ambient of $-10\,°\mathrm{C}$. Under practical conditions, the cooling rate of an individual box could be considerably slower, as it will usually be surrounded by other boxes.

Depending on the quality of the butter, the maximum time of storage will depend on the temperature of storage, and Schulz (1964) summarised this relationship as shown in Table III.

Due to the slow cooling rate of butter, bacterial counts will often increase, especially in unsalted butter, during the first few days of storage, and then decrease. Coliform organisms usually die out during cold storage. Little growth will occur in butter held below 0 °C, and

TABLE III
KEEPING QUALITY OF BUTTER

Temperature of storage (°C)	Very good keeping quality	Good keeping quality	Poor keeping quality
20	3 weeks	10 days	3 days
15	5 weeks	20 days	3 days
10	2 months	4 weeks	1 week
0	3 months	6 weeks	1–4 weeks
−12	9 months	6 months	1–3 months
−25	12 months	9 months	3–6 months

none would be expected at −15 °C, at which temperature, bacterial counts would be expected to decrease somewhat, especially in unsalted butter.

OFF-FLAVOURS

The flavour of butter may be regarded as a composite effect of the flavour of the fat, the aqueous serum and salt. The characteristic flavour of sweet cream and ripened cream butters are very different. However, it is generally agreed that butter should have a fresh, clean flavour. The term 'nutty' is often used to describe high quality sweet cream butter. The flavour of freshly made butter disappears gradually on storage and, after long-term cold storage, it is often the absence of objectionable flavours which is the criterion of evaluation. There are many undesirable off-flavours, and the more common forms have been described. It is important to distinguish between defects in freshly made butter and those in cold stored butter, since a slight defect in freshly made butter may be indicative that possibly very serious defects will arise if the butter were to be cold stored, whereas the same defect is not quite so serious after cold storage.

A flat or insipid flavour can be present in freshly made butter, and may be caused by: (1) excessive washing of the butter grains during manufacture, (2) dilution of the cream with water, or (3) the initial stages of bacterial deterioration. The appearance of a flat flavour a short time after manufacture indicates serious bacterial contamination, and the possibility of serious defects on storage. Good quality butter, after long-term storage, may also acquire a flat flavour, but this is not considered unusual.

A medicinal flavour in butter may be caused directly by the use of medicaments in treating the cow, but chlorine compounds added, either intentionally or not, to the milk or cream, are a more likely cause of this defect.

The main defects which develop in butter during storage are; (1) oxidative rancidity, (2) hydrolytic rancidity, and (3) putrefactive taint. The first two of these defects are due to the break-down of fat. The former defect is often called 'tallowy flavour', and is due to the oxidation of double bonds in the unsaturated fatty acids. The defect is catalysed by light and heavy metals, especially copper. Hydrolytic rancidity is caused by lipases, but as the lipases of animal origin are destroyed by pasteurisation, the defect is usually attributed to post-pasteurisation bacterial growth producing lipase which, in turn, hydrolyses the fat giving free fatty acids. The short-chain fatty acids (C_4, C_6 and C_8) will contribute most to the 'off-flavour' which hydrolysis produces. The prevention of hydrolytic rancidity in butter depends on

(1) low bacterial counts in the raw materials,
(2) minimal contamination,
(3) proper dispersion of moisture and salt in the butter and,
(4) low-temperature storage.

The breakdown of protein by putrefactive organisms gives rise to the so-called putrefactive taint. The presence of this defect indicates poor sanitation and manufacturing conditions. The organisms responsible (coliforms and *Pseudomonas* spp.) are readily killed by pasteurisation, and their presence in butter is usually traceable to unchlorinated water supplies.

REFERENCES

DAVIS, J. G. (1963) *J. Soc. Dairy Technol.*, **16**, 3.
DAVIS, J. G. (1956) *Laboratory Control of Dairy Plant*, Dairy Industries, London.
MURPHY, M. F. and MULCAHY, M. J. (1965) *A Review of some Factors Affecting Butter Quality*. An Foras Taluntais, Dublin.
SCHULZ, M. E. (1964) *Milchwissenschaft*, **19**(1), 9–17.
SVEDBERG, H. (1955) *Mejeritek. Medd.*, **16**, 6.
VERINGA, H. A., VAN DEN BERG, G. and STADHOUDERS, J. (1976) *Milchwissenschaft*, **31**(11), 658.

4

Microbiology of 'Starter Cultures'

A. Y. Tamime

West of Scotland Agricultural College,
Auchincruive, UK

The preservation of food by fermentation is one of the oldest methods known to mankind. A typical example is the lactic acid fermentation which is widely used during the manufacture of fermented dairy products, e.g. cheese, fermented milks (yoghurt, acidophilus milk, Ymer, Skyr, Filmjolk, cultured butter milk, etc.), sour cream, and cultured butter, or the lactic acid/alcoholic fermentation as in Kefir and Kumiss. Such a fermentation process is the result of the presence of micro-organisms (bacteria, moulds, yeasts or combinations of these) and their enzymes in milk. In the dairy industry, these organisms are known as starter cultures, and their essential roles are summarised as follows. First, for the production of lactic acid as a result of lactose fermentation; the lactic acid imparts a distinctive and fresh, acidic flavour during the manufacture of fermented milks; however, in cheesemaking, lactic acid is important during the coagulation and texturising of the curd. Secondly, for the production of volatile compounds (e.g. diacetyl and acetaldehyde) which contribute towards the flavour of these dairy products. Thirdly, the starter cultures may possess a proteolytic or lipolytic activity which may be desirable, especially during the maturation of some types of cheese. Fourthly, other compounds may be produced, for example alcohol, which is essential during the manufacture of Kefir and Kumiss. Fifthly, the acidic condition in these dairy products prevents the growth of pathogens, as well as many spoilage organisms.

Traditionally, milk was left to sour naturally prior to the manufacture of cheese and fermented milks. Such a method of production is not reliable, is prone to failure, may promote undesirable side effects, and the quality of the end product can vary tremendously. These drawbacks are basically due to a lack of scientific knowledge in the field of microbiology, i.e. starter cultures, but since the turn of the century, starter cultures have been widely studied, and their behaviour and metabolism are well established. Hence, the selection of starter cultures has become feasible, and their activity more predictable. Such an approach is important in factories handling large volumes of milk and, furthermore, greater uniformity in the quality of the end product can be expected.

Although the traditional method of manufacture was not scientifically controlled, it has provided the industry with the basic technology required today. At present, the manufacture of fermented dairy products is more centralised, and hence, starter cultures have become an integral part of a successful industry; their relevance is reflected by the economic value of the end-products. Bronn (1976) valued the world output of fermented dairy products at £7500 million. In view of the economic importance of cheese and fermented milks, the classification, main-tenance, preservation and propagation of starter cultures is of paramount importance.

ANNUAL UTILISATION OF STARTER CULTURES

There is no data available in the dairy industry concerning the actual amounts or types of starter culture utilised and/or produced every year. It is possible, however, to estimate the volume from world production figures of fermented dairy products.

According to the FAO (1979), world production of cheese was estimated at 10·5 million tonnes in 1978 (see Table I), of which 90% could be of hard cheese varieties and 10% of soft cheese; requiring approximately 94·5 and 15·75 million tonnes of liquid milk, respectively. Thus, a total of 110·25 million tonnes of milk was used for the manufacture of cheese and, on average, cheese starter cultures are inoculated at a rate of 1–1·5% (w/w) to milk. Hence, in 1978 it could be estimated that world production of cheese bulk starter cultures was in the region of 1·1–1·65 million tonnes.

World production figures of fermented milks, Kefir, Kumiss, sour cream and cultured butter are not available. However, Table II indicates the total consumption of yoghurt and other fermented dairy products in Western Europe, North America and the Soviet Union, which is

TABLE I
WORLD PRODUCTION OF CHEESE[a]

Continent	Production year		
	1976	*1977*	*1978*
Africa	339 686	344 387	352 056
North and Central America	2 140 455	2 165 850	2 109 758
South America	440 513	452 332	459 614
Asia	637 923	655 941	666 665
Europe	6 242 780	6 504 629	6 699 130
Oceania	213 360	187 292	196 490
Total	10 019 717	10 310 431	10 483 713

[a] All types of cheese.
Figures are in tonnes.
Adapted from FAO (1979).

in the region of 4·6 million tonnes. These fermented dairy products have a very short shelf-life, and they are not normally exported; hence, it is safe to assume that the data in Table II represent entirely the annual total production figures of these products in the countries concerned. If such an approach is accepted, then it is possible to estimate the volumes of different bulk starter cultures produced in 1977 as follows.

(1) Kefir and Kumiss are mainly manufactured in the Soviet Union and East European countries. Total production of these products could be estimated around 2 million tonnes (see Table II). The starter culture inoculation of Kefir and Kumiss is at a rate of 5 % and 10–30 %, respectively. Therefore, if an average of 17·5 % is considered, then 350 000 tonnes of bulk starter cultures were produced in 1977.

(2) The data in Table II under the column headed 'Others', but excluding the figures for the USSR, may represent the production of acidophilus milk, Skyr, Ymer, Filmjolk, etc. In 1977, the production was 926 000 tonnes requiring around 18 500 tonnes of bulk starter cultures, if the inoculation rate is estimated at 2 %.

(3) Nearly 2 million tonnes of yoghurt were manufactured in 1977, and under commercial practice, starter cultures are used at a rate of 2 %; hence, requiring 40 000 tonnes of yoghurt bulk starter.

A. Y. Tamime

TABLE II
TOTAL CONSUMPTION OF FERMENTED MILKS FROM HOME PRODUCED MILK
(1 000 tonnes)

Country	1975 Yoghurt	Others	1976 Yoghurt	Others	1977 Yoghurt	Others
Austria	25·0	29·2	28·2	16·6	30·8	15·7
Australia	13·9	—	13·9	—	20·0	—
Belgium	37·0	—	36·0	—	25·0	—
Brazil	67·0	—	67·0	—	69·0	—
Canada	16·2	—	20·5	—	28·5	—
Czechoslovakia	20·1	25·1	21·8	31·8	24·2	35·4
Denmark	29·9	35·8	33·4	44·8	38·7	44·8
Finland	29·5	136·8	30·4	132·0	31·8	131·6
France	(414·2)		(416·7)		(422·5)	
Germany	285·0	259·0	326·0	272·0	348·0	108·0
Ireland	(55·0)		(62·0)		(55·0)	
Israel	11·9	36·8	16·2	39·1	16·5	40·6
Japan	91·5	189·9	70·4	180·8	81·4	195·8
Luxemburg	1·2	0·1	1·3	—	1·4	—
Netherlands	193·4	—	204·8	—	204·5	—
Norway	4·9	31·6	5·6	30·1	7·0	30·7
Poland	(110·0)		(99·0)		(93·0)	
Spain	122·4	—	124·0	—	(158·6)	
Sweden	19·1	144·1	20·4	145·4	23·8	149·4
Switzerland	71·0	—	77·4	—	80·9	—
UK	67·3	—	68·0	—	75·0	—
USA	191·0	—	231·0	—	265·0	—
USSR	6·2	1 838·5	3·8	1 755·1	4·3	1 759·9
Total	4 609·6		4 625·5		4 616·8	

Figures in parentheses represent total consumption of all types of fermented milks.
After: IDF (1977, 1978, 1979).

The total volume of bulk starter culture production in the countries listed in Table II could, therefore, be estimated at 408 500 tonnes, and due to lack of data, the above estimates do not include countries like Bulgaria, Turkey and the Middle East, where yoghurt is very popular. However, *per capita* annual consumption of yoghurt in 1975 in Bulgaria was 31·5 kg (IDF, 1977), and in view of such evidence, it is safe to estimate that, in 1977, the world production of bulk starter cultures for fermented milks could be in excess of 0·5 million tonnes.

MICROBIOLOGICAL ASPECTS

Classification of Starter Organisms

As mentioned earlier, several micro-organisms (bacteria, yeasts, moulds or combinations of these) are employed in the fermentation process of milk during the manufacture of cheese and other fermented milk products. Such broad divisions could be used to classify the dairy starter cultures, and by far the most important group is the bacteria (i.e. lactic acid bacteria (LAB)) which include the genera *Streptococcus*, *Leuconostoc* and *Lactobacillus*. The classification of the LAB by Orla-Jensen (1931) is still considered the most universal method of classifying these organisms. According to *Bergey's Manual of Determinative Bacteriology, 7th edn* (Breed *et al.*, 1957), the LAB were grouped into one family, the *Lactobacillaceae*, which in turn was divided into two tribes: Tribe I, *Streptococceae* (spherical or ovoid in shape) and Tribe II, *Lactobacilleae* (rod-shaped bacteria). Now in the latest classification given in *Bergey's Manual of Determinative Bacteriology, 8th edn* (Buchanan and Gibbons, 1974), the LAB are divided into two separate families, i.e. the *Streptococcaceae* and the *Lactobacillaceae*.

In the dairy industry, interest is mainly focused on the genera *Streptococcus* (referred to as lactic streptococci), *Lactobacillus* and *Leuconostoc*. The relevance of these organisms as dairy starter cultures, including the lactobacilli, are as follows.

1. The genus *Streptococcus* is widely used in the cheese industry, and examples from this genus are: *Str. lactis*, *Str. lactis* sub-sp. *diacetylactis*, *Str. cremoris* and *Str. thermophilus*. All these organisms are homofermentative producing only lactic acid from glucose and, with the exception of *Str. thermophilus*, are classified as mesophilic bacteria. *Str. thermophilus* is a thermoduric organism, and it is used symbiotically with *Lactobacillus bulgaricus* as a yoghurt and cheese starter culture (Sharpe, 1979).

2. Of the genus *Leuconostoc*, only *L. cremoris* (previously known as *L. citrovorum*) and *L. dextranicum* are associated with dairy starter cultures. They are heterofermentative organisms capable of producing lactic acid, carbon dioxide and aroma compounds (e.g. ethanol and acetic acid) from glucose (Sharpe, 1979).

3. The genus *Lactobacillus* is divided into three main categories: *Thermobacterium*, *Streptobacterium* and *Betabacterium*. The former two sub-groups are homofermentative, and the species,

which are used in the dairy industry, are *Lac . bulgaricus, Lac. lactis, Lac. acidophilus, Lac. helveticus, Lac. casei* and *Lac. plantarum* (Sharpe, 1979). It is now accepted that *Lac. jugurti* is a mutant variant of *Lac. helveticus*, only lacking the ability to ferment maltose. However, *Lac. caucasicus*, which is sometimes reported to be among the other organisms in a Kefir grain, is now proposed as a rejected name (Rogosa and Hansen, 1971; Buchanan and Gibbons, 1974). The Betabacteria (*Lac. brevis* and *Lac. fermentum*) are heterofermentative, and they are not important as dairy starter cultures (Sharpe, 1979).

Other bacterial species are sometimes incorporated into a dairy starter culture, and these are as follows. First, *Str. faecium* which is used during the manufacture of modified Cheddar cheese in the USA (Sellars, 1967). Secondly, *Brevibacterium linens* (now accepted as closely related to *Arthrobacter globiformis*—Buchanan and Gibbons, 1974; Dagonneau and Kuzdzal-Savoie, 1978) which imparts a distinctive, reddish-orange colour to the rind of Brick and Limburger cheese (Olson, 1969). Thirdly, *Propionibacterium freudenreichii* sub-sp. *shermanii* is widely used in Swiss cheese varieties (Emmenthal and Gruyère), mainly for its ability to produce large gas holes in the cheese during the curing period. Fourthly, *Bifidobacterium bifidium* (used to be known as *Lactobacillus bifidus* (Buchanan and Gibbons, 1974)), which is found in the gut of infants, and is sometimes employed, with a yoghurt or acidophilus milk starter culture, for the manufacture of Bioghurt (a therapeutic fermented milk (Rasic and Kurmann, 1978)).

Moulds are mainly used in the cheese industry for the manufacture of some semi-soft cheese varieties. Their major role is to enhance the flavour and aroma, and to modify, slightly, the body and texture of the curd. The moulds can be divided into two types, taking into consideration their colour and growth characteristics, namely the white moulds (*Penicillium camemberti*, *P. caseiocolum* and *P. candidum*) which grow externally on the cheese, e.g. Camembert and Brie, and the blue mould, *P. roqueforti*, which grows internally in the cheese; examples of these 'blue cheeses' are Roquefort, Blue Stilton, Danish Blue, Gorgonzola, and Mycella. Other genera of moulds which have very limited application, or are traditionally used in some parts of the world, are *Mucor rasmusen*, used in Norway for the manufacture of ripened skim-milk cheese, and *Aspergillus oryzae*, used in Japan for the production of soya milk cheese varieties (Kosikowski, 1977).

The presence of yeasts in milk, beside the LAB cultures, results in a lactic acid/alcohol fermentation. This type of fermentation is limited to the manufacture of Kefir and Kumiss in the dairy industry. In the Kefir starter culture, often known as Kefir grains, the dominant yeasts are *Saccharomyces kefir* and *Torulopsis kefir* (Kosikowski, 1977; Manus, 1979). Other yeast species have been isolated from Kefir grains during the past half century, such as *Mycotorula kefir*, *Cryptococcus kefir*, *Mycotorula lactosa*, *Candida pseudotropicalis* var. *lactosa* and *Mycotorula lactis*; however, Uden and Buckley (1971) classified these organisms as synonyms for *Candida kefir*. On the other hand, the Kumiss starter culture, as reported by Kosikowski (1977), consists of *Torulopsis* spp. and *Kluyveromyces lactis* in conjunction with LAB.

It can be observed, however, that many varieties of micro-organism can be used as starter cultures in the dairy industry, and the overall classification is illustrated schematically in Fig. 1.

Terminology of Starter Cultures

The micro-organisms, which are used in the dairy industry, are used either singly, in pairs or in a mixture; thus, giving the industry the opportunity to manufacture different types of fermented dairy products. Table III illustrates the application and combinations of these micro-organisms which are required for the production of cheese, fermented milks, sour cream, cultured butter, Kefir and Kumiss.

Mesophilic lactic starter cultures (optimum temperature 20–30 °C) are widely used. In the cheese industry, for example, they are divided into three categories: single, multiple or mixed strains. In theory, a single strain starter consists only of one type of organism but, in practice, it is rarely used. However, single strains can be paired to safeguard against bacteriophage (phages) attack, intolerance of salt or cooking temperatures, and variation in the quality of the end product. An example of such a pairing is the New Zealand 'single strains system' where *Str. lactis* and *Str. cremoris* are used. The pairings are mainly based on phage relationships, and the rate of acid development, i.e. fast and slow. Crawford (1972) referred to the pairing of *Str. lactis* and *Str. cremoris* as 'the non-gas and non-aroma producing cheese lactic starters'. Multiple strain starter cultures consist of known numbers of single strains, so that the starter can be used for an extended period of time during the cheesemaking season. This type of starter type has been developed, tried and widely used in New Zealand (Limsowtin *et al.*, 1977; Turner *et al.*, 1979). A mixed strain starter is a combination

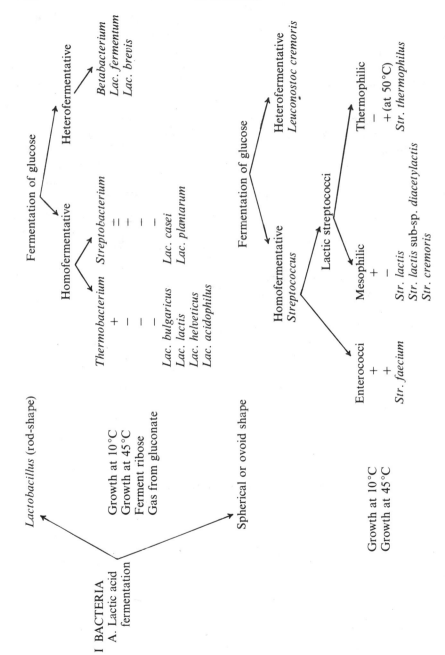

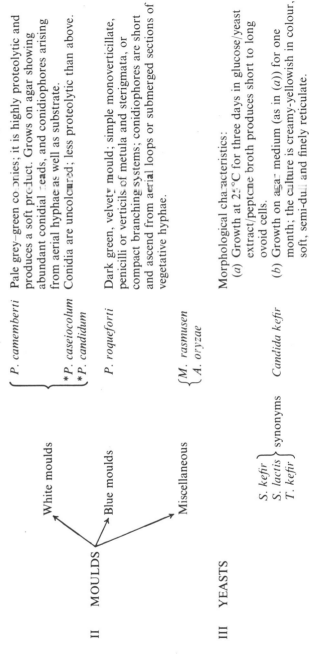

B. Miscellaneous
  Bre. linens
  P. freudenreichii sub-sp. *shermanii*

II MOULDS

White moulds ⟶ {
  P. camemberti Pale grey–green colonies; it is highly proteolytic and produces a soft product. Grows on agar showing abundant conidial heads, and conidiophores arising from aerial hyphae as well as substrate.
  *P. caseiocolum
  *P. candidum Conidia are uncoloured; less proteolytic than above.
}

Blue moulds ⟶ *P. roqueforti* Dark green, velvety mould; simple monoverticillate, penicilli or verticils of metula and sterigmata, or compact branching systems; conidiophores are short and ascend from aerial loops or submerged sections of vegetative hyphae.

Miscellaneous ⟶ {*M. rasmusen*
  A. oryzae}

III YEASTS

{*S. kefir*
S. lactis
T. kefir} synonyms *Candida kefir*

Morphological characteristics:
(*a*) Growth at 25°C for three days in glucose/yeast extract/peptone broth produces short to long ovoid cells.
(*b*) Growth on agar medium (as in (*a*)) for one month; the culture is creamy-yellowish in colour, soft, semi-dull and finely reticulate.
(*c*) Growth on corn meal agar; the pseudomycelium appears as coarse and slightly curved branches. Formation of blastospores is poor.

* *Penicillium caseiocolum* and *P. candidum* are probably biotypes of *P. camemberti*, and hence the name *P. camemberti* is the designation usually applied to the white moulds employed in the manufacture of Camembert and similar cheeses.

Fig. 1. Classification and differentiation of dairy starter cultures. Adapted from: Raper and Thom (1968); Uden and Buckley (1971); Sharpe (1979).

A. Y. Tamime

TABLE III
THE APPLICATION OF STARTER CULTURES AND ASSOCIATED ORGANISMS IN THE DAIRY INDUSTRY

Micro-organism	Product	Organisms present
Lactic starters		
Mesophilic	Hard-pressed cheese (Cheddar, Cheshire, Dunlop, Derby, Double Gloucester, Leicester, Gouda, Edam)	1, 2, 3, 4, 5[a] (refer to text for cheese starter terminology)
1. *Str. lactis*		
2. *Str. lactis* sub-sp. *diacetylactis*		
3. *Str. cremoris*		
4. *L. cremoris*[b]	Semi-hard cheese (Caerphilly, Lancashire)	1, 2, 3, 4
Thermophilic		
5. *Str. faecium*	Soft cheese (Quarg, Feta, Cottage, Cream cheese, Lactic curd)	1, 2, 3, 4
6. *Str. thermophilus*		
7. *Lac. bulgaricus*		
8. *Lac. lactis*		
9. *Lac. helveticus*	External mould (Camembert, Brie, Pont l'Eveque, Coulommiers)	1, 2, 3, 4, 13, 14, 15
10. *Lac. casei*		
11. *Lac. acidophilus*		
12. *Lac. plantarum*	Internal mould (Roquefort, Blue Stilton, Danish Blue, Gorgonzola, Mycella, Gammelöst)	1, 2, 3, 4, 16

Moulds and yeasts

13. *P. camemberti*		
14. *P. caseiocolum*		
15. *P. candidum*		
16. *P. roqueforti*		
17. *S. kefir*		
18. *S. lactis* } *C. kefir*		
19. *T. kefir*		

High temperature scalded cheese		
(i) Parmesan, Romano, Grana	6, 7, 8, 9, 10, 11, 12	
(ii) Emmenthal, Gruyère	6, 7, 8, 9, 22	
Surface slime (Limberger, Brick)	1, 2, 3, 20	
Sour cream, cultured butter, buttermilk	2, 3, 4	

Miscellaneous

20. *Bre. linens*	
21. *Bif. bifidium*	
22. *P. freudenreichii* sub-sp. *shermanii*	

Yoghurt and Skyr	6, 7
Therapeutic yoghurt	6, 7, 11, 21
Kefir	1, 4, 7, 10, 11, 17, 18, 19 and *Lac. brevis*
Kumiss	1, 6, 7, 18, plus *Torulopsis* sp.
Yemer	1, 2, 3, 4
Yakult	10
Filmjolk, Laktofil and Graddfil	1, 2, 3, 4

[a] For the production of modified Cheddar.
[b] *Leuconostoc dextranicum* could be also present.
Adapted from: Sellars (1967); Poulsen (1970); Cox (1977); Carini and Lodi (1978); Law and Sharpe (1978); Lawrence and Thomas (1979); Sharpe (1979).

TABLE IV
SELECTED CHARACTERISTICS OF DAIRY STARTER CULTURES

Micro-organisms	G+C[a] (mean %)	Acid from glucose	Gas from glucose	Gas from gluconate	Aldolase	Lactic acid configuration	% Acid in milk	NH₃ from arginine	Serological group	Growth 10°C	Growth 45°C	Requirement for (Thiamine, Riboflavine, Pyridoxal, Folic acid, Thymidine, Vitamin B₁₂)						Carbohydrate utilisation (Aesculin, Amygdalin, Arabinose, Cellobiose, Fructose, Galactose, Gelatin, Lactose, Hippurate, Malose, Mannitol, Mannose, Melezitose, Melibiose, Raffinose, Rhamnose, Ribose, Salicin, Starch, Sorbitol, Sucrose, Trehalose, Xylose)																							
Bacteria												Thi	Rib	Pyr	Fol	Thy	B₁₂	Aes	Amy	Ara	Cel	Fru	Gal	Gel	Lac	Hip	Mal	Mni	Man	Mlz	Mlb	Raf	Rha	Rib	Sal	Sta	Sor	Suc	Tre	Xyl
Str. lactis	38·5	+	−	−	−	L(+)		+	N	+	−	+						V	V			V	V			+	+	V	+						V	V	V		−	−	V
Str. lactis sub-sp. *diacetylactis*[b]	38·5	+	−	−	−	L(+)		+	N	+	−							V	V			V	V			+	+	V	+						V	V	V		−	−	V
Str. cremoris	38–40	+	−	−	−	L(+)		−	N	+	−							V	V			−	V		−	+	+	−	V						V	V	V		−	−	−
Str. thermophilus	34–38	+	−	−	−	L(+)		+	X	−	+							−				+	+		+	−	+	−	−						−	−	+	+	+	V	−
Str. faecium	39–42	+	−	−	+	D(+)		+	D	+	+	+	+	+				+	+	+	+		−	−	+	+	+	+	+	+	+		+	−	+	+	+	+	+	+	+
L. cremoris	38–39	+	−	−	+	D(−)		−		+	−		+	+				V	V	−		+	V		V	+	−	V	−	V	−	+	V	−	−	−	−	−	−	−	−
L. dextranicum			−	−	+	D(−)		−	E	−	−		V	V	V		−	V	V	−	V	+	V		V	+	+	V	V	V	V	+	V	V		−	−	−	+	+	+
Lac. bulgaricus	50·3		−	+		D(−)	1·7	−	E	−	+	−				−		−	−		+	+	+		+	+	−	−	−	−	−	−	−	−	+	−	−	−	−	−	−
Lac. lactis	50·2		−	+		DL	1·4–1·7	−	A	−	+	−				−		−	−		+	+	+		+	+	V	−	−	V	−	−	−	−	+	−	−	+	−	+	−
Lac. helveticus	39·3		−	+		DL	2·7	−		−	+	−				−		−	−		−	+	+		+	+	−	−	−	+	−	−	−	−	−	−	−	+	−	+	−
Lac. acidophilus	36·7		−	−		DL	0·8	−		−	+	−				+		+	−		+	+	+		+	+	+	+	+	+		−	V	−	+	+	−	−	+	+	−
Lac. casei	46·4	−	+		L(+)	1·2–1·5	−	BC	+	V					+		+	+	+	−	+	+		V	−	+	+	+	+		V	−	+	V	+	+	+	+	+	+	+
Lac. plantarum	45·0	−	+		DL	0·1–0·28	−	D	+	−					+		+	+		+	+	+		+	−	+	+	+	+		+	+	+	+	+	+	+	+	+	+
Bre. linens[c]	60–64·4			−				+		+	−						+	−		−	−	+	+	+		−	−	W	−	+	V	+	−	−	V	−	−	−	−	−	−	W
Bif. bifidium						1·3		−		+	−	+						+	−	+	+		+		−	−	+						+	−	−	−	−	+	+	−	+
P. freudenreichii sub-sp. *shermanii*	64–67									−	+									−	−	+	+	+		+	−		+				+	−	−	−	−	+	+	−	W

Yeasts		
S. kefir		
S. lactis } *C. kefir*	+	DL
T. kefir		

Opt. 37–42°C pantatonate, niacin — Biotin, niacin

+ −

[a] Mean % of Guanine and Cytosine in DNA.
[b] Same characteristics as *Str. lactis* with the exception of citrate utilisation and the production of carbon dioxide, acetoin and diacetyl.
[c] Information related to *A. globiformis*.
+ Positive reaction by 90% or more strains. − Negative reaction by 90% or more strains.
V Variable, slow or weak reaction. W Weak or delayed reaction. X Ungrouped.

After: Hansen (1968); Fowler (1969); Rogosa and Hansen (1971); Uden and Buckley (1971); Buchanan and Gibbons (1974); Steffen *et al.* (1974).

of *Str. lactis*, *Str. cremoris*, and the gas and aroma producing mesophilic LAB (*Str. lactis* sub-sp. *diacetylactis*, *L. cremoris* and/or *L. dextranicum*) (Crawford, 1972; Cox *et al.*, 1978; Sharpe, 1979). Incidentally, the aroma producing lactic starters are essential for the production of buttermilk, sour cream, cultured butter and some fermented milk products.

Thermophilic LAB (optimum temperature 37–45 °C) are used for manufacture of yoghurt, acidophilus milk and high-temperature scalded cheese (e.g. Swiss varieties). Examples of these organisms are *Str. thermophilus* and all the lactobacilli spp. (see Table III). These starters produce lactic acid rapidly at high temperature, and the rate of acid production is further enhanced if symbiotic relationships exist between different species, as in the case between *Str. thermophilus* and *Lac. bulgaricus*. Some strains of yoghurt starter cultures are slime producers, possibly glucan (Galesloot and Hassing, 1973; Tamime, 1977), and the use of such strains can improve the viscosity of yoghurt.

Finally, the combined activity of mesophilic and thermophilic lactic acid bacteria and yeast yields a lactic acid/alcohol fermentation in milk (e.g. Kefir, and Kumiss). Ethyl alcohol is mainly produced, and the level can reach as high as 1·5 %; the flavour components are due to acetaldehyde, diacetyl and lactic acid. The Kefir grains are irregular and whitish in colour, and the bacteria and yeasts are held together due to the formation of a glucose–galactose polymer produced by *Lac. acidophilus* (Ottogalli *et al.*, 1973).

Identification and Differentiating Characteristics
The classification of the dairy starter cultures, as illustrated in Fig. 1, is a general approach towards the identification of these micro-organisms. However, successful production of cheese and other fermented dairy products relies completely on choosing the right organism for the specific type of fermentation required. The general differentiating characteristics of these starter cultures are shown in Table IV.

STARTER CULTURE TECHNOLOGY

The fermentation process of any cultured dairy product relies entirely on the purity and activity of the starter culture, provided that the milk or the growth medium is free from any inhibitory agent, e.g. antibiotics or bacteriophage. The traditional method for the production of bulk starter is illustrated in Fig. 2 (system 1), and although the

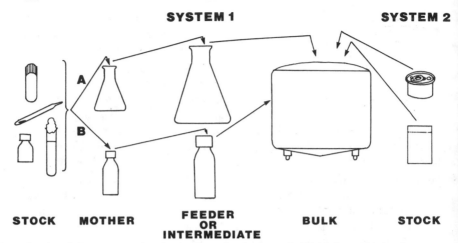

FIG. 2.   Culture preparation. System 1: stock culture may be liquid, freeze-dried or frozen at −196°C. System 2: stock culture may be freeze-dried or frozen at −196°C.

propagation procedure is time consuming, requiring skilled operators and may lead to contamination by bacteriophage, which is one of the major hazards in the industry, it is widely used. Nevertheless, research work has been intensified in this area to overcome various problems, and developments in this field have focused on the areas of starter preservation and concentration. Developments in starter technology until the early 1970s have been reviewed by Lloyd (1971), and only subsequent developments will be discussed in this section.

The starter culture must contain the maximum number of viable organisms, must be highly active under production conditions in the dairy, and must be free from contaminants. Provided that culture inoculation is carried out under aseptic conditions and growth is initiated in a sterile medium, Foster (1962) suggested that one of the following principles must be adopted in order to maintain an active starter culture:

(1)   reducing and/or controlling the metabolic activity of the micro-organisms;
(2)   separating the organisms from their waste products.

The former principle is evident in refrigeration, while the latter approach is mainly used during the concentration and/or preservation of the starter, i.e. during the production of a concentrated active bulk starter, either in a continuous fermentor or in a batch process, for direct inoculation of the milk in the vat.

Methods of Preservation

It is essential that starter cultures are preserved in order to maintain an available stock of organisms, especially in the case of a starter failure. Also, successive sub-culturing can induce mutant strains which may alter the overall behaviour and general characteristics of the starter. Dairy cultures may be obtained from research establishments, educational colleges, culture bank organisations or commercial manufacturers, and starter culture bacteria may be preserved by one of the following methods:

(1) liquid starter,
(2) dried starter:

    (i) spray-dried,
    (ii) freeze-dried or lyophilised,
    (iii) concentrated freeze-dried,

(3) frozen starter:

    (i) deep frozen at $-40\,°C$,
    (ii) low-temperature freezing at $-196\,°C$ in liquid nitrogen.

Liquid Starter

This is the most popular and widely used form in which starter cultures are handled in the dairy. Starters are normally preserved in small quantities, but to meet the required volume for any production line, a scale-up system of propagation is required.

Stock culture → Mother → Feeder or *Intermediate → Bulk*

For example, processing 10 000 litre of milk into cheese per day with a rate of inoculation of 2%, would require a scale-up propagation as follows:

$$Stock \xrightarrow{1\%} Mother \xrightarrow{1\%} Feeder \xrightarrow{2\%} Bulk$$
$$0\cdot4\,ml \qquad 40\,ml \qquad 4\,litre \qquad 200\,litre$$

The working stock cultures are maintained in autoclaved ($15\,lb\,in^{-2}$ for 15 min) reconstituted antibiotic-free skim-milk powder (10–12% SNF), with either weekly or daily sub-culturing. Cheese cultures (*Str. lactis*, *Str. cremoris*, *L. cremoris*) can be propagated up to 50 times without any fear of mutation, and the sterilised medium is inoculated at a rate of 1% and incubated at 22 or 30°C for 18 or 6 h, respectively (Walker, 1980). However, yoghurt starter cultures (*Str. thermophilus* and *Lac. bulgaricus*) are normally sub-cultured only 15–20 times as

TABLE V

GROWTH MEDIUM FOR THE PRESERVATION OF LIQUID
STOCK CULTURES

Skim-milk	10–12 % SNF
Litmus solution (5 %)	2 %
'Yeast extract	0·3 %
Dextrose/lactose	1·0 %
Chalk	Enough calcium carbonate to cover the bottom of the test tube
Panmede (adjust to pH 7)	0·25 %
Lecithin (adjust to pH 7)	1·0 %

Autoclave 10 lb in^{-2} for 10 min and incubate for
1 week at 30 °C to check sterility.
After: Shankar (1975).

a safe-guard against mutation, and to retain the ratio of cocci:rods
as 1:1 (Sellars, 1975). Incubation is carried out at 42 °C for 3–4 h,
or at 30 °C for 16–18 h using 2 or 1 % inocula, respectively.

Starter culture activity is affected by the rate of cooling after incubation,
level of acidity at the end of the incubation period, and the temperature
and duration of storage. Cooling is important to control the metabolic
activity of the starter; in practice, however, a warm starter (freshly
incubated and uncooled) is sometimes used in cheese factories and,
to some extent, in the yoghurt industry.

The reserve stock culture can be maintained in a liquid form, and
a typical growth medium for most LAB is illustrated in Table V.
The inoculated medium is incubated for a short period of time and
stored under ordinary refrigeration. Reactivation is only necessary once
every 3 months.

Dried Starter

The preservation of starter cultures by drying techniques is an alternative
method for culture retention. The development of such processes seeks
to overcome the work involved in maintaining liquid stock cultures.
It also facilitates the dispatch of dried cultures by post without any
loss in their activity. Prior to 1950, vacuum drying was the normal
practice (Tofte-Jespersen, 1974a,b; 1976). The process consists of mixing
a liquid culture with lactose, and then neutralising the excess acid
with calcium carbonate. The mixture is partially concentrated by separation and/or expressing the whey, so yielding granules which are dried

Hydrolyse milk protein with trypsin
for 4 h at 37 °C

↓

Steam the milk

↓

Cool the milk, inoculate with starter culture and
incubate at 20 °C with pH control
(calcium hydroxide is used as the neutralising
agent)

↓

Evaporate at 27 °C to 22 % dry matter after obtaining maximum
cell concentration

↓

Spray-dry at air temperature 70 °C to 9 % moisture
(powder temperature does not exceed 42 °C)

↓

Vacuum-dry at 27 °C and 1–2 mm Hg (dried culture
has 5–6 % moisture)

FIG. 3.   The Dutch process for the production of spray-dried starter cultures. Adapted from
Stadhouders *et al.* (1969).

under vacuum. The dried starter contains only 1–2% viable bacteria,
and may require several sub-cultures before regaining its maximum
activity.

Higher survival rates in dried starters can be achieved by spray-drying,
a method which was evolved in Holland (Stadhouders *et al.*, 1969).
The process is illustrated in Fig. 3, and the dried starter is claimed
to be as active as a 24-h liquid starter. Although this development
in starter technology proved promising, this system has not been developed
commercially. The reason could be the usually low survival rates of
the dried cultures, i.e. 10% for the majority of mesophilic LAB and
44% for *Str. lactis* sub-sp. *diacetylactis*. However, the addition of
mono-Na-glutamate and ascorbic acid to a starter culture propagated
in a buffered medium did protect the bacterial cells to some extent,
and the spray-dried culture retained its activity after storage for
6 months at 21 °C (Porubcan and Sellars, 1975*a*). Anderson (1975)
claimed in a Swedish Patent that yoghurt starter cultures (ratio of
cocci:rods—40:60 to 60:40) can be obtained in the spray-dried form
when the starter is propagated in concentrated skim-milk (18–24% TS)
fortified with cyanocobalamin, lysine and cystine. The drying temperature

TABLE VI

EFFECT OF SELECTED CRYOGENIC AGENTS[a] ON THE SURVIVAL RATE OF FREEZE-DRIED LACTIC ACID BACTERIA (ALL FIGURES AS %) OF ORIGINAL CELL NUMBER)

Micro-organism	L-Glutamic acid	L-Arginine	L-Lysine	DL-Threonine	DL-Pyrrolidone-carboxylic acid	Acetyl glycine	DL-Malic acid
Str. cremoris	40–60	42–60	1–16	20–39	53–67	48–59	23–57
Str. lactis	31–74	36–53	0–7	8–39	19–81	10–56	26–44
Str. lactis sub-sp. diacetylactis	44–53	48–54	4	14–37	47–60	38–43	20–33
Str. thermophilus	35–40	21–40	6–7	7–11	24–48	29–44	52–59
Lac. bulgaricus	16–21	20–35	1–10	6–10	9–11	7–23	6–15
Lac. acidophilus	42–63	39–57	4–38	6–21	24–56	3–35	28–66
Lac. helveticus	48	35	23	14	23	32	35
Lac. casei	42–49	28–40	6–10	6–9	13–29	4–18	6–22
Lac. plantarum	57	44	5	10	48	20	46

The range of survival (%) is due to different strains tested.
[a] Suspending medium (0·06 M solution of agent adjusted to pH 7).
Adapted from Morichi (1972).

can be as high as 75–80 °C without causing any bacterial damage. Despite the advantages claimed for spray-dried cultures, it seems that this system of preservation is not widely used.

Freeze-dried cultures are produced when the starter is dried in the frozen state. This method of starter preservation improves the survival rate of the dried culture, and good results have been achieved as compared with spray-dried starters. It has been observed that the process of freezing and drying can damage the bacterial cell membrane, but the damage is minimised with the addition of certain cryogenic agents/compounds prior to freezing and drying. These protective solutes are hydrogen bonding and/or ionising groups which help to prevent cellular injury by stabilising the cell membrane constituents during the preservation procedures (Morichi, 1972, 1974). Table VI illustrates the effect of such solutes on the survival rates of the preserved LAB.

Many different media, additives or techniques have been studied to determine the optimum conditions for the production of freeze-dried cultures, and a review of such work is illustrated in Table VII. Based on the work of Morichi (1972, 1974) and the information presented in Table VII, it is possible to summarise the factors that can affect the survival rate of freeze-dried dairy starter cultures as follows.

1. Most LAB preserve well with the exception of *Lac. bulgaricus* which is sensitive to freezing and drying.
2. Propagation of the starter culture in milk fortified with yeast extract and hydrolysed protein improves the survival rate. However, raising the cell concentration of a culture, e.g. $> 10^{10}$ cells ml^{-1}, can increase the viable number of bacteria in the dried culture.
3. Starter cultures are less sensitive to freezing and drying if the cells are harvested towards the latter part of the exponential phase, with the exception of *Lac. bulgaricus* and *Str. cremoris* where the cells are harvested in the early stages of the stationary phase.
4. Media in a pH range of 5–6 are more favourable for higher survival rates, and neutralisation of the growth and the suspending medium is essential.
5. The suspending medium, which ensures the best survival rate, can vary with different types of organism. For example, skim-milk plus Na-malate proved suitable for *Str. thermophilus*; a solution of lactose and arginine hydrochloride gave higher protective activity to *Lac. bulgaricus*; and glutamic acid for the *Leuconostoc* spp.

TABLE VII

A SELECTION OF DIFFERENT CRYOGENIC COMPOUNDS EMPLOYED DURING THE
PRODUCTION OF FREEZE-DRIED STARTERS

Method of preparation	Test organism	Reference
Skim-milk + 3% lactose Horse serum + 8% glucose Naylor and Smith (1946) reducing medium	*Lac. bulgaricus, Lac.* *acidophilus, Lac.* *casei*	Briggs *et al.* (1955)
Skim-milk + 1% peptonised milk + 10% saccharose Concentrated milk (20–25% TS) Starter + milk + rennet Biomass in whey + skim-milk + 10% saccharose + 1–2% Na-glutamate	*Str. thermophilus,* *Lac. bulgaricus*	Gavin (1968)
Suspend concentrated mixed strain cheese starter cultures in 7% lactose or 0·06 M mono-Na- glutamate or 0·06 M arginine- mono-hydrochloride	*Str. lactis/Str.* *cremoris, Str.* *lactis* sub-sp. *diacetylactis,* *L. cremoris*	Pettersson (1975)
Propagate dairy starter in milk- based medium (pH adjusted to 6·0–6·5) + additives (ascorbic acid, mono-Na-glutamate, aspartate compound) + cryoprotective agents (inositol, sorbitol, mannitol, glucose, sucrose, corn syrup, DMSO, starches, PVP, maltose, mono- or disaccharides)	Patent for the production of dairy starter cultures	Porubcan and Sellars (1975*b*)
Suspend concentrated culture in GCGS (5% gelatin + 5% Na- citrate + 2% mono-Na- glutamate + 10% sucrose)	*Str. lactis, Str.* *cremoris, Str.* *lactis* sub-sp. *diacetylactis,* *Leuconostoc* spp., *Propionibacterium* spp.	Speckman (1975)
Propagate yoghurt starter in skim- milk (11% TS) + 0·1% Tween 80 Suspend biomass in 6% malt extract	*Str. thermophilus,* *Lac. bulgaricus*	
Grow lactobacilli in MRS broth + 10% (w/v), cryogenic agent (casitone, lactose, malt extract, milk solids, mono-Na-glutamate, Myvacet, whey powder, or peptonised milk)	*Str. lactis, Str.* *thermophilus, Lac.* *bulgaricus, Lac.* *acidophilus*	Kilara *et al.* (1976)

TABLE VII—*contd.*

Method of preparation	Test organism	Reference
Grow streptococci in all purpose Tween + 10 % (w/v) cryogenic agent (casitone, dimethyl sulphoxide, glycerol, lactose, malt extract, milk solids, mono-Na-glutamate, pectin, peptonised milk, whey powder)		
Propagate LAB in low lactose medium fortified with soya and casein protein (0·13–0·26 % w/w). The mixture is continuously buffered, and freeze-drying commences at a certain cell biomass concentration	Lactic starters	Hup and Stadhouders (1977)
Suspend washed cells in skim-milk (10 % TS) + 0·5 % ascorbic acid + 0·5 % thiourea + 0·5 % ammonium chloride	*Str. lactis, Str. cremoris, Str. lactis* sub-sp. *diacetylactis, Str. thermophilus, L. cremoris, L. dextranicum, Lac. bulgaricus, Lac. lactis, Lac. casei, Bif. bifidium*	Sinha *et al.* (1970); Sinha *et al.* (1974)
Suspend concentrated culture in sucrose solution + buffering salt + gelatin + Na-glutamate	Mixed strain cheese and butter cultures	Bannikova and Lagoda (1970)
Propagate culture in hydrolysed milk enriched with 5 % Difco yeast extract or 3 % aqueous extract from maize + 0·012 % MnSO$_4$, MgSO$_4$ and ZnSo$_4$ + glucose or β-D-galactosidase Continuous neutralisation Suspend cell biomass in skim-milk (10 % TS)	*Str. lactis, Str. cremoris, Lac. casei, Str. lactis* sub-sp. *diacetylactis*	Kornacki *et al.* (1974)
Cultures grown in reconstituted skim-milk powder (10 % TS), buffered to pH 6·9 or neutralised with marble For streptococci: + Yeast extract and Vit. E.	*Str. lactis, Str. thermophilus, Lac. bulgaricus*	Naghmoush *et al.* (1978)

(continued)

TABLE VII—contd.

Method of preparation	Test organism	Reference
For lactobacilli: + Yeast extract, Tween 80 and sheep serum		
Propagate culture in skim-milk (11% TS) plus 15% β-glycerophosphate	Str. lactis, Str. cremoris	Yang and Sandine (1978, 1979)
Grow culture in skim-milk + 1·5% Na-glutamate and 8% sucrose	Str. thermophilus, Lac. bulgaricus	Ozlap and Ozlap (1979)
Propagate culture in protein digested milk + yeast extract + lactose and constant buffering with ammonia Concentrate by centrifugation Bacterial cells + protective medium	Cheese starters	Bouillanne and Auclair (1978)
Mix starter culture + 1'0% sugar solution, or 5–10% peptone solution or polymer 1500	Str. thermophilus, Lac. bulgaricus	Nikolova (1978)
Cultivate butter culture in skim-milk + proteolytic enzymes + buffering salts + stimulants Concentrate by centrifugation Suspend biomass in sucrose and Na-citrate	Str. lactis, Str. cremoris, Str. lactis sub-sp. diacetylactis	Chuzhova and Belova (1978)

6. Moisture content in the dried culture must be less than 3%.
7. Dried cultures stored at 5–10 °C showed higher rates of survival during prolonged storage than those stored at room temperature (Nikolova, 1975).
8. Vacuum packaging of the dried cultures is highly recommended because the preserved organisms are sensitive to oxygen.
9. The most popular type of packaging material for dried cultures is the glass vial, followed by the laminated, aluminium foil sachet. The feasibility of using a nylon sachet has been studied by Tsvetkov and Nikolova (1975) and Nikolova (1974), and results indicate a deterioration in survival rates due to the permeation of oxygen through the packaging material.
10. Freezing cultures at −20 to −30 °C and drying at temperatures between −10 and 30 °C results in high bacterial activity of the dried culture.

11. Carbonyl compounds, such as pyruvate and diacetyl, which can react with the amino groups within the preserved cells can accelerate their death. It is recommended that these compounds should be separated from the harvested cells. For long-term preservation of freeze-dried cultures, the suspending medium must be fortified with non-reducing sugars, amino acids and/or semicarbazide.

12. Rehydration temperature can affect the leakage of cellular ribonucleotides from the damaged cells. Although *Str. thermophilus*, *Str. lactis* and *Str. cremoris* showed little response to different rehydration temperatures, it is recommended that *Lac. bulgaricus* is rehydrated at 20–25 °C (Morichi *et al.*, 1967).

Starter cultures preserved by freeze-drying tend to have a prolonged lag phase, and are mainly used as inoculants for the propagation of mother cultures (see Fig. 2, system 1). Larger quantities are needed for direct inoculation of the bulk starter, and an extended incubation time may be required (Sellars and Babel, 1978). Developments in the last few years have made it feasible to produce concentrated freeze-dried cultures (CFDC) for direct inoculation of the bulk starter (see Fig. 2, system 2) or direct-in-vat inoculation of milk for the manufacture of cheese and other fermented dairy products (see Table VIII).

Frozen Starters

Liquid starter cultures (mother or feeder) can be preserved by freezing at -20 to -40 °C for a few months. This method of starter preservation is found in centralised laboratories, and the frozen cultures are dispatched to a dairy, whenever needed, as direct inoculants for bulk starters. Freezing and prolonged storage at -40 °C can lead to a deterioration in starter culture activity, and can damage certain lactobacilli, but the use of a medium containing 10 % skim-milk, 5 % sucrose, fresh cream, 0·9 % sodium chloride or 1 % gelatin can improve survival rates (Imai and Kato, 1975). In addition, concentrated cells (10^{10}–10^{12} colony forming units (cfu) ml^{-1}) frozen at -30 °C, and in the presence of certain cryogenic compounds (Na-citrates, glycerol or Na-β-glycerophosphate), have been retained as active mesophilic, lactobacilli or propionic acid starter cultures (Accolas *et al.*, 1970, 1978; Lloyd and Pont, 1973; Libudzisz *et al.*, 1977; Jackubowska *et al.*, 1978; Oberman *et al.*, 1978; Piatkiewicz *et al.*, 1978).

Although freezing at -40 °C proved to be a successful process to

A. Y. Tamime

TABLE VIII

THE MANUFACTURE OF BULK STARTER AND FERMENTED DAIRY PRODUCTS USING CFDC FOR DIRECT-IN-VAT INOCULATION

Country	Product	Inoculation rate	References
UK	Cheddar cheese	4·4 units per 100 litres milk[a]	Chapman (1978); Chapman et al. (1978)
France	Carré de L'Est cheese	2 units per 100 litres milk[a]	Mietton et al. (1978, 1979)
France	Camembert cheese	0·5 units per 1 000 litres milk[a]	Vassal et al. (1978)
France	Saint Paulin cheese	4 units per 100 litres milk[a]	Lablee and Perrot (1978)
France	Tome De Savoie cheese	4 units per 100 litres milk[a]	Bouty et al. (1978)
USSR	Cultured butter	0·3–60 g per ton cream	Chuzhova (1977a,b)
Germany	Cheese and butter bulk starter	25 g per 1 000 litres skim (11% TS)[b]	Pallasdies (1977)
Poland	Edam and Tilsit cheese bulk starter	1 g per 100 litres milk	Kornacki et al. (1978)
UK/Denmark	Yoghurt or bulk starter	One sachet per 500–1 000 litres milk[c]	Anon. (1980)
	Cheese starter	One sachet per 200–1 000 litres milk (red or blue)[c]	

[a] CFDC manufactured by Vitex in France known as ICF. The unit is equivalent to the quantity of CFDC used to produce 150 m mol of lactic acid in reconstituted skim-milk (10% TS) at 30°C for 4 h.
[b] Visbyvac Instant CFDC.
[c] Chr. Hansen's Freeze-Dried Redi-Set.

preserve starter cultures, freezing in liquid nitrogen at − 196 °C is by far the best method. The reviews by Gilliland and Speck (1974) and Hurst (1977) show the extent of the work which has been carried out in this field, and the freezing and thawing cycle is still regarded as the important factor in the successful use of frozen cultures in the dairy industry. An organism which is highly susceptible to damage during freezing is *Lac. bulgaricus*, but it was found that the presence of Tween 80 and Na-oleate improved cell stability (Smittle *et al.*, 1972; Smittle, 1973; Smittle *et al.*, 1974).

Freezing cultures in liquid nitrogen has made possible the direct-in-vat inoculation of milk for cheese and yoghurt production, or direct inoculation of the bulk starter (see Fig. 2, system 2). The advantages of such an approach are as follows: convenience, culture reliability, improved daily performance and strain balance, greater flexibility, better control of phage, and possible improvement in quality. However, the disadvantages are: difficulties in providing liquid nitrogen facilities, higher cost, greater dependence on starter suppliers, and apportioning of responsibility in case of starter failure (Tamime and Robinson, 1976; Wigley, 1977). More recently, Wigley (1980) has discussed the successful use of liquid nitrogen frozen cultures for the manufacture of Cheddar cheese.

Concentration of Cells

It can be observed from the information above that the survival rate of the preserved starter culture is dependent on the processing conditions (growth medium, presence of cryogenic compounds, freezing and drying), and on the method of cell concentration. In brief, the different systems which are used for concentration of cell biomass are as follows. Firstly, mechanical means (e.g. Sharples separators, ultra-centrifuge (15 000–20 000 × g)) which can cause some physical damage to the bacterial cells. Secondly, continuous neutralisation of the growth medium at around pH 6 can be used to produce a culture with high cell numbers; the formation of lactate can be inhibitory, and hence limit the degree of concentration. Thirdly, the diffusion culture technique, a method developed for removing the lactate from the growth medium, so that the system is capable of achieving 10^{11} cfu ml^{-1} in the concentrated culture (Osborne, 1977).

Preservation of Moulds and Kefir Grains

Blue and white moulds are normally preserved by freeze-drying, or with a fluid-bed drier as reported by Hylmar and Teply (1970) for the preservation of P. roqueforti. The dried cultures are resuspended in sterile or boiled water, and the resultant preparation is referred to as the 'working culture'. The rehydrated spores can remain active for a week at 5°C. The application of these moulds in the dairy industry depends on the end-product, and according to Galloway (1980), white moulds (working culture) can be used in one of the following ways during the manufacture of Camembert and Brie:

(1) direct inoculation of the milk with the lactic starter, prior to the addition of rennet;

(2) spraying the mould solution on the cheese curd before salting;

(3) coating the surface of cheese with a special mixture of salt and dried mould spores.

However, the application of blue mould is as follows:

(i) using the same methods as (1) and (3) mentioned above;

(ii) adding the mould culture to the curd immediately after filling the cheese moulds;

(iii) growing the moulds on brown bread crumbs, shaping the mycelial mass into balls, which are then wrapped in muslin cloth and rubbed by hand into the milk after the addition of the lactic starter.

Ottogalli and Rondinini (1976) reported on the preservation of the moulds and LAB used in the production of Gorgonzola cheese. The method comprised growing the cultures in milk at pH 5·5–6·5 to give 10^8–10^9 cfu ml^{-1}, followed by freezing at $-18°C$. The survival rates for the particular organisms were as follows: Str. thermophilus, 18%; Str. lactis, 38%; L. cremoris, 80%; Lac. bulgaricus, 47%; Torula spp., 50% and P. roqueforti, 100%. From such results it is safe to assume that blue moulds can be preserved by freezing without any loss in their activity.

Kefir grains are either preserved dry or wet. One simple method consists of washing the excess grains with water and drying at room temperature. This crude method of preservation can lead to contamination of the dried grains, and it is possible that the symbiotic microflora may be altered. The preservation of the grains by freeze-drying is a better system. The preserved culture is in the form of a powder or small crystals, and after two to three sub-cultures, the grains start to form in the growth medium. The production of freeze-dried Kefir grains is discussed in a Russian Patent described by Lagoda et al.

(1979). Alternatively, the washed grains are suspended in a sterile medium which can be stored for a few months at ordinary refrigeration temperature without any appreciable loss in activity; the grains retain their normal shape and form.

FACTORS CAUSING INHIBITION OF STARTER CULTURES

There are many factors that can cause an inhibition of, and/or a reduction in, starter culture activity, and either event can lead to poor quality fermented dairy products reaching the consumer, and financial loss to the manufacturer. It is recommended, therefore, that milk intended for bulk starter culture production, or the manufacture of these dairy products, should be free from these factors. The causes of inhibition of starter cultures are summarised in the following paragraphs.

Antibiotics
Residues of antibiotics in milk result from mastitis therapy in the dairy cow. There are various types of antibiotics used, and starter cultures are susceptible to very low concentrations (Cogan, 1972). Reinbold and Reddy (1974) surveyed 30 different antibiotics and antimicrobial agents, and concluded that lactic streptococci and yoghurt starter cultures were very sensitive to all the antibiotics tested; however, strains of *Leuconostoc* spp., *Str. faecium*, and *Bre. linens* were less sensitive. Teply and Cerminova (1974) tested the effect of eight antibiotics under processing conditions, and their results are illustrated in Table IX. It can be observed that the inhibitory levels of streptomycin, chloramphenicol, oxytetracycline and tetracycline seem rather high, and this apparent resistance could be attributed to culture strain variation, variation in the commercial preparations of antibiotics used, or some feature of the test method used to detect the level of antibiotics.

Bacteriophages
The occurrence of bacteriophages (phages) in dairy starter cultures was first reported by Whitehead and Cox (1935). Since that period, research work has been intensified in this area in order to know more about phages and the possibility of controlling them. Recent reviews on cheese starter cultures have been reported by Reiter (1973), Lawrence *et al.* (1976), Lawrence (1978), and Lawrence and Thomas (1979), while the yoghurt cultures have been studied by Reinbold and Reddy (1973), Peake and Stanley (1978*a,b*) and Accolas and Spillmann (1979*a,b*).

A. Y. Tamime

TABLE IX

SENSITIVITY OF DAIRY STARTER CULTURES TO VARIOUS ANTIBIOTICS UNDER
PROCESSING CONDITIONS

Inhibitory level ($IU\,ml^{-1}$)

Starter cultures	Penicillin	Streptomycin	Chloramphenicol	Erythromycin	Chlortetracycline	Bacitracin	Oxytetracycline	Tetracycline
Yoghurt	0·01	1·0	0·5	0·1	0·1	0·04	0·4	1·0
Cream	0·2	2·5	1·0	0·5	0·05	0·1	0·4	1·0
Acidophilus	0·07	2·5	2·0	0·07	0·5	—	2·5	2·0
Emmenthal cheese	0·2	2·5	2·0	0·08	0·1	0·03	—	—
Kefir	2·5	2·5	3·0	0·5	2·0	—	1·5	1·0
Mesophilic lactic starter	0·01	2·5	2·0	0·05	0·5	0·12	2·0	—

Adapted from Teply and Cerminova (1974).

Phages are viruses that can attack and destroy starter cultures. The result is a failure of lactic acid production. Lactic streptococci and lactobacilli are the most vulnerable organisms of the dairy starter cultures to phage attack. The morphology of the majority of the phages shows a head and a tail made up of a nucleic acid core (deoxyribonucleic acid (DNA) and ribonucleic acid (RNA)) which is protected with a protein layer (Sandine, 1979). It is possible to classify starter cultures into three main groups based on sensitivity: phase insensitive (not affected), phage carriers (slight reduction in the rate of lactic acid being produced), or phage-sensitive (the bacteria undergo complete lysis). In addition, the phage–host interactions have been divided into three categories (Sandine, 1979): virulent phages (which can cause lysis of the cells, or alternatively partial lysis and the survivors become phage resistant), temperate phages (four different types of interaction may occur) and, finally, carrier phages (phage is present in the host cell and can be removed by phage antiserum).

It is evident that the existence of phage is a problem in the dairy industry, and the following precautionary measures may be practised.

1. The propagation of the starter culture must be carried out employing aseptic techniques (refer to later section).

2. Heat treatment of bulk starter milk ensures the destruction of these viruses, but it is vital that the starter tank should be filled to its maximum capacity, otherwise prolonged heat treatment is necessary to eliminate the phages in the airspace (Pearce and Brice, 1973).
3. Daily rotation of phage-unrelated starter strains, or phage-resistant strains, must be used in the dairy.
4. Effective filtration of the air in the starter room and the production area can help to control the phage problem.
5. The equipment must be properly sanitised, i.e. by heat or chemicals (Ciblis, 1966; Sandine, 1979).
6. Location of the starter room far away from the production area and whey handling department reduces the possibility of air-borne infection.
7. Ensure that plant personnel, particularly those from the cheese room, are not allowed into the starter handling area.
8. Propagate starter culture in phage inhibitory medium (refer to section on Production Systems of Bulk Starter Cultures for further details).
9. Do not use starter culture strains sensitive to those phages categorised by having short latent periods and large burst sizes (Sandine, 1979).
10. 'Fogging' the airspace of the starter preparation room with a solution of hypochlorite, or alternatively, the use of UV light can control phage in the atmosphere.
11. 'Fogging' the cheese production area, although this approach is regarded by some as an extremely dubious practice.
12. Development of phage-resistant strains (Thomas and Lowrie, 1975; Heap and Lawrence, 1976; Marshall and Berridge, 1976; Czulak *et al.*, 1979; Erickson, 1980).
13. Use of mixed strain starter cultures (Reiter and Kirikova, 1976).

Detergent and Disinfectant Residues

Detergents and disinfectants are used for cleaning and sanitation purposes for dairy equipment. Residues of these compounds (alkaline detergents, chlorine-based materials, iodophors, quaternary ammonium compounds (QAC) and ampholytes) can affect the starter culture activity. Pearce (1978) reported that strains of *Str. lactis* and *Str. cremoris* showed inhibition at 2–4 mg QAC litre^{-1}, except one strain which tolerated up to 12 mg QAC litre^{-1}. Yoghurt cultures are more tolerant, and the inhibitory levels (mg litre^{-1}) of chlorine compounds, QAC and iodophors for *Str. thermophilus* and *Lac. bulgaricus* are 100, 100–500 and 60; 100, 50–100 and 60, respectively (Bester and Lombard, 1974).

Contamination of starter milk with these compounds is due to human error, or a break-down in the automatic cleaning cycle. In practice, it is necessary to ensure that the rinsing cycle is long enough to wash down these chemicals from the bulk starter tank.

Miscellaneous Inhibitors

Natural antibodies (lactenins/agglutinins) are present in milk and can inhibit the growth of the starter cultures. These antibodies are heat-sensitive, and heat treatment of bulk starter milk ensures their destruction. Leucocytes in mastitis milk can cause phagocytosis of the starter organisms, and heating results in no significant improvement (Sellars and Babel, 1978). However, Gajduseck and Sebela (1973) observed that while yoghurt starter cultures are inhibited by up to 35 % in milk containing high somatic cell counts, boiling the milk for 2 min, or heating it at 90 °C for 20 min, inactivated the inhibitory materials and allowed normal acid production.

Late lactation milk and spring milk have some effect on starter culture activity. The reason(s) has not been elucidated yet. It is possible to speculate that thiocyanates may be responsible (see Lawrence, 1970; Rutkowski *et al.*, 1973).

Other inhibitors in milk could be due to environmental pollution, such as insecticides, which can inhibit the starter organisms (Deane and Patten, 1971; Gajduskova and Lat, 1972). Volatile and non-volatile compounds (fatty acids, formic acid, formaldehyde, acetonitrile, chloroform, ether, etc.) in concentrations up to 100 ppm inhibit the growth of the *Streptococcus* spp. and *L. cremoris* (Kulshrestha and Marth, 1974*a,b,c*).

PRODUCTION SYSTEMS FOR BULK STARTER CULTURES

The production of a bulk starter culture, i.e. the culture used directly in the fermentation process, necessitates several stages of sub-culturing in order to meet the quantity required. The most important aspect of starter preparation is the protection of the culture from phage attack, and the methods used may be divided into two main systems: first, the use of mechanically protected tanks, and secondly, the propagation of starter in phage-resistant/inhibitory medium (PRM/PIM). The former approach is widely used in the UK, Australia and New Zealand, while the latter system is popular in the United States.

Mechanically Protected Systems

Examples of such methods for starter production are the Lewis and the Jones methods. The development of the former type is well documented by Lewis (1956) and Cox and Lewis (1972). The technique consists of using re-usable polythene bottles (115 and 850 g capacity) for mother and feeder cultures, respectively. These bottles are fitted with Astell rubber seals, and the growth medium, i.e. 10–12 % reconstituted antibiotic-free skim-milk powder, is sterilised in these bottles. The starter culture transfers are carried out by means of two-way hypodermic needles, and the overall technique is illustrated schematically in Fig. 4. The Lewis system requires a pressurised bulk starter tank where the growth medium is heat treated inside the sealed vessel. It is worth while pointing

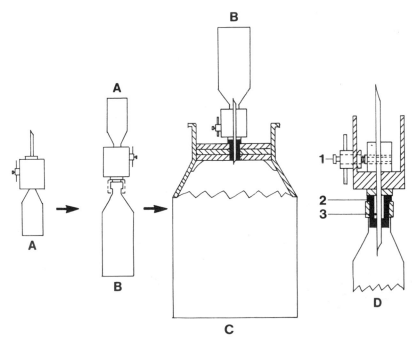

FIG. 4. Schematic illustration of the Lewis system for starter culture transfers. A, mother culture; B, feeder/intermediate culture; C, bulk culture; D, detail of needle assembly: (1) tap, (2) Astell seal, (3) hypochlorite solution.

(a)

(b)

FIG. 5.   Illustrations of some Lewis-type bulk starter tanks: (a) side view of starter tanks; (b) general view of the tank; (c) submerged inoculation orifice. Reproduced by courtesy of Dairy Cultures Ltd, Sherborne, UK.

(c)

Fig. 5—*contd.*

out that, during the heating or cooling of the milk, no air escapes from or enters the tank. The top of the tank is flooded with sodium hypochlorite solution (100 mg litre^{-1}), so that the transfer of the starter inoculum from feeder to bulk starter milk medium is through a sterile barrier (see Fig. 5). Figure 6 provides an example of such a tank (1136–2273 litre capacity), and Fig. 7 shows how inoculation is carried out. Incidentally, special lids can be fitted to 22·7–45·5 litre churns, so that the Lewis system can be used successfully for the production of smaller volumes of starter culture.

In the Lewis system, the starter transfer from one container to another relies on squeezing the polythene bottle to eject the culture. However, the Alfa-Laval system (Anon., 1975) uses filter-sterilised air, under pressure, for transferring the culture. The filter consists of a special hydrophobic fibre paper fitted with a prefilter on each side, and the whole unit is enclosed in a protective casing.

The other protective method is the Jones system, and the specification of the tank is illustrated in Fig. 8. During the heating or cooling of the milk, air leaves or enters the starter vessel, i.e. the tank is not pressurised. The manway cover is water sealed, and the air is

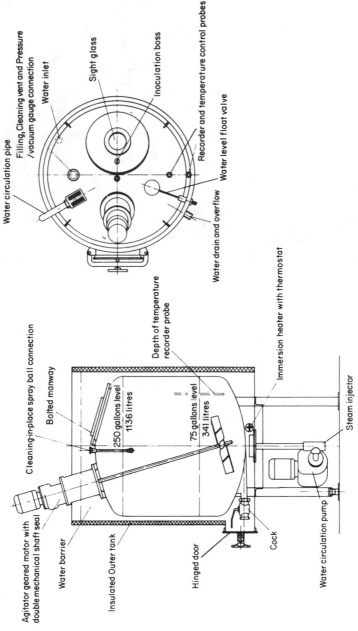

Fig. 6. Wincanton bulk starter tank designed for the Lewis system of starter culture preparation. Reproduced by courtesy of Wincanton Engineering Co. Ltd, Sherborne, UK.

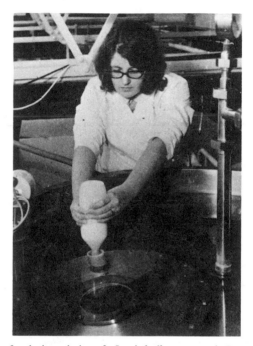

FIG. 7.   Procedure for the inoculation of a Lewis bulk starter tank. Reproduced by courtesy
of Dairy Cultures Ltd, Sherborne, UK.

sterilised using a combination of heat and cotton wool filtration. The starter inoculum is poured into the tank through a special narrow opening, and a 'ring of flame' or steam is used to provide a sterile point of entry. The historical background to, and development of, this system has been reported by Whitehead (1956) and Robertson (1966a,b). More recently Tofte-Jespersen (1979) discussed the principles of a similar tank with a different air filtration system (see Fig. 9).

PRM/PIM Method

The proliferation of phage in a dairy starter culture is dependent on the presence of free calcium ions in the growth medium. Reiter (1956) observed the inhibition of phage of lactic streptococci in a milk medium lacking in calcium. The medium was referred to as a 'phage-resistant medium' (PRM), but sometimes it is also called a 'phage-inhibitory medium' (PIM). The use of phosphates to sequester the free calcium ions in bulk starter skim-milk was reported by Hargrove (1959), and

148   A. Y. Tamime

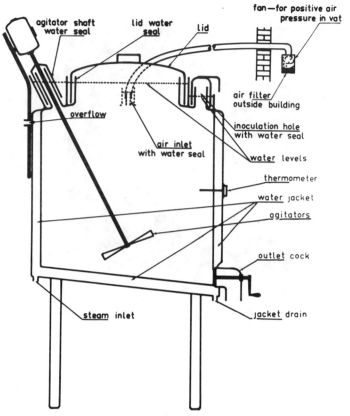

FIG. 8. Schematic drawing of the Jones-type bulk starter tank (130 or 250 gal. bulk starter unit).

this initiated the commercial production of PRM/PIM. Since 1960 there have been tremendous developments in PRM/PIM composition, and the existing PRM/PIM on the market consists mainly of milk solids, sugar, growth stimulatory factor(s) and buffering agents, e.g. phosphates and citrates. However, the effectiveness of PRM/PIM to protect and stimulate growth of starters is limited; a view based on the following research works.

1.   Sozzi (1972) and Sozzi et al. (1978) observed that some phages can still destroy starter cultures despite the absence of calcium, and that phage activity is maximised at the optimum growth temperatures of the starter organisms.

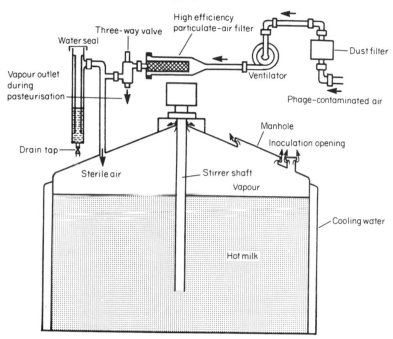

FIG. 9. Bulk starter tank incorporating a special air filtration unit. After: Tofte-Jespersen (1979); reprinted with permission of the *Journal of the Society of Dairy Technology.*

2. Henning *et al.* (1965) reported that *Leuconostoc* spp. do not grow well in PRM/PIM.

3. Ledford and Speck (1979) observed that phosphates in PRM/PIM cause metabolic injury to starter cultures, but that this effect was least pronounced in cultures and growth medium obtained from the same commercial source.

4. Gulstrom *et al.* (1979) evaluated seven commercial brands of PRM/PIM, and concluded that the bulk starter media were ineffective against phage, and failed to provide adequate growth. Best results were obtained in a medium containing hydrolysed cereal solids and buffered with citrates.

The inhibitory effect of phosphates can be reduced by using ammonia, and the development of a whey-based PIM/PRM was first reported by Ausavanodom *et al.* (1973). More recently, their PIM/PRM method has been replaced by the production of an active, concentrated starter culture in whey buffered with ammonia (Richardson, 1978; Richardson *et al.*, 1979). Some of the advantages claimed for such starter cultures are:

(1) less expensive as compared with PIM/PRM method;
(2) an active, highly concentrated, culture that never sours, and requires no ripening time in the cheese vat;
(3) daily or weekly propagation is not required;
(4) growth medium is cheap, readily available and consists of whey, water, lactose and stimulants.

As an alternative, Willrett *et al.* (1979) developed a new blend for bulk starter propagation. Cultures grown in this medium were more active and more resistant to phage inhibition than starters propagated in PIM/PRM, reconstituted skim-milk powder, or the Richardson's system mentioned above.

CLEANING AND SANITISATION OF EQUIPMENT

The mother and feeder equipment are cleaned by hand using a suitable detergent, and sanitisation, after cleaning, can be achieved using either steam or chemicals (e.g. a solution of hypochlorite containing 200 mg litre^{-1} of available chlorine) followed by a cold water rinse.

The majority of the bulk starter vessels are equipped for cleaning-in-place (CIP). The cleaning and sanitisation procedures are: cold water rinse to remove milk residues, hot detergent wash to remove denatured milk from the surface of the tank, water rinse followed by heat sanitation. If chemical sterilants are used, the starter vessel is rinsed with water to remove residues. In view of the fact that the starter milk is heat treated above 88 °C in the tank, sanitation of the tank after cleaning is not necessary (BSI, 1977).

QUALITY CONTROL

The ability of a starter culture to perform its function during the manufacture of fermented dairy products depends on its purity and activity. Some of the recommended routine quality control tests on starter cultures are as follows:

1. Lactic and other bacterial starter cultures.
    (i) Microscopic examination using Gram's stain, and/or Newman's staining method. The starter bacteria are Gram-positive, and

the latter method is used to monitor the ratio of cocci to rods in yoghurt cultures.

(ii) Detection of contaminants:
  (*a*) purity using the catalase test; LAB are catalase-negative, and a positive reaction is due to contamination;
  (*b*) a positive coliform test indicates gross contamination;
  (*c*) yeast and moulds must not be present in LAB cultures;
  (*d*) detection of phages.

(iii) An activity or vitality test helps to determine the rate of acid development by a starter culture prior to its use in the processing vat. For example, a simulated cheesemaking process in the laboratory.

(iv) Tolerance to scalding temperatures is an important test for cheese cultures.

(v) The Voges–Proskauer or Creatine test is a biochemical test to check aroma-producing cultures, and this test is sometimes useful to detect gas-producing organisms in a non-gas-forming culture.

2. Moulds
  (i) The activity is determined by counting the spores in a sterile suspension.
  (ii)* Sometimes a coliform test is carried out to detect gross contamination.

3. Kefir grains.
  (i) Activity test on the dried grains.
  (ii) Two possible contaminants may be present in the grains, i.e. coliforms and white moulds, e.g. *Geotrichum candidum*.

4. Swab tests on starter culture equipment and propagation vessels are essential to check on the efficiency of cleaning, and the sanitary condition of the plant.

ACKNOWLEDGEMENTS

The author acknowledges Dr R. J. M. Crawford (WSAC) for his critical comments during the preparation of this chapter, Mr J. E. Lewis (Dairy Cultures Ltd) for providing the photographs, Mr T. Rothwell (Wincanton Engineering Ltd) for providing schematic drawings of bulk starter tank, and Mr S. Crawford (WSAC) for the preparation of the schematic diagrams.

REFERENCES

ACCOLAS, J. P. and SPILLMANN, H. (1979a) *J. Appl. Bacteriol.*, **47**, 135.

ACCOLAS, J. P. and SPILLMANN, H. (1979b) *J. Appl. Bacteriol.*, **47**, 309.

ACCOLAS, J. P., AUCLAIR, J., BOUILLANNE, C., MOCQUOT, G., ROUSSEAUX, P., VALLES, E. and VASSAL, L. (1970) *XVIIIth Internat. Dairy Congr.*, 1E, 275.

ACCOLAS, J. P., VEAUX, M., ROUSSEAUX, P., SPILLMANN, H. and VASSAL, L. (1978) *XXth Internat. Dairy Congr.*, E, 597.

ANDERSON, L. (1975) *Dairy Sci. Abstr.*, **37**, 310. (*Swedish Patent* (1974) No. 369 470).

ANON. (1975) In: *Culture Preparation*, Alfa-Laval Technical Bulletin No. TB 60 738E. Alfa-Laval Co. Ltd, London.

ANON. (1980) *Chr. Hansen's Sales Information Bulletin*. Chr. Hansen's Lab. Ltd, Reading, UK.

AUSAVANODOM, N., WHITE, R. S. and RICHARDSON, G. H. (1973) *J. Dairy Sci.*, **56**, 637.

BANNIKOVA, L. A. and LAGODA, I. V. (1970) *XVIIIth Internat. Dairy Congr.*, 1E, 277.

BESTER, B. H. and LOMBARD, S. H. (1974) *S. African J. Dairy Technol.*, **6**, 47.

BOUILLANNE, C. and AUCLAIR, J. (1978) *XXth Internat. Dairy Congr.*, E, 583.

BOUTY, J. L., CINQUIN, J. L. and MOUCHOT, J. C. (1978) *XXth Internat. Dairy Congr.*, E, 589.

BREED, R. S., MURRAY, E. G. D. and SMITH, N. R. (Eds.) (1957) *Bergey's Manual of Determinative Bacteriology*, 7th edn. Williams and Wilkins Co., Baltimore, USA.

BRIGGS, M., TULL, G., NEWLAND, L. G. M. and BRIGGS, C. A. E. (1955) *J. Gen. Microbiol.*, **12**, 503.

BRONN, W. K. (1976) Cited by Sharpe, M. E. (1979).

BSI (1977) *Recommendations for Sterilization of Plant and Equipment Used in the Dairy Industry*, BS-5305. British Standards Institution, London.

BUCHANAN, R. E. and GIBBONS, N. E. (Eds.) (1974) *Bergey's Manual of Determinative Bacteriology*, 8th edn. Williams & Wilkins Co., Baltimore, USA.

CARINI, S. and LODI, R. (1978) *XXth Internat. Dairy Congr.*, E, 529.

CHAPMAN, H. R. (1978) *J. Soc. Dairy Technol.*, **31**, 99.

CHAPMAN, H. R., MCINTYRE, H. and SCURLOCK, E. M. W. (1978) *XXth Internat. Dairy Congr.*, E, 590.

CHUZHOVA, Z. P. (1977a) *Dairy Sci. Abstr.*, **39**, 670.

CHUZHOVA, Z. P. (1977b) *Dairy Sci. Abstr.*, **39**, 670.

CHUZHOVA, Z. P. and BELOVA, G. A. (1978). *XXth Internat. Dairy Congr.*, E, 591.

CIBLIS, E. (1966) *XVIIth Internat. Dairy Congr.*, C, 395.

COGAN, T. M. (1972) *J. Appl. Microbiol.*, **23**, 960.

COX, W. A. (1977) *J. Soc. Dairy Technol.*, **30**, 5.

COX, W. A. and LEWIS, J. E. (1972) In: *Safety in Microbiology*, Society for Applied Bacteriology—Technical Series No. 6, Eds. Shapton, D. A. and Board, R. G. Academic Press Inc. (London) Ltd, London.

COX, W. A., STANLEY, G. and LEWIS, J. E. (1978) In: *Streptococci*, Society for Applied Bacteriology—Technical Series No. 7, Eds. Skinner, F. A. and Quesnel, L. B. Academic Press Inc. (London) Ltd, London.

CRAWFORD, R. J. M. (1972) *Dairy Ind.*, **37**, 648.
CZULAK, J., BANT, D. J., BLYTH, S. C. and CRACE, J. B. (1979) *Dairy Ind. Internat.*, **44**(2), 17.
DAGONNEAU, H. and KUZDZAL-SAVOIE, S. (1978) *XXth Internat. Dairy Congr.*, E, 594.
DEANE, D. D. and PATTEN, M. M. VAN (1971) *J. Milk Fd Technol.*, **34**, 16.
ERICKSON, R. J. (1980) *Dairy Ind. Internat.*, **45**(3), 37.
FAO (1979) In: *Production Year Book*, Vol. 33. Food and Agriculture Organization of the United Nations, Rome.
FOSTER, E. M. (1962) *J. Dairy Sci.*, **45**, 1290.
FOWLER, G. G. (1969) *Milchwissenschaft*, **24**, 211.
GAJDUSEK, S. and SEBELA, F. (1973) *Dairy Sci. Abstr.*, **35**, 364.
GAJDUSKOVA, V. and LAT, J. (1972) *Acta Veterinaria—Brno*, **41**, 447.
GALESLOOT, TH. E. and HASSING, F. (1973) Cited by Sharpe, M. E. (1979).
GALLOWAY, J. (1980) Personal Communication.
GAVIN, M. (1968) In: *La Lyophilisation de Cultures de Yoghourt*, Thesis No. 4227. Ecole Polytechnique Federale, Zurich, Switzerland.
GILLILAND, S. E. and SPECK, M. L. (1974) *J. Milk Fd Technol.*, **37**, 107.
GULSTROM, T. J., PEARCE, L. E., SANDINE, W. E. and ELLIKER, P. R. (1979) *J. Dairy Sci.*, **62**, 208.
HANSEN, P. A. (1968) In: *Type Strains of* Lactobacillus *Species—A Report by the Taxonomic Subcommittee on Lactobacilli and Closely Related Organisms.* American Type Culture Collection, Rockville, Maryland, USA.
HARGROVE, R. E. (1959) *J. Dairy Sci.*, **42**, 906.
HEAP, H. A. and LAWRENCE, R. C. (1976) *New Zealand J. Dairy Sci. Technol.*, **11**, 16.
HENNING, D. R., SANDINE, W. E., ELLIKER, P. R. and HAYS, H. A. (1965) *J. Milk Fd Technol.*, **28**, 273.
HUP, G. and STADHOUDERS, J. J. (1977) *US Patent No.* 4 053 642.
HURST, A. (1977). *Canadian J. Microbiol.*, **23**, 936.
HYLMAR, B. and TEPLY, M. (1970) *XVIIIth Internat. Dairy Congr.*, 1E, 127.
IDF (1977) In: *Consumption Statistics for Milk and Milk Products*, Annual Bulletin Doc. No. 93. International Dairy Federation, Bruxelles, Belgium.
IDF (1978) In: *Consumption Statistics for Milk and Milk Products*, Annual Bulletin Doc. No. 103. International Dairy Federation, Bruxelles, Belgium.
IDF (1979) In: *Consumption Statistics for Milk and Milk Products*, Annual Bulletin Doc. No. 109. International Dairy Federation, Bruxelles, Belgium.
IMAI, M. and KATO, M. (1975) *J. Agric. Chem. Soc. Japan*, **49**, 93.
JAKUBOWSKA, J., LIBUDZISZ, Z. and PIATKIEWICZ, A. (1978) *XXth Internat. Dairy Congr.*, E, 578.
KILARA, A., SHAHANI, K. M. and DAS, N. K. (1976) *Cultured Dairy Products J.*, **11**(2), 8.
KORNACKI, K., RYMAZEWSKI, J., OZIMEK, G. and SMIETANA, Z. (1974) *XIXth Internat. Dairy Congr.*, 1E, 436.
KORNACKI, K., KAMKOWSKA, A., STEPANIAK, E., BARTOSZEWICZ, E., RYMASZEWSKI, J., POZNANSKI, S. and PSZCZOIKOWSKA, G. (1978) *XXth Internat. Dairy Congr.*, E, 590.
KOSIKOWSKI, F. V. (1977) In: *Cheese and Fermented Milk Foods*, 2nd edn. Edwards Brothers Inc., Michigan, USA.

KULSHRESTHA, D. C. and MARTH, E. H. (1974a) *J. Milk Fd Technol.*, **37**, 593.
KULSHRESTHA, D. C. and MARTH, E. H. (1974b) *J. Milk Fd Technol.*, **37**, 600.
KULSHRESTHA, D. C. and MARTH, E. H. (1974c) *J. Milk Fd Technol.*, **37**, 606.
LABLEE, J. and PERROT, C. (1978) *XXth Internat. Dairy Congr.*, E, 588.
LAGODA, I. V., BANNIKOVA, L. A., BAVINA, N. A., ROZHKOVA, I. V. and PRUSAKOVA, N. V. (1979) *Dairy Sci. Abstr.*, **41**, 109.
LAW, B. A. and SHARPE, M. E. (1978) In: *Streptococci*, Society for Applied Bacteriology—Technical Series No. 7, Eds. Skinner, F. A. and Quesnel, L. B. Academic Press Inc. (London) Ltd, London.
LAWRENCE, R. C. (1970) *XVIIIth Internat. Dairy Congr.*, 1E, 99.
LAWRENCE, R. C. (1978) *Paper '2ST'* read at *XXth Internat. Dairy Congr.*, Paris, France.
LAWRENCE, R. C. and THOMAS, T. D. (1979) In: *Microbial Technology: Current State, Future Prospect*, 29th Symposium of the Society of General Microbiology, Eds. Ball, A. T., Ellwood, D. C. and Ratledge, C. Cambridge University Press, London.
LAWRENCE, R. C., THOMAS, T. D. and TERZAGHI, B. E. (1976) *J. Dairy Res.*, **43**, 141.
LEDFORD, R. A. and SPECK, M. L. (1979). *J. Dairy Sci.*, **62**, 781.
LEWIS, J. E. (1956) *J. Soc. Dairy Technol.*, **9**, 123.
LIBUDZISZ, Z., PIATKIEWICZ, A. and JAKUBOWSKA, J. (1977) *Acta Alimentaria Polonica*, **3**, 433.
LIMSOWTIN, G. K. Y., HEAP, H. A. and LAWRENCE, R. C. (1977) *New Zealand J. Dairy Sci. Technol.*, **12**, 101.
LLOYD, G. T. (1971) *Dairy Sci. Abstr.*, *Review Article No. 163*, **33**, 411.
LLOYD, G. T. and PONT, E. G. (1973) *Australian J. Dairy Technol.*, **28**, 104.
MANUS, L. J. (1979) *Cultured Dairy Products J.*, **14**(1), 9.
MARSHALL, R. S. and BERRIDGE, N. J. (1976) *J. Dairy Res.*, **43**, 449.
MIETTON, B., PERNODET, G., SAGEAUX, R. and TINGUELY, P. (1978) *XXth Internat. Dairy Congr.*, E, 587.
MIETTON, B., HARRY, C., PERNODET, G. and TINGUELY, P. (1979) *Dairy Sci. Abstr.*, **41**, 39.
MORICHI, T. (1972) In: *Mechanism and Prevention of Cellular Injury in Lactic Acid Bacteria, Subjected to Freezing and Drying*. IDF Scientific Conference—Japan, 56th Annual Session.
MORICHI, T. (1974) *Japan Agric. Res. Quarterly*, **8**, 171.
MORICHI, T., IRIE, R., YANO, N. and KEMBO, H. (1967) *Agric. Biol. Chem.*, **31**, 137.
NAGHMOUSH, M. R., GIRGIS, E. S., GUIRGUIS, A. H. and FAHMI, A. H. (1978) *Egyptian J. Dairy Sci.*, **6**, 39.
NAYLOR, H. B. and SMITH, P. A. (1946) *J. Bacteriol.*, **52**, 565.
NIKOLOVA, N. (1974) *Dairy Sci. Abstr.*, **36**, 596.
NIKOLOVA, N. (1975) *Dairy Sci. Abstr.*, **37**, 414.
NIKOLOVA, N. (1978) *XXth Internat. Dairy Congr.*, E, 584.
OBERMAN, H., PIATKIEWICZ, A. and LIBUDZISZ, Z. (1978) *XXth Internat. Dairy Congr.*, E, 577.
OLSON, N. F. (1969) In: *Ripened Semisoft Cheeses*, Pfizer Cheese Monograph Vol. IV. Chas. Pfizer & Co. Inc., New York.
ORLA-JENSEN, S. (1931) In: *Dairy Bacteriology*, 2nd edn., Translated by Arup, P. S. J. & A. Churchill, London.

OSBORNE, R. J. W. (1977) *J. Soc. Dairy Technol.*, **30**, 40.
OTTOGALLI, G. and RONDININI, G. (1976) *Dairy Sci. Abstr.*, **38**, 109.
OTTOGALLI, G., GALLI, A., REMINI, P. and VOLONTERIO, G. (1973) Cited by Sharpe, M. E. (1979).
OZLAP, E. and OZLAP, G. (1979) *Dairy Sci. Abstr.*, **41**, 871.
PALLASDIES, K. (1977) *Dairy Sci. Abstr.*, **39**, 298.
PEAKE, S. E. and STANLEY, G. (1978a) *J. Appl. Bacteriol.*, **44**, 321.
PEAKE, S. E. and STANLEY, G. (1978b) *XXth Internat. Dairy Congr.*, E, 833.
PEARCE, L. E. (1978) *New Zealand J. Dairy Sci. Technol.*, **13**, 56.
PEARCE, L. E. and BRICE, S. A. (1973) *New Zealand J. Dairy Sci. Technol.*, **8**, 17.
PETTERSSON, H. E. (1975) *Milchwissenschaft*, **30**, 539.
PIATKIEWICZ, A., LUBUDZISZ, Z. and OBERMAN, H. (1978) *XXth Internat. Dairy Congr.*, E, 582.
PORUBCAN, R. S. and SELLARS, R. L. (1975a) *J. Dairy Sci.*, **58**, 787.
PORUBCAN, R. S. and SELLARS, R. L. (1975b) *US Patent No.* 3 897 307.
POULSEN, P. R. (1970) *XVIIIth Internat. Dairy Congr.*, 1E, 409.
RAPER, K. B. and THOM, C. (1968) In: *A Manual of the Penicillia*. Hafner Pub. Co., New York.
RASIC, J. L. and KURMANN, J. A. (1978) In: *Yoghurt—Scientific Grounds, Technology, Manufacture and Preparations*. Technical Dairy Pub. House, Copenhagen, Denmark.
REINBOLD, G. W. and REDDY, M. S. (1973) *Dairy Ind.*, **38**, 413.
REINBOLD, G. W. and REDDY, M. S. (1974) *J. Milk Fd Technol.*, **37**, 517.
REITER, B. (1956) *Dairy Ind.*, **21**, 877.
REITER, B. (1973) *J. Soc. Dairy Technol.*, **26**, 3.
REITER, B. and KIRIKOVA, M. (1976) *J. Soc. Dairy Technol.*, **29**, 221.
RICHARDSON, G. H. (1978) *Dairy and Ice Cream Field*, **161**(9), 80A.
RICHARDSON, G. H., HONG, G. L. and ERNSTROM, C. A. (1979) In: *Lactic Cultures*, Utah Agricultural Experimental Station, Research Report No. 42. Utah State University, USA.
ROBERTSON, P. S. (1966a) *Dairy Ind.*, **31**, 805.
ROBERTSON, P. S. (1966b) *XVIIth Internat. Dairy Congr.*, D, 439.
ROGOSA, M. and HANSEN, P. A. (1971) *Internat. J. Systematic Bacteriol.*, **21**, 177.
RUTKOWSKI, A., KOZLOWSKA, H. and HOPPE, K. (1973) *Dairy Sci. Abstr.*, **35**, 243.
SANDINE, W. E. (1979) In: *Lactic Starter Culture Technology*, Pfizer Cheese Monographs Vol. VI. Chas. Pfizer & Co. Inc., New York.
SELLARS, R. L. (1967) In: *Microbial Technology*, Ed. Peppler, H. J. Reinbold Pub. Corp., New York.
SELLARS, R. L. (1975) Personal communication.
SELLARS, R. L. and BABEL, F. J. (1978) In: *Cultures for the Manufacture of Dairy Products*, Revised edn. Chr. Hansen's Lab. Inc., Wisconsin, USA.
SHANKAR, P. A. (1975) Personal communication.
SHARPE, M. E. (1979) *J. Soc. Dairy Technol.*, **32**, 9.
SINHA, R. N., NAMBUDRIPAD, V. K. N. and DUDANI, A. T. (1970) *XVIIIth Internat. Dairy Congr.*, 1E, 128.
SINHA, R. N., DUDANI, A. T. and RANGANATHAN, B. (1974) *J. Fd Sci.*, **39**, 641.
SMITTLE, R. B. (1973) *Dissertation Abstr. Internat. B*, **34**, 2446.

156   *A. Y. Tamime*

SMITTLE, R. B., GILLILAND, S. E. and SPECK, M. L. (1972) *J. Appl. Microbiol.*, **24**, 551.
SMITTLE, R. B., GILLILAND, S. E., SPECK, M. L. and WALTER JR, W. M. (1974) *J. Appl. Microbiol.*, **27**, 738.
SOZZI, T. (1972) *Milchwissenschaft*, **27**, 503.
SOZZI, T., POULIN, J. M. and MARET, R. (1978) *J. Dairy Res.*, **45**, 259.
SPECKMAN, C. A. (1975) *Dissertation Abstr. Internat. B*, **35**, 3474.
STADHOUDERS, J., JANSEN, L. A. and HUP, G. (1969) *Netherland Milk and Dairy J.*, **23**, 182.
STEFFEN, C., NICK, B. and BLANC, B. H. (1974) *XIXth Internat. Dairy Congr.*, 1E, 404.
TAMIME, A. Y. (1977) In: *Some Aspects of the Production of Yoghurt and Condensed Yoghurt*, Ph.D. Thesis. University of Reading, England, UK.
TAMIME, A. Y. and ROBINSON, R. K. (1976) *Dairy Ind.*, **41**, 408.
TEPLY, M. and CERMINOVA, N. (1974) *XIXth Internat. Dairy Congr.*, 1E, 428.
THOMAS, T. D. and LOWRIE, R. J. (1975) *J. Milk Fd Technol.*, **38**, 6.
TOFTE-JESPERSEN, N. J. (1974a) *S. African J. Dairy Technol.*, **6**, 63.
TOFTE-JESPERSEN, N. J. (1974b) In: *A New View of International Cheese Production*. Published by Chr. Hansen's Lab. A/S, Copenhagen, Denmark.
TOFTE-JESPERSEN, N. J. (1976) *Dairy and Ice Cream Field*, **159**(5), 58 A.
TOFTE-JESPERSEN, N. J. (1979) *J. Soc. Dairy Technol.*, **32**, 190.
TSVETKOV, TS. and NIKOLOVA, N. (1975) *Dairy Sci. Abstr.*, **37**, 618.
TURNER, K. W., DAVEY, G. P., RICHARDSON, G. H. and PEARCE, L. E. (1979). *New Zealand J. Dairy Sci. Technol.*, **14**, 16.
UDEN, N. VAN and BUCKLEY, H. (1971) In: *The Yeast—A Taxonomic Study*, Ed. Lodder, J. North Holland Pub. Co., Amsterdam, Holland.
VASSAL, L., BOUILLANNE, C. and AUCLAIR, J. (1978) *XXth Internat. Dairy Congr.*, E, 585.
WALKER, A. (1980) Personal communication.
WHITEHEAD, H. R. (1956) *Dairy Eng.*, **73**, 159.
WHITEHEAD, H. R. and COX, G. A. (1935) Cited by Reiter and Kirikova (1976).
WIGLEY, R. C. (1977) *J. Soc. Dairy Technol.*, **30**, 45.
WIGLEY, R. C. (1980) *J. Soc. Dairy Technol.*, **33**, 24.
WILLRETT, D. L., SANDINE, W. E. and AYRES, J. W. (1979) *The Cheese Reporter*, **103**(18), 8.
YANG, N. L. and SANDINE, W. E. (1978) *J. Dairy Sci. (Sup. 1)*, **61**, 121.
YANG, N. L. and SANDINE, W. E. (1979) *J. Dairy Sci.*, **62**, 908.

5

Microbiology of Cheese

HELEN R. CHAPMAN and M. ELISABETH SHARPE

*National Institute for Research in Dairying,
Shinfield, UK*

Made with the biological materials milk, rennet and micro-organisms, cheese is a continually developing substrate, requiring time to reach maturity, and affected greatly by the conditions under which it is produced and stored. An understanding of the properties and behaviour of the materials used in the manufacture of cheese is essential for those concerned with its production, ripening and marketing.

HISTORICAL

There is no real knowledge of the origins of cheese or cheese making, but the earliest records of human activities refer to cows and milk. These may be found in Sanskrit writings of the Sumarians, *c.* 4000 BC, in Babylonian records of 2000 BC, and in the Vedic hymns. The preparation of cheese probably dates back many centuries to the time when nomadic tribes of eastern Mediterranean countries carried milk of domesticated mammals in sacks, made from animal skins, or gourds, or in vessels such as stomachs or bladders. If kept warm the milk rapidly became sour and separated into curds and whey. If the whey was drained from the curds, the latter could be dried to form a firm, cheesy mass that could be eaten fresh, or stored and eaten over long periods. In this way much of the food value of milk could be preserved for use when supplies of liquid milk were not available. This primitive system survives today in the preparation of 'Kishk', the dried curd of Arabia, which according to some reports consists of soured curds separated from whey, which are pressed, salted and left to dry on the goat-hair roof of the nomadic tents.

In time it was found that the secretion from the stomach of a young ruminant had the power to coagulate milk, which reduced the time required to drain the whey from the curds. This eventually led to the use of rennet, an extracted enzymic secretion of the fourth stomach of a young calf, lamb or kid, to bring about the coagulation of milk, which is the first step in the process of cheesemaking as it is practised today.

Aristotle (384–322 BC) speaks of 'Phrygian' cheese made from ass' milk and mare's milk, and he was aware that cheese consisted of fat, casein and water, and the Greeks and Romans certainly recognised its nutritional value. Their methods of making cheese were simple, but they had progressed far beyond the primitive stages of drying an acid curd in the sun. The cheesemaking process described by Columella (*De Re Rustica, c.* 50) in the first century AD shows that there had been a gradual evolution from the acidic curdling of milk by natural fermentation to the controlled production of a form of curd which could be preserved. This knowledge spread through the countries of the Roman Empire. In Britain there is no positive evidence of cheesemaking before the Roman conquest, though presumably some form of milk curds were used. During the ensuing years of Roman occupation, however, cheese became a well known food. Palladius, who wrote a treatise on '*Agriculture in Britain*' 300 years after the conquest advocated that cheesemaking should take place in early summer, that milk should be curdled with rennet obtained from the stomachs of the kid, the lamb or the calf or, alternatively, by the milk from the fig tree or teasel flowers. This is direct evidence of the coagulation of milk by agents other than natural acidity, and by rennets of vegetable as well as animal origin.

Different ways of making cheese were developed in different countries, and in different areas within a country, as the outcome of experience and to suit local and market demands. These cheeses, whatever the country of origin, were the forerunners of varieties which have been stabilised and named, and have assumed local, national and often international importance.

Whilst there are over 400 varieties of cheese, there are only about 18 distinctly different types. Many varieties are named after their place of origin and differ from one another only in shape and method of packaging; their method of manufacture and general characteristics being very similar.

TYPES OF CHEESE

All varieties of natural cheeses are made from milk. They can be divided into three main classes, viz., soft, blue-veined and hard-pressed cheese. These vary widely in moisture content and, therefore, in keeping quality and method of ripening.

Soft cheese curd retains a high proportion of moisture (whey) (55–80%). Some varieties are eaten fresh (Cambridge, Coulommier, Bondon, etc.), whilst others are ripened, usually by the growth of surface moulds (Brie, Camembert, Pont l'Eveque, etc.). Semi-soft cheeses, such as Limburger, Tilsit and Brie, are made from slightly firmer curds (45–55% moisture) and are ripened by the surface growth of micro-organisms, particularly *Brevibacterium linens*. These are the smear-ripened cheeses.

Blue-veined cheeses, such as Stilton, Roquefort and Gorgonzola, are made from semi-soft/semi-hard curd with 42–52% moisture, and are ripened by species of *Penicillium* moulds which grow within the cheese.

The semi-hard cheeses, such as Edam and Gouda, are made from firmer curd with a moisture content within the range of 45–50%. The cheeses are ripened by bacteria and are consumed within 2–3 months.

The hard-pressed cheeses are made from firm, relatively dry curd (35–45% moisture). They are ripened by bacteria and mature slowly over a period varying from 3–12 months. In some varieties (e.g. Cheddar and Cheshire) acid is developed in the curd before it is salted and pressed. In other varieties (e.g. Emmenthal and Gruyère) acid is developed whilst the curd is draining and being pressed, but before it is salted.

The very hard, grating cheeses, such as Parmesan, Romano and Asiago, are made from very firm curd. They are low-moisture cheeses (26–34%), made from partly skimmed milk and are ripened by bacteria, slowly, over a period of 1–2 years.

The consistency of a cheese, its firmness or body, is determined by certain basic factors, and control of these is essential to ensure that the properties are characteristic of the variety and will provide suitable conditions for correct ripening. Softness is favoured by high moisture content, high fat content and extensive proteolysis. The opposite of these features characterise the hard varieties of cheese with firm body.

The variety of cheese to be produced in any class is determined by the type of milk used, the preparation of the young curds, and the inclusion in the milk or curds of certain micro-organisms responsible for the development of acidity during manufacture, and the development of characteristic features and flavours during ripening. The types of bacteria or moulds which, by their growth during cheesemaking or cheese-ripening, participate in the process, are determined by deliberate inoculation of specific organisms, conditions of cheesemaking and environmental factors.

MICRO-ORGANISMS IN CHEESE

In order to study the microbiology of cheese it is necessary to enumerate and isolate the main groups of bacteria present in the milk, the cheese curd, and in the ripening cheese. To assess their role in cheese-ripening, it may be necessary to follow the survival or multiplication of these groups over a long period. For other purposes it may be necessary to detect potentially toxin forming, pathogenic or spoilage organisms. To determine the total numbers and main groups of micro-organisms present in cheese milk and cheese, both general purpose and selective plating media are used. Most of these media are described by Law et al. (1973).

General Purpose Media
Total Bacterial Counts

Milk agar is used for total counts in milk. Yeast glucose agar is used for total counts in curd and young cheese, i.e. where starter streptococci predominate.

Selective and Diagnostic Media
Starter Streptococci

A β-glycerophosphate medium is used for enumeration of the total starter organisms present, but to distinguish and enumerate the different types of lactic streptococci in a multiple strain starter, diagnostic media are used. The M16 BCP medium distinguishes *Streptococcus cremoris* by the production of yellow colonies; those of *Str. lactis* remain white. A calcium citrate medium differentiates aroma-producing streptococci and leuconostocs by a clearing zone round the colonies. These media can be used to follow the survival rates of different starter components in the cheese (see Sharpe, 1978).

Non-Starter Lactic Acid Bacteria

Rogosa's acetate (SL) medium is used for the isolation of lactobacilli, pediococci and leuconostocs. Its selective action, due to a low pH (5·4) and a high concentration of acetate ions, suppresses the growth of all other organisms including starter streptococci.

Staphylococci and Micrococci

Chapman's salt mannitol medium is used. The high salt content inhibits other organisms, and mannitol fermentation differentiates the mannitol-positive species which are likely to be *Staphylococcus aureus*. Coagulase-positive strains can be differentiated on Baird-Parker medium if necessary.

Gram-Negative Rods

Crystal violet (2 ppm) nutrient agar is used for Gram-negative rods, including *Pseudomonas*, *Achromobacter* and coliforms. Violet red bile agar further selects the coliforms in this group.

Group D Streptococci

Thallous acetate tetrazolium glucose agar is used to enumerate Group D streptococci, including enterococci. Incubation of this medium at 45 °C instead of 37 °C inhibits the growth of starter streptococci.

Lipolytic Bacteria

These may include organisms over a wide range of species, and are detected on Victoria blue butter-fat agar (Fryer *et al.*, 1967). Hydrolysis of the butter-fat to free fatty acids is detected by the indicator Victoria blue changing from pink to blue.

Proteolytic (Caseinolytic) Bacteria

These may also cover a wide range of species. They are enumerated on a milk agar, containing 15% skim-milk, where a clearing zone, brought about by hydrolysis of the milk casein, develops round the colonies.

Corynebacteria

These organisms usually occur as orange or yellow pigmented colonies on nutrient agar. They can only be detected, however, if the total number of *non*-lactic acid bacteria, which also grow on this medium, is small enough not to overgrow them.

Moulds

These can be cultivated on yeast maltose agar, a similar medium to yeast glucose agar but with the glucose replaced by maltose, and the pH adjusted to 5·5.

MILK FOR CHEESEMAKING

Cheese is generally made from cow's milk but in some countries, and for making certain varieties of cheese, milk of other mammals is used. Sheep's milk is used for making Roquefort cheese, goat's milk for many varieties of cheese in Italy and Greece, and buffalo's milk is used in India and Egypt. Whatever the source of the milk it is essential that it is produced by healthy animals. Mastitis milk is pathogenically abnormal, and although it may not affect factory cheesemaking because of the dilution effect, the pathogens present are a potential danger to the farmhouse cheesemaker, and may well contaminate the product itself. The raw milk should be of good general bacteriological quality to avoid undesirable fermentations and enzymic reactions, and should be free from inhibitory substances, such as residual antibiotics, which interfere with the growth of the starter bacteria.

Raw milk is used in the UK by some farmhouse cheesemakers who are able to exercise personal control over the production of their milk supply, in much small-scale cheesemaking in the rest of Europe, and largely in Asia. Untreated milk is still considered by many cheesemakers to produce more flavourful cheese.

Milk Storage

Modern cheesemaking plants, designed to handle daily throughputs of up to 250 000 gal. of milk by mechanised and semi-continuous systems, can only operate effectively if there is an adequate supply of milk at the start of the making process. Thus it has become general practice in many cheese factories to collect and store the whole of the incoming milk for use the following day, or on subsequent days. Developments in methods of cooling, transport and storage of milk have made this possible. At the farm, milk is cooled to 4 °C and stored in refrigerated bulk tanks. It is collected in refrigerated or insulated road tankers, and transported to the cheese factory where it is stored in insulated silos until it is used (Flückiger *et al.*, 1980). At this stage the milk could be between 24 and 72 h old, and may contain over 10^6 bacteria ml^{-1}, mainly psychrotrophs. Raw milk on arrival at the creamery will have total counts of about 10^3–10^7 ml^{-1}, depending on the levels of hygiene at the farms. The organisms present consist of psychrotrophs, mostly *Pseudomonas, Aeromonas, Alcaligenes,* small numbers of lactic acid bacteria, spore-forming Gram-positive rods, coryneform bacteria, micrococci and coliforms. Of these, only the psychrotrophs will multiply during transport and storage, particularly if the temperature in the insulated tankers and milk silos is allowed to rise. This growth leads to the production of extracellular lipases and proteinases, particularly by pseudomonads, *Achromobacter, Acinetobacter* and *Aeromonas.* Efficient pasteurisation of the milk for cheesemaking (72°C/15–17 s) greatly reduces the total count of micro-organisms in the cheese milk, but their enzymes may survive the heat treatment and give rise to off-flavours in the cheese (Law *et al.,* 1976).

Effect on Cheese Yields

These enzymes have the potential to affect adversely the processing properties of the milk, and to impair the quality of milk products (Law, 1979). Lipases may cause rancidity in Cheddar cheese (Law *et al.,* 1976); in Dutch cheese (Driessen and Stadhouders, 1971); in Swiss cheese (Pinheiro *et al.,* 1965) and in Camembert (Dumont *et al.,* 1977). Proteinases may degrade casein in raw milk during storage producing peptides, which sequentially are lost in the whey and may affect the yields of cheese. Excessive amounts of nitrogen were lost into the whey of soft cheese made from milk with psychrotrophic counts of 10^6 (Feuillat *et al.,* 1976), whereas with Cheddar cheese, losses of nitrogen into the whey, and therefore the yield of cheese, were not influenced by psychrotrophic counts of 10^7 cfu ml^{-1} in the raw milk (Law *et al.,* 1979).

Standardisation

To meet a standard fat content in the cheese, the milk may be 'standardised', that is the composition of the milk may be adjusted for fat by adding cream or by removing part of the fat, or by adding skim-milk or non-fat milk solids (skim-milk powder). A reduced fat content is necessary in making the very hard varieties of cheese, whereas other varieties require a more balanced ratio of fat to casein, and cream cheeses are made from cream or milk with a high fat content (Chapman *et al.,* 1974).

Clarification

This is a process for the removal of extraneous matter from milk by centrifugal force in a clarifier, a machine similar to a cream separator. It is sometimes used as a more effective alternative to filtration, and is carried out at 32·2–37·8 °C as the milk is being pasteurised. The influence of clarification on the quality of most cheese is problematic, but for Swiss-type cheese it has a beneficial effect on eye-formation, and some makers of Cheddar cheese consider that it improves the quality of their cheese, possibly by removing leucocytes and some bacteria. Clarification breaks up clumps of bacteria which increases the bacterial count of the milk, but not the total bacterial content.

Bactofugation

Bactofugation, or bacterial centrifugation, is a process for the removal of bacteria from milk in a high speed centrifuge. It is particularly effective in removing the relatively heavier spore-forming bacteria. The bactofuge is a modified hermetic clarifier, in which the heavier milk components (bacterial concentrate or bactofugate) are separated from the bactofugated milk and discharged separately. The bactofugate amounts to 2–3 % of the total milk flow, and contains 80–90 % of the bacteria present in the original milk. This is sterilised, while the bulk of the milk is pasteurised normally, and is then re-united with the main flow of bactofugated milk. Bactofugation increases the quality of Cheddar cheese made from low-grade milk (Kosikowski and O'Sullivan, 1966).

Hydrogen Peroxide–Catalase Treatment

This treatment of milk for cheesemaking with hydrogen peroxide (0·04–0·08 % H_2O_2, at 52–53 °C for 30 min), effectively reduces the total count of bacteria present in the milk, and the numbers of coliform organisms. It does increase the cost of cheesemaking, and it may impart a 'chemical-like' flavour to the cheese. It does not ensure destruction of all pathogenic bacteria (Keogh, 1964).

Homogenisation

This is a process for pumping milk under pressure. It disrupts the fat globules, dispersing them throughout the milk and thereby increasing its viscosity. At low pressures (500–1000 lb in^{-2}), the treatment can be beneficial in the manufacture of cream cheese, soft cheese and blue-veined cheese, where smooth curds are required. The smaller fat globules increase lipolysis in the cheese which accelerates flavour development in certain varieties of blue-veined cheese. The process is carried out as the milk is being pasteurised.

Pasteurisation
Where large volumes of milk are processed into cheese daily, the presence of potential spoilage and pathogenic organisms dictate that some form of heat treatment be used to destroy most of these organisms. In the UK, milk for cheesemaking in factories, and in some large-scale farmhouse enterprises, is given a form of heat treatment sufficient to destroy undesirable bacteria without affecting the physical and chemical properties of the milk. The temperature–time conditions of pasteurisation by the HTST method (72 °C for 15 s) meet those requirements, and the minimum heat treatment recommended is 68 °C for 15 s. All the vegetative bacteria (including pathogens) are killed except for heat-resistant micrococci, Group D streptococci and coryneform bacteria. Spores of *Bacillus* and *Clostridium* species also survive.

Thermisation
This term is used to describe a low heat treatment in which the conditions of temperature and time vary considerably with the use of the treatment. Milk, which must be stored at the factory before it can be processed, may be subjected to thermisation (63 °C/10–15 s) and cooled to 5 °C prior to storage in insulated silos. Milk for making cheese in certain parts of Europe and America may be thermised or heat treated at various temperature and time combinations, such as 63–68 °C for 25 s, 68 °C for 40 s, 70 °C for 15 s, 60 °C for 16 s or 65 °C for 2 s; not quite sufficient to give a negative phosphatase reaction. Thermisation is not a legal process and it is not a substitute for pasteurisation.

Inhibitory Salts
In some countries it is permissible to add sodium nitrate ($NaNO_3$) to the milk for making some of the less acid varieties of cheese, such as Edam and Gouda, to prevent the growth of gas-producing clostridia which cause 'blowing' of cheese. Up to 20 g per 100 litres are added to raw or pasteurised milk in the vat to prevent the growth of *Cl. tyrobutyricum* in the curd until the salt concentration reaches the desired level in the cheese after brining. Lactic acid bacteria and propionic acid bacteria are not affected.

Additional Salts
Where soluble calcium salts have been precipitated by an excessively high temperature treatment of the milk, calcium chloride, added at the rate of 0·01–0·03 %, improves the firmness of the rennet coagulum.

THE NORMAL FLORA OF CHEESE MILK

The bacterial flora of pasteurised cheese milk finally consists of the thermoduric organisms which have survived pasteurisation, namely some corynebacteria, micrococci, enterococci and spores of *Bacillus* and *Clostridium*; and post-pasteurisation contaminants such as other micrococci, occasionally coagulase-positive staphylococci, coliforms, lactic acid bacteria including lactobacilli, pediococci and sometimes leuconostocs and enterococci. These organisms are derived from the pipe-lines, cheese vats and utensils, the air, the washing water and from the dairy personnel. Where cheese is manufactured on a large scale and closed vats and mechanised processes are used, the level of contamination will be much less than where open vats and manual processes are used. The level of hygiene in the creamery plays a large part in influencing the level of post-pasteurisation contamination. The necessity to control phage proliferation, especially in large factories where the vats are re-used at least once during the day, means that strict hygiene must be observed.

In addition to the micro-organisms present, proteolytic and lipolytic heat-resistant enzymes will also be present. These enzymes, derived from psychrotrophs which are themselves killed by pasteurisation, survive and remain active in the milk. All the organisms present in the milk become part of the fresh curd flora, being concentrated about 10 times in the curd. The level of all these non-starter bacteria present in the curd is related to the history of the raw milk (e.g. cooling, refrigeration, age, etc.), the heat treatment it receives, and to the level of hygiene and cleanliness in the creamery.

CHEESE STARTERS

Production of cheese depends on fermentation of the lactose by lactic acid bacteria (LAB) to form mainly lactic acid. This imparts a fresh, acid flavour to curd cheeses, assists in the formation of the rennet coagulum and, by causing shrinkage of the curd and moisture expulsion, promotes characteristic texture formation during cheesemaking. The low pH of fresh cheese curd (5·0–5·2) helps to suppress the growth of pathogenic and spoilage bacteria, and thus preserves the product. Lactic acid bacteria also produce traces of flavourful aroma compounds, and their proteolytic and, to a lesser degree, lipolytic activity aids the maturation of cheese. In addition, growth of lactic acid bacteria produces the low oxidation–reduction potential (E_h), necessary for the production of reduced sulphur compounds, such as methanethiol, which may contribute to the aroma of Cheddar cheese (Manning *et al.*, 1976). At one time, the fermentation of lactose in milk for cheesemaking depended on natural contamination of the milk with LAB from the environment. In the dairy industry of today, where large volumes of milk, sometimes up to 250 000 gal. day^{-1}, are processed into cheese, a rapid and reliable fermentation is essential. Naturally occurring LAB have been destroyed by the pasteurisation process, and are replaced by LAB deliberately inoculated into the cheese milk. Selected strains, with predictable acid development and production of flavourful products, are used as starters to obtain a steady rate of acidity throughout the curd-making process, greater regularity of acid production in the day-to-day cheesemaking, and a uniform product.

Types of Starters
Lactic acid bacteria used as starters in cheesemaking include streptococci, leuconostocs and lactobacilli. Selected species of these genera are used as combined cultures, or as single strain cultures, or as mixtures of single strain cultures (Sharpe, 1979). Mesophilic starters (optimum temperature 20–30 °C) are used to produce a wide variety of cheese (Table I). For hard cheese, selected strains of *Str. cremoris* consistently give cheese of good flavour and free from off-flavours, but pairing with selected strains of *Str. lactis* can shorten the manufacturing time. *Streptococcus lactis* sub-sp. *diacetylactis* and/or leuconostocs, which produce CO_2, give the desired open texture for mould-ripened cheese, or for slight eye formation as in Dutch cheese. These species also produce flavourful diacetyl. Thermophilic starters (optimum temperature 37–45 °C) are used for production of cooked cheese varieties (e.g. Swiss and Parmesan), where the starter must be able to withstand a high cooking temperature of ~ 45 °C and grow at relatively high temperatures (Table I).

It is essential to have a satisfactory starter which produces sufficient acid at a rate required for the particular making process, does not produce off-flavours or bitterness, and provides conditions in the curd which are suitable for typical flavour development. Starters multiply during cheesemaking from about 10^7 cfu ml^{-1} in milk to 10^8–10^9 cfu g^{-1} of curd. Their growth is checked by the salting stage.

All strains of starter bacteria are inhibited by antibiotics such as penicillin, aureomycin, etc., which may be present in milk following

TABLE I
LACTIC ACID BACTERIA EMPLOYED AS STARTER CULTURES

Bacteria	Examples of usage
Mesophilic starters[a]	
Str. cremoris, Str. lactis or Str. cremoris Str. lactis Str. lactis sub-sp. diacetylactis Leuconostoc spp.	Hard-pressed cheese, e.g. Cheddar, Gouda; Mould-ripened, e.g. Stilton; Soft ripened, e.g. Camembert, Feta, many other types
Str. cremoris Str. lactis sub-sp. diacetylactis L. cremoris	Soft, unripened, e.g. Coulommier, Cottage cheese, Quarg, Cream cheese
Thermophilic starters[b]	
Str. thermophilus with Lac. helveticus, Lac. lactis or Lac. bulgaricus	Swiss-type, e.g. Emmenthal; Italian, very hard, e.g. Parmesan; Semi-soft, smear types, e.g. Limburger
Mixed starters	
Str. lactis, Str. thermophilus or Str. faecalis and Lac. bulgaricus	Italian pasta filata type, e.g. Mozzarella, Provolone

[a] Optimum temperature 20–30 °C.
[b] Optimum temperature 37–45 °C.
For further details see Chapter 4.

the treatment of cows for mastitis. Penicillin is the most powerful, and 0·05 to 0·10 unit ml^{-1} milk affects the growth of starter bacteria. These antibiotics can be detected by assay tests (Davis, 1965a; Berridge, 1956).

Many strains of starter bacteria are inhibited by agglutinins and peroxidases which occur naturally in some milks. Such strains grow better in pasteurised milk as the inhibitory substances are inactivated by the heat treatment.

CURD MAKING

Addition of Starter
The milk for cheesemaking is inoculated with starters consisting of mixtures of known and compatible strains and species, or mixed strain starters containing many different strains of the different species, or in the case of some hard cheeses, pairs of known strains of mesophilic

starters used in strict rotation. They are added to the vat milk or injected into the milk line by means of a metering pump. The milk may or may not be ripened by the starter to a desired degree of acidity before rennet is added, but ripening increases the risk of attack by bacteriophage (Chapman and Harrison, 1963).

Coagulation of Milk

The digestive enzyme rennin, derived from the stomachs of sucking calves, lambs and goats, has been used for many centuries to coagulate milk for cheesemaking, and commercial preparations of the enzyme became available in 1874 (Bille Brahe, 1974). These are called rennet, and are prepared by extracting the enzyme from the abomasum, or fourth stomach, or vel of the young ruminant by soaking the vels for several days in a 10% solution of NaCl. Liquid commercial rennet contains 18–20% NaCl and a small amount of preservative, sodium benzoate. The rennet extract contains two principal enzymes, rennin and pepsin. Rennin is probably the more active in the coagulation of casein during curd-making, and both assist in the proteolysis of casein during cheese ripening. Rennet prepared from older ruminants contains a higher proportion of pepsin, which increases with the age of the animal. Most animal rennet nowadays contains a certain proportion of bovine pepsin (Green, 1972).

During the past decade, a world-wide increase in the amount of milk available for cheesemaking and a concomitant decrease in the number of calves being slaughtered has led to fears of a possible shortage of calf-rennet, and has revived interest in other milk-coagulating enzymes to replace rennet (Green, 1977). Whilst such rennets may affect the cheese flavour, they have little effect on the growth or acid production of the starter cultures.

Rennet is a very powerful milk coagulant, and its effect in cheesemaking takes place in three phases: (1) an enzymic, destabilising phase, where the protective colloidal nature of k-casein is destroyed and *para*-k-casein is formed within the casein micelles; this change can proceed at low temperatures and is the basis of continuous methods of curd-making (Berridge, 1972); (2) a non-enzymic, coagulating phase which can proceed only at higher temperatures; and (3) a mainly proteolytic phase which takes place around pH 5·2–5·8 and includes the breakdown of milk proteins to peptides. This last phase is essential for cheese ripening, and the peptides formed are utilised and further degraded by the starter cultures (Law and Sharpe, 1978*b*).

The amount of rennet used, the temperature and acidity of the milk, and the previous treatment of the milk all affect the firmness of the coagulum and the texture of the curd formed for the manufacture of different varieties of cheese.

Expulsion of Moisture from Curd

As the rigidity of the rennet coagulum increases, its water holding capacity decreases. Shrinkage, or syneresis, of the coagulum, causing it to lose more water and become firmer, is facilitated by cutting; by development of acidity (the pH falling from 6·6 to 5·0 as a result of lactic acid produced by the starter bacteria; and by warmth, the temperature rising from 28–33 °C at renneting to 38–45 °C or more when the curd is cooked. The extent of shrinkage governs the moisture content and firmness of the curd, and the consistency or 'body' of the cheese; the amount of moisture retained in the curd governs the lactose content and, therefore, the acidity of the fresh/young cheese.

Cutting the rennet coagulum has the greatest effect on moisture expulsion. It releases a large volume of whey, establishes the size of the curd particles, and increases the area of curd from which whey can escape. Stirring maintains the curd in particulate form in the whey, making it possible to heat (cook) it uniformly and to control the expulsion of moisture.

Low cooking temperatures leave more moisture, and therefore more lactose, in the curd and so produce a softer type of cheese in which the starter bacteria develop acid rapidly in the early stages of a relatively short maturation period. High cooking temperatures produce a drier, harder curd suitable for a long-keeping, slow-maturing cheese. These high temperatures also affect the protein, resulting in a more elastic, springy type of curd, such as that required for Emmenthal cheese.

The starter bacteria are trapped in the coagulum at renneting, and the majority are retained in the curd particles after cutting. These ferment lactose in the curd moisture producing lactic acid which cannot immediately escape into the whey, so the concentration of lactic acid is greater within the curd particles. The acid brings about certain chemical changes in the casein, and some calcium phosphate is dissolved. Simultaneously, whey constituents, including lactose, diffuse back into the curd particles to replace constituents which have been used up. This development of acidity in and around the curd particles causes further contraction, and the more active the starter culture, the faster the acid development and the greater the contraction of the curd particles, and the quicker the expulsion of moisture. Thus the growth of the starter bacteria helps to determine the rate of moisture expulsion from the curd, the diffusion of lactose back into it, and the rate of acid production during the later stages of the making process. In Cheddar cheesemaking, the most rapid increase in numbers of starter bacteria occurs during curd-making. The average count of *Str. cremoris* (NCDO 924) in 15 Cheddar cheeses made over a 12 month period rose from 2×10^7 cfu ml^{-1} in the milk 20 min after inoculation to 4×10^8 cfu g^{-1} in the curd at maximum scald (39 °C) 2 h later, and to $1 \cdot 4 \times 10^9$ cfu g^{-1} in the curd at separation from the whey 1 h 20 min later.

The rate of heating the curds and whey also helps to determine the rate of syneresis, and the final temperature of the scald/cook depends on the required moisture content of the cheese and the heat-tolerance of the starter bacteria (Table I).

When the required state of acid development and curd firmness is reached, the curd is separated from the whey. This is a crucial stage in the manufacture of all varieties of cheese as it relates to the physical condition of the curd. In types of cheese where little starter activity is required during curd-making, e.g. Emmenthal, the curd is dipped out of the whey, placed in hoops and pressed. The starter bacteria begin to multiply and produce lactic acid when the curd temperature falls to suitable levels. In other types of cheese where high levels of lactic acid are required in the curd at pressing, the curd particles are separated from the whey, allowed to mat together and undergo a texturing process, such as 'cheddaring', during which time the starter bacteria continue to multiply. This type of curd is eventually cut into portions which are placed in moulds and subjected to light pressure (e.g. Edam and Tilsit), or the curd is passed through a tearing or slicing mill, salted, placed in moulds and subjected to high pressure (e.g. Cheddar).

The extent of acid production by the starter bacteria during cheesemaking has a marked effect on the chemical and physical composition of the curd, and control of acid development is important (Davis, 1949). This is achieved by reducing further the moisture content of the curd and maintaining the growth of the starter bacteria, so that fermentation of the lactose remaining in the curd moisture will yield the required pH. Casein exists in rennet curd as calcium paracaseinate, and when lactic acid is produced in appreciable amounts, it reacts gradually to remove the calcium and the curd begins to show elasticity.

If acid production continues, however, entirely protonated paracasein is formed. The amount of lactic acid formed by the starter bacteria in fresh cheese curd determines the form in which the casein exists. In high acid varieties of cheese (e.g. Camembert, pH 4·6–4·8), all the casein appears as free paracasein. In Cheddar, Emmenthal and related types (pH 5·0–5·2), the casein is partly in the calcium form. Whether this affects the action of the ripening organisms and proteolytic enzymes is not known, but the form in which the protein exists affects the elasticity of the curd, and this affects the characteristics of the ripening cheese.

SALTING THE CURD OR CHEESE

Sodium chloride is added, in some form, to all cheeses, including lactic cheese and Cottage cheese, at some stage in their manufacture to suppress the growth of unwanted bacteria, to control the growth of wanted micro-organisms and thus the rate of ripening, to assist the physico-chemical changes in the curd, and to give flavour to the cheese. The whole, fresh cheese may be floated in brine (10% solution), or the surface of the cheese may be rubbed with dry salt, or a definite amount of salt may be added to the milled curd, mixed with it and allowed to dissolve before the cheese is pressed.

The amount of salt taken up by cheese salted in brine, or dry salted on the surface, depends on the concentration of the brine, the moisture content of the cheese, the length and the temperature of exposure, and the ratio of surface area to volume of the cheese. Following initial concentration at the surface of the cheese, salt gradually diffuses through it and the final distribution is fairly uniform. Where salt is sprinkled onto milled curd and mixed with it, the salt is rapidly distributed throughout the cheese. Salt dissolves in the water content of the cheese forming a weak brine, the actual concentration of which helps to control the ripening and preservation of the cheese.

Mesophilic lactic acid bacteria vary in their tolerance to salt. Most strains of *Str. cremoris* are inhibited by 2% salt, but *Str. lactis* can withstand up to 4% salt. The growth of unwanted bacteria is suppressed by the salt concentration, although *Escherichia coli* requires almost 12% salt for inhibition and may even be stimulated by concentrations of 3% (Davis, 1965b). Suppression by salt of the growth of undesirable proteolytic and lipolytic bacteria, such as butyric acid bacteria (Goudkov and Sharpe, 1965), is of special importance in varieties such as Edam and Gouda, where only low levels of lactic acid are produced during curd-making.

RIPENING OF CHEESE

Ripening involves changes in the chemical and physical properties of the cheese accompanied by the development of characteristic flavour.

Fresh, young or 'green' cheese curd is tough and sometimes rubbery. It consists mainly of protein, fat and moisture, in varying proportions depending on the type of cheese, together with small amounts of salt, lactose, lactic acid, whey proteins and minerals. In ripening, this curd is gradually digested by enzymes, and the mature cheese acquires the firm, or plastic, or soft body characteristic of the particular variety. The chemical changes responsible for ripening cheese are: (1) fermentation of lactose to lactic acid, small amounts of acetic and propionic acid, CO_2 and diacetyl, (2) proteolysis, and (3) lipolysis. These changes are brought about by enzymes from (i) the lactic acid bacteria of the starter culture, (ii) miscellaneous, non-starter bacteria in the milk, (iii) the rennet, rennet paste or rennet substitute used to coagulate the milk, (iv) the milk itself, and (v) other micro-organisms growing within or on the surface of the cheese. These metabolic changes are accompanied by the development of characteristic flavour. They are affected by the size and composition of the young cheese, and are controlled by the conditions of temperature and humidity at which the cheese is ripened and stored. Block stacking of warm cheese on pallets and block stacking of pallets can influence temperature and flavour differences between blocks of cheese from the same making vat (Miah *et al.*, 1974). Some varieties (e.g. Emmenthal, Camembert and Stilton) require special periods of controlled temperature and humidity for the ripening process, during which bacterial or fungal activity produces specific changes in the body, texture and flavour of the cheese. Ripening is then followed by storage until the cheese is ready for sale. Other varieties, particularly the hard cheeses without eyes (e.g. Cheddar and Parmesan) are stored at constant temperature throughout the ripening period, and maturation may extend over many months (Table II).

During ripening, characteristic changes take place in the body, texture and flavour of the cheese.

TABLE II

CHARACTERISTICS OF CHEESE VARIETIES

Cheese	Country of origin	Moisture (% max.)	Fat in dry matter (% max.)	Size (cm) Diameter	Height	Weight (kg)	Ripening (months)
A. Hard cheese (26–50% moisture)—ripened by bacteria							
Very hard (grating)		26–34					
Parmesan	Italy	34	32	35–45	18–24	22–40	24–48
Romano	Italy	34	38	25	15	9	5+
Asiago (old)	Italy	32–45	30–40	30–45	10–15	8–18	12+
Sbrinz	Switzerland	28	47–50	45–50	10–15	13–20	24–36
Hard (with eyes)		36–45					
Emmenthal	Switzerland	41	43	68–78	13–25	70–100	3–12
Gruyère	France	39	45	40–63	8–13	30–40	3+
Herrgård	Sweden	39	30–45	38	10–15	12–18	3–6
Hard (without eyes)							
Cheddar	UK	39	48	28–35	12–15[T]	20–30[T]	4–12
Double Gloucester	UK	44	48	35	22[T]	13–16[T]	4–6
Derby	UK	42	48	35	10–13[T]	13–16[T]	
Cantal	France	42	45	30–50	40	19–45	1–3
Cheshire	UK	44	48	30	30[T]	21[T]	1–3
Leicester	UK	42	48	35	13[T]	15–16[T]	2–12
Provolone	Italy	45	45	Pear	36–45	4–5	2–3
Cacciocavallo	Italy	40	42			2	2–4
Semi-hard		45–50					
Caerphilly	UK	46	48	25	7·5	4	0·5–1
Lancashire	UK	48	48	36	25–31	20–22	3–12
Edam	Holland	45	40	12–15	Sphere	1–5	1·5–2
Gouda	Holland	45	48	24–51	6–15	3·5–25	2–5

B. Bacterial surface-ripened cheese							
Semi-soft		45–55					
Limburger	Belgium	52	40	(Block)	15 × 15 × 8	1·0	1+
Münster	France	46	45	11–18	3–5	0·3–0·9	
Port du Salut	France	56	42	20	4–6	2·0	1–2
Bel Paesa	Italy	47	49–55	25	13	5·0	4–5
Taleggio	Italy	47		(Block)	20 × 20 × 5	2·0	2
Tilsit	Germany	45–55	49–55	25	13	5·0	2–5
Romadour	Germany	49–61	20–50	(Block)	5 × 5 × 10	0·5	0·5–1
Brick	USA	44	50	(Block)	25 × 12 × 13	2·5	1·5–2
Monterey (Jack)	USA	44	50	24	10	3·0–4·0	0·5–1·5
C. Internal mould-ripened cheese							
Semi-hard		42–52					
Stilton	UK	42	48	20	30	5·5–6·5	4–6
Wensleydale	UK	46	48	20	30	4·5–5·5	6
Roquefort	France	45	50	25	10	2·0	3+
Bleu d'Auvergne	France	50	40	23	10	2·5	2+
Gorgonzola	Italy	42	48	20–30	16–20	6–13	3–6
Blue	USA	46	50	20	15	2·0	6+
Gammelöst	Norway	46–52	0·5–1·0[a]	15	10–15	3–4	1+
D. Soft cheese (48–55% moisture)—surface mould-ripened							
Brie	France	56	40	14–16, 22–24, 32–36	2·5	0·5, 1·5, 3·0	4–6
Camembert	France	48	45	10–11	3	0·210–0·260	4–6
Carre'd L'Est	France	55	45–50	10	2·5	0·225	2–3
Neufchatel	France	52	51	5	5–7	0·100	1+

(continued)

TABLE II—contd.

Cheese	Country of origin	Moisture (% max.)	Fat in dry matter (% max.)	Size (cm) Diameter	Height	Weight (kg)	Ripening (months)
E. Soft cheese (50–80% moisture)—unripened							
Coulommier	France	58^b	22^a				
Gervaise (carre)	France	55^b	60				
Cream	UK	55^b	45^a				
Cambridge	UK	58^b	22^a				
Lactic	UK						
Cottage	USA	80^b	—^{c}				
Mozzarella	Italy	53^b	18^a				
Ricotta (whole milk)	Italy	72^b	12^a				
F. Miscellaneous							
White brined cheese							
Feta	Greece						4–5 days
Whey cheese							
Ricotta	Italy	82^b	0.5^a				
Mysöst	Norway and Sweden	13–18^b	10–20				
Processed cheese	Many^d	Variable^d	Variable^d				

T Traditional sizes and weights.
^{a} % Fat.
^{b} % Moisture.
^{c} Creamed Cottage cheese minimum 4% fat.
^{d} Regulations for moisture and fat in dry matter vary in the different countries of production.

Changes in Body, Texture and Flavour

The term 'body' is used to describe the consistency of cheese, and includes such attributes as firmness, elasticity, plasticity and cohesiveness. Transformation of the tough, rubbery curd is brought about by enzymic digestion of the casein. The cheese becomes softer, and if the moisture content is low, more crumbly (e.g. Cheshire). Large amounts of acid developed during manufacture may produce a crumbly, brittle, curd, and the body of such cheese is described as 'short', meaning that its lacks elasticity. This condition is a defect in some varieties (e.g. Cheddar and Emmenthal), where acid development is carefully controlled to avoid it. In other cheeses (e.g. Stilton and Roquefort), it is a desirable feature.

Texture describes the structure or presence of 'holes' within the cheese. Close texture refers to cheese with no holes, whilst cheese with holes is said to have an open texture. Where this arises from a failure of the curd particles to fuse together, it is described as mechanical openness and, if more than slight, it is a defect. With blue-veined cheese, however, an open texture is necessary to allow the growth of moulds throughout the cheese. Where openness arises from unacceptable gas production it is a defect, but moderate openness, due to gas production, is characteristic of some types of cheese. It is acceptable in cheeses such as Limburger and Gouda, and is essential in cheeses such as Emmenthal and Gruyère.

Primarily, however, the ripening of cheese involves development of the desired, characteristic flavour and aroma compounds by the action of micro-organisms and enzymes which break down proteins, fats and carbohydrates and, in some cases, metabolise lactic acid, lactate and citrate. Various products of protein hydrolysis, as well as fatty acids and their esters or ketones, may be present in varying amounts in the cheese. This produces a complex mixture of components which give the required balance of flavour (Fryer, 1969) characteristic for the variety. The starter bacteria die out during ripening, as do most other organisms present in the curd, including enterococci and leuconostocs. Only the lactobacilli, which may be present in fresh curd in small numbers, multiply, and these may reach levels of 10^6 to $10^8 \, g^{-1}$ in cheese in 3–6 weeks (Sharpe, 1979).

Chemical and Biochemical Changes

Most of the lactose disappears from hard cheese (30–40% moisture) within the first few days of manufacture, but this time may be longer in the case of soft, high moisture (50 %) types of cheese. Fermentation of the lactose by the starter bacteria produces mainly lactic acid, with some volatile acids, ethanol and small amounts of other by-products. Some lactic acid combines with basic radicals in the cheese to form salts, and in certain varieties of cheese (e.g. Emmenthal), where the starter includes propionic acid bacteria, there is secondary fermentation of the lactic acid with the production of propionic acid, acetic acid and carbon dioxide.

The greater part of the nitrogenous material in the young cheese is present as water-insoluble protein, but as ripening proceeds, part, or all, of this is hydrolysed by enzyme action to more simple, soluble compounds.

$$\text{Protein} \xrightarrow{+H_2O} \text{Proteoses} \xrightarrow{+H_2O} \text{Peptones} \xrightarrow{+H_2O} \text{Peptides} \xrightarrow{+H_2O} \text{Amino acids}$$

(Insoluble)   ($\leftarrow$ - - - - - - - - - - - - - - - Soluble - - - - - - - - - - - - - - - - - - $\rightarrow$)

Micro-organisms may reduce amino acids to ammonia and organic acids, or oxidise them to form carbon dioxide and amines.

The extent of proteolysis and the formation of resulting compounds helps to establish the characteristics of the mature cheese. Some soft cheeses, such as Camembert, Brie and Limburger, undergo extensive proteolysis; and the formation of water-soluble compounds, including peptides, amino acids and ammonia, together with the high moisture content of these varieties, is responsible for the soft, 'velvety' body and texture of the mature cheese. Hard cheeses, on the other hand (e.g. Cheddar and Emmenthal) undergo much less proteolysis, and only some 25–35 % of the protein is made soluble; in a well-matured cheese, a high proportion of the breakdown products is in the form of peptides and amino acids.

Rennin can break down casein to water-soluble compounds (i.e. mostly proteoses and peptones), but microbial enzymes bring about further breakdown with the formation of amino acids and even ammonia. In hard cheese, the micro-organisms present are mainly lactic acid forming cocci and rods which are dispersed throughout the whole of the cheese mass, and these produce only small amounts of extracellular proteinases. When these cells die and autolyse, however, they release intracellular enzymes, and ripening progresses evenly through the whole cheese. In soft, quick-ripening cheeses, most of the proteolysis is brought about by extracellular proteinases released by various micro-organisms which grow on the surface of the cheese. Moulds play an important part in the proteolysis of some cheeses, e.g. blue-veined cheese (Stilton) and surface mould-ripened cheese (e.g. Brie and Camembert).

Decomposition of the fat in cheese is not extensive, but some hydrolysis of fat occurs during cheese ripening, and the products of greatest importance are volatile lower fatty acids, including butyric, caproic, caprillic and capric. Lipolytic enzymes in cheese may have come from the milk, from micro-organisms, or from enzyme preparations added to the milk (e.g. coagulants of bacterial or fungal origin). Milk lipase is active only in cheese made from raw milk. Certain strains of lactobacilli and possibly other bacteria liberate, upon autolysis, intra-cellular lipases, and these are thought to account for much of the lipolytic activity in hard cheese (e.g. Cheddar and Emmenthal). Moulds, growing in, or on the surface of, certain types of cheese (e.g. Stilton and Camembert) are sources of lipolytic enzymes. Rennet extract has little lipolytic activity, but rennet pastes are actively lipolytic, and are used in the manufacture of Italian hard cheese (e.g. Parmesan).

Methods of Ripening and Storing
There are two distinct and basically different ways of ripening cheese. The very hard cheeses (e.g. Parmesan), hard cheeses (e.g. Cheddar) and some semi-hard cheeses (e.g. Caerphilly and Lancashire) are stored under conditions which discourage surface growth of micro-organisms, and so restrict metabolic activity to micro-organisms and enzymes within the cheese. Ripening proceeds slowly and uniformly throughout the whole mass of cheese, and the process is not affected by the size of the cheese. Hard cheeses vary in size from Caerphilly, 25 cm diameter and 6 cm high and weighing 3·5 kg, to Emmenthal, 70 cm in diameter and 20 cm high and weighing 70 kg. They may be made in the traditional, cylindrical shape, or in modern, rectangular blocks. All soft cheeses, however, and some semi-soft cheeses (e.g. Limburger and Brie) are stored under conditions that promote the growth of surface organisms. These may be moulds, e.g. *Penicillium camemberti* in the case of the surface mould-ripened soft cheeses (Camembert, Brie, etc.) or bacteria, e.g. *Brevibacterium linens* on the bacterial surface-ripened cheeses such as Limburger. Enzymes produced by surface organisms diffuse into the cheese bringing about changes, associated with ripening, which progress from the surface towards the centre. Hence the relatively small size and flat shape of these types of cheese (Table II). The blue-veined cheeses (e.g. Stilton and Roquefort) are ripened by a combination

of both of these methods. Initially micro-organisms and enzymes are responsible for ripening changes within the cheese, then air is admitted to the interior of the cheese and moulds, such as *P. roqueforti* (or *P. glaucum*) introduced naturally from the air in the store or deliberately during the manufacturing process, develop in the cheese, bringing about proteolytic and lipolytic changes which characterise the flavour and appearance of the blue-veined cheeses.

Size and Protection

Generally speaking the size and shape of a cheese is associated with the type of ripening that it is to undergo, and with the conditions of temperature and humidity under which the cheese is to be matured and stored. Hard cheeses ripen slowly during periods of several months to several years, and ripening proceeds evenly throughout the cheese mass whatever the size or shape of the particular variety (see section on Cheddar). The cheeses are held at temperatures ranging from 4–14 °C, and where relative humidity has to be controlled, this is kept fairly low (86–88 %) to discourage mould growth, but sufficiently high to prevent excessive evaporation of moisture. Some varieties are coated with paraffin wax or plastic emulsions to minimise the loss of moisture. Modern methods of packaging enable many varieties of hard cheese to be packed in special films which may be heat-sealed, sealed under vacuum, or vacuum-packed in a shrink-type film (e.g. Cryovac). All these methods exclude air, thus preventing mould growth on the cheese surface and preventing evaporation from it. Where cheeses are packaged by these methods they are not exposed to the atmosphere in the store, and therefore control of humidity in the cheese store is not necessary. The cheeses may be stacked one on top of another to maximise the utilisation of floor space, but it is necessary to hold these cheeses at lower temperatures (6 °C) to discourage gas production. This use of low temperature slows down the ripening process, so these cheeses require a longer period to mature. Thus, a Cheddar cheese ripened at 6 °C may require 9 months to reach the same degree of maturity as can be reached by a cheese made from the same batch of curd in 6 months with ripening at 13 °C (Law *et al.*, 1979).

Where surface growth of micro-organisms is required for ripening, as is the case for soft and semi-soft varieties of cheese, the extent of surface area in relation to cheese mass is important; it helps to govern the rate of ripening. The surface-ripened cheeses are dry-salted on the surface, or have their surfaces rubbed with a brine-soaked cloth, for several successive days which helps to control the surface flora. They are then held under carefully controlled conditions of temperature (15–20°C) and at relatively high humidity (90–95%) to encourage the growth of desired micro-organisms. This method of ripening allows a regular succession of organisms to develop. At first only certain salt-tolerant yeasts, and moulds such as *Geotrichum candidum*, are capable of growth on the surface of these cheeses with their high acid and salt content. These fungi utilise the lactate present, and by increasing the pH, allow salt-tolerant bacteria (e.g. *Bre. linens*) or other moulds (e.g. *P. camemberti*) to grow.

In stores where bacterial surface-ripened cheeses are held, *Bre. linens* is indigenous. Its growth is characterised by a reddish-brown colour, and the bacteria produce enzymes which break down cheese protein to amino nitrogen. Each day a cloth, dipped in mild, warm brine is rubbed gently over the surface of the cheeses, 'smearing' the bacterial growth all over it. Some 2–14 days later, the cheeses are wiped dry and stored on mats or shelves until packaged for sale.

Surface mould-ripened cheeses are sprinkled with an aqueous spore suspension of *P. camemberti*; the characteristic white 'velvety' growth appears within 2–14 days. This changes to a very pale, somewhat translucent, grey as proteolysis progresses, and touches of reddish-brown appear where *Bre. linens* has grown with the mould. At 14 days old, the cheeses are wrapped in tinfoil and stored in a cool room (10°C) for 7 days, and then prepared for distribution.

PRESERVATION OF CHEESE

Cheesemaking evolved as a means of preserving milk, and in general, the lower the moisture content the longer the keeping quality of the cheese. The hard cheeses, with low moisture and high acid contents, mature slowly and may be kept for long periods; semi-hard and semi-soft cheeses keep for a few months; soft, ripened cheeses for a few weeks, and the soft, unripened cheeses and Cottage cheese with their very high moisture contents keep for only a matter of days without refrigerated storage. Other factors which control the keeping quality of cheese include its content of lactose, oxygen and salt, as well as its pH and the temperature of storage.

In ripened cheese, the lactose has been used by the starter bacteria, and none is available for spoilage organisms which might be present.

This limits the spoilage of ripened cheese to organisms that can oxidise lactate or utilise protein decomposition products. These include the film yeasts, moulds and certain anaerobic spore-forming bacteria such as *Clostridium tyrobutyricum.*

The essentially anaerobic conditions which prevail within all cheese limits the activity of film yeasts and moulds to the surface. Here their growth is controlled by cleaning the surface (e.g. Emmenthal and Gruyère), or by packaging the cheese in film (e.g. Cheddar), or coating cheese with paraffin wax (e.g. Edam) or vegetable oil (e.g. Parmesan). In the case of surface-ripened cheese, it is essential to suppress the growth of unwanted micro-organisms such as wild yeasts and moulds (e.g. *Mucor* spp.) in the ripening rooms.

Whatever the method of salting cheese or the amount of salt added, the preservative effect of salt depends on the concentration of NaCl in the water phase of the cheese. Thus, 1.60% NaCl in a hard cheese containing 36% water gives an actual brine concentration of 4.4% ($1.6/36 \times 100$) which is sufficient to inhibit potential spoilage bacteria.

The minimum pH of most cheese is 5.3, and in many varieties it reaches 4.5. Acidity does not prevent the growth of film yeasts and moulds, but it does inhibit growth of spoilage bacteria within the cheese, particularly if associated with low moisture, lack of oxygen and high salt concentration. Combinations of two or more of these preservative factors effectively prevents spoilage by micro-organisms at normal, low levels of contamination. High levels of contaminants, however, may result in serious spoilage even at satisfactory concentrations of moisture, salt and acidity.

Spoilage of Cheese
Mould Growth

Certain mould species are essential for ripening specific varieties of cheese, but mould growth on most cheese is undesirable. It spoils the appearance, can impart musty flavours, and may produce mycotoxins. The mould most commonly found in ripening rooms include species of *Alternaria, Aspergillus, Cladosporium, Monilia, Mucor* and *Penicillium.* In addition, the high moisture soft cheeses, Cottage cheese and cream cheese, are subject to spoilage by species of *Geotrichum.* Yeasts and moulds develop on walls and shelves of ripening rooms, and strict cleanliness is the most important factor in controlling their growth (see section on Dairy Hygiene).

Attempts have been made to control mould growth on cheese surfaces by impregnating the wrapping or packaging material with fungicides or fungistatic chemicals. Most of these have not been approved by regulatory authorities, but sorbic acid has been given tentative approval and pimaracin, once considered undesirable because of its antibiotic properties, may be reconsidered as a fungicide in view of the equally undesirable aflatoxin produced by moulds.

Gas Formation

Unwanted gas production can occur during the manufacture or ripening of cheese depending on the number and kind of gas-producing micro-organisms in the milk, or contaminating the curd. Raw milk heavily contaminated with coliform bacteria may form so much gas that the curd floats in the vat. This defect is rare in modern dairies, where milk is efficiently pasteurised and good methods of manufacture are used, but it can occur in Cottage cheese curd during the long setting period (4–5 h at 31–32 °C) if coliforms are present, and it is possible that leuconostocs in the starter can contribute to this tendency for curd to float (Sandine *et al.*, 1957).

Generally, however, early blowing occurs in the cheese during the first few days of ripening. This 'early gas' is produced by coliform bacteria, *Aerobacter* species being more active than species of *Escherichia*. The extent of spoilage depends on the number of coliform organisms, the activity of the starter bacteria, the amount of lactose available and the temperature at which the cheeses are held. Coliform bacteria are able to tolerate the acid and salt conditions of most cheese, they are not inhibited by the starter bacteria, they ferment lactose readily, grow well at the temperatures used in making most varieties of cheese, and in the early days of ripening whilst the cheeses are adjusting to the storage temperature. They are best controlled by using milk of good bacteriological quality, and by using adequate conditions of heat treatment.

Early blowing of cheese may also be caused by lactose-fermenting yeasts, which also produce fruity flavours. Aerobic spore-formers (e.g. *Bacillus subtilis*) also ferment lactose producing carbon dioxide, hydrogen, acetic acid and ethyl alcohol, but they are not of great importance in hard cheeses because of the low oxygen and low pH.

Late blowing of cheese, or gas production which occurs some weeks after manufacture, is caused by species of clostridia (e.g. *Cl. sporogenes*, *Cl. butyricum* and *Cl. tyrobutyricum*). Small numbers of these anaerobic spore-forming rods are present in most milk samples, but they are sensitive to acid and salt and the ripening conditions of most varieties are not conducive to their growth. They can cause serious problems in Swiss-type cheese, where the extent of blowing can vary from a few small holes to large holes, cracks and fissures depending on the number of organisms, the amount of gas and rate of production, and the body (elasticity) of the cheese curd.

Rind Rot

Failure to keep the surface of a hard cheese dry allows moisture to accumulate and micro-organisms, such as film yeasts, moulds and proteolytic bacteria, to grow. This growth causes softening and discolouration of the cheese, and even the production of undesirable odours. This condition is prevented by turning the cheese are regular intervals and by keeping the surface dry.

Discolouration

Surface discolouration of cheese may result from mould growth: (1) black spots of *Aspergillus niger* on hard cheese, (2) red spots of *Sporendonema casei* on blue cheese. Coloured spots within the cheese are rare, but 'rusty spots' in Cheddar and Cheshire cheese and Herrgärd and Svecia cheese have been attributed to pigmented variants of *Lac. plantarum* and *Lac. brevis*, known as sub-sp. *rudensis* (Sharpe, 1962), and pigmented species of propionibacteria; *Propionibacterium rubrum* may cause coloured spots in Swiss cheese (Langsrud and Reinbold, 1974).

PRACTICAL ASPECTS

Equipment

Equipment used for cheesemaking should be of a simple design, easy to clean and sterilise, and to maintain in working order. Large-scale cheese manufacture calls for large capacity, specialised items of equipment, but their construction should be as simple as possible. They include milk storage silos (25 000 gal. capacity), cheese vats of 1000 to 3000 gal. capacity, cutters, agitators and curd drainers, curd conveyors and cheese presses holding 1 ton of curd and more.

Stainless steel is widely used in the manufacture of all large items of equipment in direct contact with milk, curd and whey, and plastics materials may be used for curd conveyors.

During the last twenty years, all operations in the cheesemaking process have been mechanised, and this permits the use of automatic control where it is required. The extent to which mechanisation is employed will depend on the economics of production, and will range from simple mechanical handling aids to automated systems controlled manually or programmed and supervised from a central computerised control station.

Cheese vats, curd conveyors and curd texturing equipment can all be covered, and curds and whey can be transported by pipeline systems so that the whole cheese manufacturing process can be carried out in an enclosed system. This provides a very hygienic system of processing, and permits the use of in-place cleaning systems.

Dairy Hygiene

Equipment used in the cheese manufacturing process must be maintained in a clean and bacteriologically satisfactory condition. Failure to do so can lead to problems with starter failure, due to bacteriophage, during processing, and with gas production, mould growth and unclean flavours in the cheese. Effective cleaning is essential. It should be carried out immediately after equipment is used, and should be followed by the application of heat or a suitable chemical disinfectant (sanitiser) such as chlorine, iodine or a quaternary ammonium compound. Where cheese vats are refilled with milk during the day, they should be cleaned and sanitised with hypochlorite before re-use.

Cleaning techniques vary from manual cleaning of simple, small-scale equipment to cleaning-in-place (CIP) systems operated by automatic or manually controlled programmes (British Standards Institution, 1977). Cleaning should extend to the manufacturing area, to press rooms and cheese stores, and is particularly important in areas used for dressing, cutting and packaging cheese.

Floors should be constructed of an impermeable material and be kept clean at all times. Walls and ceilings should be covered with smooth-surface materials and should be cleaned regularly. Temperature and humidity control in curing rooms assists with the control of mould growth. Clean air systems may be installed with advantage, particularly in manufacturing and packaging areas.

Brine tanks should be lined with ceramic tiles or constructed from some other non-corrosive material such as plastics. They should be emptied and cleaned regularly. Brines should be kept at the correct strength and should be renewed regularly, otherwise halophilic, often slime-forming, organisms can grow, especially in the weaker brines.

In ripening rooms and cheese stores the walls, floors and ceilings should be washed with a fungicide solution. Shelves should be washed with warm detergent and treated with a fungicide (e.g. 10% solution of formalin or 5000 ppm chlorine).

Fungistatic paints have been used with some success. Ultra-violet ray lamps have had limited success, partly due to the difficulty of exposing all cheese surfaces to effective amounts of the germicidal rays, and partly due to the strong resistance to them of *A. niger*, the mould which causes black discolouration on the surface of hard cheeses. Ozone generators have not given satisfactory results, because of the low concentration of ozone that can be tolerated without causing oxidation of the fat in the cheese.

THE HARD CHEESES

The hard cheeses constitute a large group of varieties comprising four distinct types (Table II). They are characterised by: (1) low moisture content, achieved by the size of cut and the use of high cooking temperatures for the curd; (2) clean rinds, or no rinds where film-wrapped cheese are made; and (3) their method of ripening which proceeds evenly throughout the whole of the cheese, and is carried out by enzymes derived from the starter and non-starter bacteria, and the rennet. Ripening is augmented and characterised in Swiss-type cheeses by the fermentation of lactate by propionic acid bacteria with the production of CO_2 which forms 'eyes'.

Very Hard Cheese

These are the very hard, grating varieties such as Grana (Parmesan), Romano, Asiago (old) and Sbrinz. They are characterised by their very low fat and moisture contents (Table II). Essential features in their manufacture include the use of low-fat milk; starter cultures of thermophilic lactic acid bacteria (Table I); very high scald temperatures; a long immersion in brine, and a long, slow maturation period. During maturation, free fatty acids in the cheese increase slowly producing the sharp, rancid flavour characteristic of the mature cheese. Some of the varieties may be eaten as fresh cheese (e.g. Asiago), but they are usually stored for long periods, grated and sold as flavouring for culinary purposes.

Grana or Parmesan

The name Grana, attributed to the granular appearance of this cheese when broken, is a general name which embraces several varieties of hard, grating cheese made in the valley of the River Po in northern Italy. Grana Padano is made in certain areas of the Lombary Plain, and Parmesan (Parmigiano Reggiano) in the particular part around Parma.

Essential items of equipment for the manufacture of Grana or Parmesan cheese include: shallow creaming pans or a centrifugal separator; cone-shaped, copper making vessels (kettles); cutting tools (spinos); dipping cloths; hoists and draining hooks; cheese moulds, presses and cloths; a brining chamber equipped with shelving and brine baths, and maturing stores fitted with atmospheric controls for temperature and humidity.

The process. Milk for cheesemaking is adjusted to a fat content of about 2%, and delivered to making vessels (kettles) which hold about 150 gal. of milk. The contents of each kettle is used to make one cheese. The milk may be used raw but pasteurisation is being adopted; the very high scalding temperatures used later in the process are said to make pasteurisation less important than for other types of cheese. The milk is heated to between 32 and 35 °C, and starter cultures consisting of *Str. thermophilus* and the strong lactic acid producer *Lac. bulgaricus* (or similar heat-resistant *Lactobacillus*) are added. Normally about 1% of starter is used, and after a short ripening period, rennet is added (25 ml per 100 litres). A special lipase-rich type of rennet or rennet paste may be used. The coagulum is cut, with special cutting tools called spinos, into fine particles about 3 mm in size. The curds and whey are stirred and heated in two stages, first to 43 °C over a period of 15 min, and then 15 min later, to 54–58 °C over some 15–30 min; the time depends on the the rate of acid development by the starter bacteria. When sufficiently firm, the curd particles are settled for 10 min and then dipped out of the whey in a special cloth. The mass of curd is lifted out of the kettle, the cloth suspended from hooks, and the curd allowed to drain for about 30 min prior to pressing overnight. On removal from press, the cheeses are held, in their moulds, in the brining chamber for 3 days at 15 °C for the starter bacteria to continue to ferment lactose. The cheeses are then immersed in brine (26% NaCl) at 8–10 °C for 14 days, then allowed to dry for several days at 15–20 °C and finally removed to the curing room.

Maturation takes place in two stages. During the first stage the cheeses are held at 10 °C and a relative humidity (RH) of 85% for 6–12 months. They are turned frequently, to maintain their shape and avoid spoilage of the rind, and are washed, scraped and rubbed with vegetable oil to inhibit mould growth. The cheeses may be sold at one year old for consumption as natural cheese, or alternatively they may be matured for a further period (2–4 years). These latter cheeses are coated with a mixture of lamp-black and vegetable oil, to prevent undue evaporation of moisture, and are stored at 10–12 °C and 85–90% RH. Proteolytic and lipolytic changes produce strong flavours, and the cheeses become drier and are eventually grated for culinary purposes.

Microbiological changes. During manufacture the starter bacteria, *Str. thermophilis* and *Lac. bulgaricus*, increase in number, and continue to multiply for a few days afterwards whilst lactose is available in the cheese. Non-starter bacteria from the raw milk have little time to grow before rennet is added, and the majority are killed by the temperature of the scald; their extracellular enzymes contribute to proteolytic and lipolytic changes in the ripening cheese. There is little inhibition of the cheese flora by salt during the first two weeks of ripening, but the temperature of the brine (10 °C) is not conducive to growth. The few days at 15–20 °C provides a brief opportunity for growth of any bacteria which have survived until then (14 days), and which can utilise the products of lactose fermentation. The low moisture, low pH, low temperature and long, slow ripening restrict further microbiological changes in the cheese, and the biochemical changes in the protein, fat and carbohydrate which constitute the ripening are brought about by enzymes produced by bacteria in the original milk or added in the starter culture, or by enzymes from the rennet paste. Bottazzi (1959) reported the isolation of *Propionibacterium shermanii* and *P. freudenreichii* from Parmesan cheese made in Italy, and on the influence of season of production and temperature of ripening on their numbers. Abnormal counts of *P. shermanii* were associated with defects. The same author (Bottazzi, 1960) reported the isolation of a micro-organism, probably a *Pediococcus* sp., producing a different chromatographic pattern from lactic streptococci or the lactobacilli, from Parmesan cheese throughout all stages of ripening. Marth (1979) discussed some of the bacteriological changes that occur in Parmesan cheese during ripening in the USA. He showed that the total numbers of bacteria in the cheese decreased markedly during the ripening period from $\geq 10^7 \, g^{-1}$ in fresh cheese to $10^5-10^4 \, g^{-1}$ after 14 months. The decreases were most pronounced after the brine treatment, or early in the ripening period. Psychrotrophic bacteria, coliform bacteria and *Staph. aureus* were regarded as unlikely to cause problems in normal cheese, and whilst propionic acid bacteria occurred, their importance was not evident. Lactic acid bacteria grew in relatively fresh cheese, but the starter strain *Lac. bulgaricus* was eventually replaced by *Lac. casei* as the cheese matured. The spores of aerobic and anaerobic bacteria generally failed to germinate.

Defects in Parmesan and other very hard, grating types of cheese may arise from enzymic changes following bacterial contamination of the cheese milk, and are usually associated with rancid, fruity and fermented flavours. Surface defects, such as mould growth and rind rot, should not occur if the cheeses receive proper care during ripening. The heat treatment received by the curds and whey, and the subsequent treatment of the cheese in brine and in store, combine to provide a substrate unconducive to the growth of spoilage micro-organisms.

Cheddar Cheese and Related Types
The varieties of cheese in this group (Table II) are not as hard as the grating cheeses. They are made from whole milk and contain more fat (48% fat in dry matter (FDM)) and more moisture (39%), and the starters used in their manufacture are mesophilic lactic acid streptococci. Lower cooking temperatures are used, and acidity is developed during the manufacturing process before the curd is salted and pressed. Moisture is expelled from the rennet coagulum by cutting, heating (cooking or scalding) and the development of acidity; and, following the removal of whey, the curd undergoes some form of 'texturing', either in the vat or in special devices (Renwick, 1978).

Curd for all varieties of cheese in this group is milled in a tearing or slicing mill, and dry salt is added to the milled curd. After thorough mixing, the curd is filled into suitable cheese moulds and pressed. Traditionally all the varieties in this group are cylindrical in shape, but differ in diameter and height (Table II). These conventional cheeses may be coated with paraffin wax to protect the rind and prevent excessive evaporation of moisture. They are matured on shelves in stores at 15°C and 88% RH. Modern methods of pressing and packaging Cheddar and related varieties in rindless blocks or wheels enable the cheese to be stacked on pallets, which in turn are stacked on top of one another, in the store. Humidity control is not necessary under these circumstances, but the temperature of the store is reduced to

FIG. 1.   A traditional cheesemaking room showing the vats for making Cheddar cheese; note the sweep-type agitators and the fact that the vats can be tilted to facilitate drainage of whey. Although this particular type of vat is rarely found nowadays, similar designs are widely employed throughout the cheese industry.

FIG. 2.   A group of modern cylindrical vats (3000 gal.) which are enclosed to reduce risk of infection by bacteriophage. Reproduced by courtesy of the Milk Marketing Board, Thames Ditton, UK.

4–6 °C. At this temperature the cheeses mature slowly over periods of 4–10 months. Ripening of these cheeses is brought about by enzyme systems derived from the starter bacteria and the rennet (Law and Sharpe, 1977). When mature, the body of the cheese is firm, but not hard, and the texture is either close (e.g. Cheddar, Cantal, Double Gloucester and Derby) or open and crumbly (e.g. Cheshire).

Cheddar

Cheddar cheese takes its name from the village of Cheddar in the South-West of England. Originally made in farmhouse dairies in the County of Somerset by methods handed down by word of mouth over the centuries, the general method of manufacture was standardised by Joseph Harding in 1875 and accepted into commercial practice. Since then its manufacture has spread throughout the world. It has excellent keeping quality and is the most popular variety of cheese produced in the UK, USA, Australia, New Zealand and Canada.

The farmhouse method of manufacture became centralised in large creameries from 1870 onwards. Modifications and developments have been introduced and the process has been mechanised, but the principles of manufacture have remained unchanged, and are the basis of the highly mechanised systems of Cheddar cheesemaking operated in farmhouses and factories today. In the UK in 1980, farmhouse manufacture accounted for about 10% of the Cheddar cheese production, but the greater proportion of cheese is produced in modern, mechanised creameries handling up to 200 000 gal. of milk per day.

Essential items of equipment for making Cheddar cheese and related varieties include: making vats, curd drainers, curd mills, cheese moulds and cheese presses. Making vats may be rectangular (see Fig. 1), rectangular with rounded ends, or cylindrical in shape (see Fig. 2), and may be open or covered. Their capacity ranges from 1000 to 4000 gal. The vats are stainless steel, jacketed vessels heated by circulation of water and/or steam, and fitted with mechanical sweep-type agitators or rotating, multiple-bladed cutters-cum-agitators. They are generally used solely for curd production, and are available for re-use immediately following the removal of curds and whey. Curd drainers or coolers are used for 'cheddaring' or texturing the curd, and these are especially designed for use in the different mechanised systems, such as the New Zealand Cheddarmaster and the Australian Lactomatic (Renwick, 1978). Curd mills may be of the tearing, slicing or chipping type. Cheese moulds may be of the conventional, cylindrical type (28 cm diameter), or rectangular for making 18–20 kg cheeses. In these moulds, the cheeses are pressed in horizontal or upright 'gang presses', often with the application of vacuum. For large capacity pressing, large hoop presses, holding a tonne of curd, may be used, and from these, 18–20 kg blocks of pressed cheese are cut and packaged. An alternative and recent development is the Wincanton 'block-former' system of continuous cheese pressing and packaging. Both systems involve the application of pressure after air has been evacuated from between the curd particles.

The process. Milk for cheesemaking may be standardised to a certain ratio of casein to fat (C:F) (0·68–0·72), or fat to solids-not-fat (F:SNF) (0·33–0·46) (Chapman, 1974). The milk is pasteurised (71 °C for 15 s), cooled to 30 °C and delivered to the making vats. It is inoculated with 1·0–2·0 % of a starter culture of mesophilic lactic acid streptococci (Table I). About 20 min after inoculation, rennet is added (22 ml per 100 litres milk), and 30–40 min later the coagulum is cut into small cubes about 5 mm in size; at this stage the acidity of the whey should be 0·10 to 0·14 % lactic acid. The curd is stirred in the whey, and heated to 39–40 °C over a period of 40–45 minutes (0·11–0·15 % lactic acid). Stirring continues until the curd is sufficiently firm (dry) and suitable for 'pitching' and separation from the whey (0·15–0·19 % lactic acid). This is the final stage in curd-making, the maximum increase in numbers of starter organisms has taken place, and the moisture content of the curd and its physical condition should be optimal. After separation from the whey, the curd (0·20–0·22 % lactic acid) is 'cheddared', either by hand in the traditional method, or by one of the mechanised cheddaring systems.

The starter bacteria continue to multiply and produce lactic acid at the rate of approximately 0·10 % every 20 min. Whey continues to drain from the curd which becomes elastic, smooth and silky and eventually exhibits the characteristic 'chicken-breast' texture (0·60–0·80 % lactic acid). The curd is then milled, and salt (2 %) is added and mixed thoroughly to ensure even distribution (see Fig. 3). Salted curd is delivered to cheese moulds and pressed hard (2–3 tonnes) for 15–18 h.

The microflora of Cheddar cheese and related varieties is composed of the starter bacteria, *Str. lactis, Str. cremoris* and *Str. lactis* sub-sp. *diacetylactis*, added deliberately to the milk, and the non-starter bacteria present in the milk, having survived pasteurisation or gained access as post-pasteurisation contaminants. These may include lactobacilli, pediococci, leuconostocs, staphylococci, micrococci, Group D streptococci and Gram-negative rods such as coliform bacteria, pseudomonads and

FIG. 3. Pieces of Cheddar curd passing along the conveyor of a Cheddarmaster prior to the pressing stages; note the automatic salt distributor in the foreground. Reproduced by courtesy of the Milk Marketing Board, Thames Ditton, UK.

achromobacteria. During cheesemaking, the starter streptococci increase in numbers from about $2\cdot0 \times 10^7$ cfu ml^{-1} in the cheese milk 20 min after inoculation, to about 2×10^9 cfu g^{-1} in the curd at pressing. The greatest increase occurs in the first $3-3\frac{1}{2}$ h of the making process, whilst the curd is in contact with the whey. In the period up to salting (2–3 h), the starter population is maintained or may decline, depending on the species and strains of starter bacteria. *Streptococcus lactis* maintains its maximum population until the cheese leaves the press, whereas *Str. cremoris* declines after reaching its maximum (Fig. 4). Survival also depends on salt tolerance, *Str. lactis* being more resistant than *Str. cremoris* (Dawson and Feagan, 1957). The fresh cheese curd contains between 10^8 and 10^9 viable starter bacteria per gram. These die out during the early stages of maturation at a rate dependent on the species; strains of *Str. cremoris* tend to die out more rapidly than

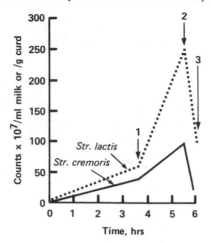

FIG. 4. Counts of *Str. lactis* (·····) and *Str. cremoris* (——) in making Cheddar cheese: (1) curd at pitching; (2) curd at milling; (3) curd at pressing after salting.

strains of *Str. lactis* (Martley and Lawrence, 1972) or *Str. lactis* sub-sp. *diacetylactis*.

Lactobacilli may be present in numbers ranging from 10 to $10^4 g^{-1}$ in the curd. They are the only lactic acid bacteria to multiply in the maturing cheese (except for the very similar pediococci), and they reach levels of 10^6 to $10^8 g^{-1}$ in 10 to 60 days, maintain these numbers then decline after 4–6 months. Species present may include *Lac. casei*, *Lac. plantarum*, the heterofermentative species *Lac. brevis* and *Lac. buchneri* and many unclassified streptobacteria. Pediococci occur much less frequently, but multiply at a similar rate and, if present, may reach levels of $10^7 g^{-1}$. The species present is *Pediococcus pentosaceus* (once known as *P. cereviseae*). Investigations, where specific strains of lactobacilli could be identified by serological tests, have shown that a dairy environment can become the permanent habitat of particular strains. This resident flora can infect each batch of cheese milk, so that a lactobacillus flora, peculiar to an individual plant, may be built up. Rigorous cleaning can remove this flora, but a new and perhaps different one is soon acquired. These stable floras may impart individual overtones of flavour to cheese from a particular factory.

Leuconostocs which may be present in the curd in small numbers, usually only $1–100 g^{-1}$, do not multiply but slowly die out during cheese ripening (Reiter *et al.*, 1967). The most common species are *Leuconostoc lactis*, *L. cremoris*, *L. dextranicus* and *L. mesenteroides*.

With Group D streptococci, the numbers present may vary considerably, from being completely absent, being present in small numbers of $10 \, g^{-1}$, or being present in high numbers of $10^4–10^6 \, g^{-1}$. They die out at a rate depending on the strain, sometimes surviving for a few weeks, and others lasting for more than 3 months. Species most frequently found include *Str. faecium* and *Str. bovis*.

Micrococci occur in the curd at levels of $10^2–10^6 \, g^{-1}$, depending partly on the heat treatment of the milk. They slowly decrease in number during ripening, so that after 6 months there are still about $10–100 \, g^{-1}$ of cheese. Coagulase-positive *Staph. aureus* strains which are potential toxin-producers are destroyed by pasteurisation. If a lower heat treatment is used, this sub-lethal heating often damages the cells so that, although still viable and able to grow on favourable nutrient media, the heat injured cells are unable to multiply on the unfavourable acid (pH 5·2) and salt containing (4·5 %) substrate of the Cheddar curd. Heat-resistant micrococci are commonly *Micrococcus varians* and other unidentified micrococci. Other species found are *M. luteus* and *M. lacticus*.

Gram-negative rods are killed by pasteurisation. Pseudomonads are only low-level post-pasteurisation contaminants. Coliforms are found much more frequently, depending on the hygiene of the dairy personnel. They are the only non-starter bacteria to multiply during actual cheese-making, as they are not inhibited by the acid produced; they die out rapidly in the cheese.

Corynebacteria and aerobic spore-formers occur in small numbers in the curd and survive for many months, without either multiplying or decreasing.

The changes which occur during the ripening of Cheddar are essentially the controlled degradation of the milk carbohydrate, protein and butter-fat. They are brought about by the action of the enzymes from the milk, the rennet, the starter and other microflora, and by chemical action such as oxidation, and yield a complex mixture of compounds which give mature cheese the required balance of flavour and aroma.

Cheddar flavour consists of taste components which are non-volatile, and aroma compounds which are volatile. Aroma compounds contribute the typical Cheddar flavour, whilst the non-volatile ones include lactic acid, amino acids, salt and fat which make up the essential background flavour.

The three basic metabolic processes, namely glycolysis, proteolysis and lipolysis (Reiter and Sharpe, 1971; Sharpe, 1972), occur as follows.

1. *Glycolysis.* Fermentation of the lactose by the starter streptococci produces lactic acid, small amounts of acetic acid, CO_2 and diacetyl; lactobacilli multiplying during cheese maturation may also produce these compounds.

2. *Lipolysis.* The levels of free fatty acids in the cheese are an important indication of the progress of maturation and of flavour production. Lactic acid bacteria are weakly lipolytic and are able to release fatty acids from milk fat. The essential, although not fully understood, role of fat and lipolysis is shown by the fact that Cheddar cheese made from skim-milk has no Cheddar flavour at all. The free fatty acids released are, at low levels, part of the background flavour.

3. *Proteolysis.* The rennet hydrolyses the casein to peptides, but cannot degrade it further. Starter streptococci, however, can produce both peptides and amino acids by their proteolytic action. These organisms have cell-bound extracellular proteinases which break down casein to peptides, or break down the large peptides formed by rennet to small peptides. These peptides are taken up by the bacterial cell and hydrolysed intracellularly to amino acids (Law and Sharpe, 1978*b*).

The nature of all these reactions, and of further minor reactions producing flavour compounds, and the way in which their relative rates of production are controlled, is not yet fully understood. This is partly due to the complexity of the cheese microflora which are the potential producers of flavour compounds.

Cheese microflora and flavour development. As cheese flavour intensity and numbers of lactobacilli increase during ripening, it was once thought that these organisms were responsible for the flavour of Cheddar cheese and related varieties. To determine the role of lactobacilli and other bacteria in the development of cheese flavour, an aseptic technique was developed (Mabbit *et al.*, 1959; Chapman *et al.*, 1966) at the NIRD, Shinfield, whereby cheese could be made under bacteriologically controlled conditions in an enclosed aseptic vat. With this technique it was shown that neither the lactobacilli nor the other non-starter bacteria present were responsible for Cheddar flavour (Reiter and Sharpe, 1971). Aseptic cheeses made with a single strain starter produced Cheddar flavour, whereas cheese made without starter, using δ-gluconic acid lactone for acidification did not, suggesting that the starter was responsible for flavour. As the starter bacteria die out rapidly during ripening, and flavour begins to develop only when levels of viable organisms have greatly decreased (see Fig. 5), Law *et al.* (1974) suggested that the action of intracellular enzymes released by the autolysed starter cells, and not metabolites of viable cells, were responsible for the production of flavour compounds; they demonstrated the release of intracellular starter enzymes which remained active at a high level during cheese ripening.

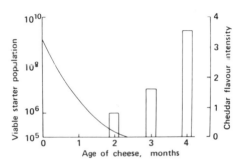

FIG. 5. Relationship between decline of the viable starter population and development of typical Cheddar cheese flavour in a cheese made with *Str. cremoris* NCDO 924.

However, more recent work indicates that Cheddar aroma is not directly due to starter activity. To release extra starter enzymes into the curd, a special technique can be used. Large amounts of pre-grown starter cells, pretreated with lysozyme to prevent acid production, are added to cheese milk, together with the normal starter. These treated cells lyse when the salt is added to the curd, and release intracellular enzymes. However, it was found that although protein breakdown was accelerated by the addition of extra starter cells, the flavour did not increase (Law and Sharpe, 1977). This finding that increased levels of starter enzymes and increased proteolysis did not affect flavour is in agreement with the observation of Lowrie *et al.* (1974), that in New Zealand cheese, more intense flavour often developed in cheeses with low curd starter populations, rather than in those with high populations. Starter enzymes are now considered not to be directly responsible for typical Cheddar flavour, and it is thought that apart from the diacetyl present, probably produced by the starter and contributing synergically to the aroma (Manning *et al.*, 1976), non-microbial chemical reactions are responsible. It is now known that methanethiol production is closely associated with the development of Cheddar flavour, and is probably one of the chief components of this flavour (Manning et al., 1976); this does not correlate with starter activity because these organisms do not produce methanethiol (Law and Sharpe, 1978a). It is likely that the starter only produces the correct conditions for the essential chemical reactions to occur, such as supply of flavour precursors derived from enzymic breakdown of milk components, a low pH which stops the reactions from proceeding too quickly, and a low E_h which keeps compounds such as methanethiol in their reduced form.

Related Types of Cheese

Modifications to the system of Cheddar cheesemaking produce the related varieties such as Double Gloucester, Derby, Cantal and Leicester (Table II). Slight differences in body, texture and moisture content (controlled by curd treatment) produce differences in the rate at which glycolysis, proteolysis and lipolysis proceed, and account for the individual characteristics of these varieties. Cheshire differs from Cheddar in its open texture, crumbly body and mild lactic flavour. These properties are brought about by differences in manufacture arising from the fast rate of acid production by very active strains of starter streptococci (mainly selected strains of Str. lactis), and the need for rapid moisture expulsion from the curd in a short making time (3 h). Raw milk is favoured by some farmhouse makers of Cheshire to preserve the natural image of their product, and the extent and rate of acid production helps to suppress the growth of unwanted bacteria. Cheshire can be made to ripen quickly (6–8 weeks), or ripen more slowly and reach a more mature stage in 3–4 months. A blue variety of Cheshire cheese is made by inoculating spores of Penicillium roqueforti into the milk or curd. The high acid curd and open texture of the cheese provide suitable conditions for mould growth, and this is encouraged by piercing the cheese during ripening.

Defects

Defects in Cheddar and related types of cheese due to micro-organisms are rare, as long as suitable starters are used which produce sufficient acidity at a satisfactory rate, and which do not give rise to fruity off-flavours or bitterness (see Table III).

Gassy curd. This and blowing of the film-wrap may arise from:

(1)   the presence of Str. lactis sub-sp. diacetylactis in the starter which utilises citrate in the cheese milk producing CO_2;

TABLE III
OFF-FLAVOUR PRODUCTION BY STARTER STREPTOCOCCI IN CHEDDAR CHEESE

Organism	Off-flavour	Cause
Str. lactis (e.g. NCDO 763)	Fruity	Ethyl butyrate
Str. lactis sub-sp. *diacetylactis*	Fruity	Ethyl hexanoate
Str. lactis (e.g. NCDO 604)	Malty	3-Methylbutanal
'Fast' starter strains (most strains of *Str. lactis* but also *Str. cremoris* NCDO 607)	Bitter	Hydrophobic oligopeptides

After: Law and Sharpe (1978*b*).

(2)   *Lac. casei* which also utilises citrate, if it is present in the curd, grows during ripening producing CO_2, and causes blowing of the pack at a later stage of maturation;

(3)   high levels of heterofermentative lactobacilli which also produce CO_2 during glycolysis, and can cause blowing of the film-wrap, but very high numbers of vigorously growing lactobacilli must be present for this to occur;

(4)   coliforms, if present in large numbers in the cheese milk, may multiply during cheesemaking producing a gassy curd, which may sometimes blow the film-wrap and give a 'faecal' off-flavour to the cheese.

Rancidity. Heat-resistant lipases, originating from psychrotrophs such as pseudomonads present in the raw milk, survive in the cheese and, if present at high enough levels, cause rancidity to develop by hydrolysing the milk fat triglycerides to release free fatty acids (FFA), such as butyric, caproic and capric. Such lipases may remain active in the cheese for more than 12 months, so that levels of FFA and intensity of rancidity continue to rise.

Mould growth. Weak bodied cheeses which develop cracks and splits, due to incorrect cheesemaking, may become infected with moulds, usually *Penicillium* spp. Aerobic conditions caused by the splits allow this growth to penetrate the cheese. The concentration of oxygen at the surface of film-wrapped cheese should be too low to support mould growth (Dolby, 1966).

Cheese with Eyes
Eye formation is characteristic of certain varieties of hard cheese, notably

the Swiss-type cheeses, Emmenthal and Gruyère. Eyes result from the production of CO_2 in the cheese by propionibacteria. The gas forms bubbles in the cheese which increase in size depending on the rate of gas production and the physical nature of the cheese. In making these varieties it is essential to produce curd which is sufficiently elastic for eye formation to occur, yet firm enough to retain its shape. Propionic acid forms salts which contribute to the flavour of these cheeses.

Emmenthal

This cheese is named after the valley of the River Emmen in the canton of Bern in Switzerland where it originated many centuries ago. The traditional Swiss cheese is made in large wheels, 68–78 cm in diameter and weighing between 70 and 80 kg (Table II), but nowadays much Emmenthal is made in rindless, rectangular blocks weighing about 40 kg. The mature cheese has a pliable body with smooth texture, many large holes (1·0–2·0 cm in diameter) which should be smooth and shiny, and a characteristic sweet, nut-like flavour.

Equipment for the manufacture of Emmenthal includes jacketed copper kettles which hold around 200 gal. of milk; cutting knives with wires about 2·5 cm apart (called harps); 'dipping cloths'; hoists; cheese moulds and presses. In addition to the making room, it is necessary to have a salting room (10 °C) in which the brine tanks are situated; a chilled room (10–12 °C and 80–85 % RH); a warm room (18–22 °C and 80–85 % RH), and a store for the mature cheeses (14–15 °C and 84–87 % RH).

During the last 20 years there has been a trend towards increasing the size of cheese factories and their daily throughputs of milk. A large factory may have a daily intake of some 15 000 gal., mostly in bulk tankers, and the cheese vats may be refilled several times in the day.

The process. Good quality raw milk is standardised to a fat content of 2·8–3·1 %, to give a cheese with 45 % FDM. The milk, sometimes clarified, is delivered to the kettles and warmed to 30–35 °C. A starter culture of thermophilic lactic acid bacteria (Table I) is then added. This generally consists of 0·03–0·10 % of a culture of *Str. thermophilus*, and up to 0·20 % of *Lac. helveticus* or *Lac. bulgaricus*, together with a few drops (1·0–2·5 ml per 1000 litres) of a culture of propionibacteria. Rennet extract (16 ml per 100 litres) is added, and 25–30 min later the coagulum is cut into strips, then cubes, and finally into particles about 3 mm in diameter. The temperature of the curd is then raised to 45 °C at the rate of 1 °C every 2 min, and then to 53–57 °C at the rate of 1 °C min^{-1}; this treatment is critical to kill-off non-thermoduric bacteria. The total heating time is about 45 min, and when the curd is judged to be sufficiently firm, it is collected in a large, coarse dipping cloth, hoisted out of the vat (kettle) and lowered into a cylindrical mould (hoop), made of wood or stainless steel, standing on a draining table with a pressing device above it. The cheese is pressed for 20–24 h, during which time the pressure is increased gradually to 15 kg per kg of cheese, and the cheese is turned at intervals to promote uniform drainage. As the curd cools, the starter bacteria multiply and produce the acidity required to cause the curd particles to mat together and form a dense, elastic mass.

After pressing, the cheese, still in its hoop, is held in the salting room, temperature 10 °C, for 1–2 days to cool. It is turned at intervals and dry salt is sprinkled on the surface. It is then removed from the hoop, and immersed for 2–3 days in brine containing about 23 % NaCl at 8–12 °C. The strength of the brine is important, as the cheese sinks in a weak brine becoming slimy and developing unpleasant flavours, and the brine becomes putrefactive. The cheese is turned once a day and salt is sprinkled over the surface. This helps to maintain the strength of the brine, replacing salt which has been absorbed into the cheese. On removal from the brine bath, the cheese is placed on a clean, dry base board and held in a chilled room (10–12 °C and 80–85 % RH) for 8–10 days. Each day it is brushed or wiped with a cloth dipped in salt water, turned and dry salted. For the next stage in maturing, the cheese is moved to a warm room (18–23 °C and 80–85 % RH) to encourage development of the eye-forming bacteria. The cheese is turned at intervals during the next 4–8 weeks, and brushed and washed with brine to prevent mould growth. The cheese acquires a slightly rounded shape, and when the eye development is considered satisfactory, it is moved to the curing room (13 °C and 80–85 % RH); here it is washed and turned at intervals throughout the remainder of the ripening period (3–9 months). At least 6 months are necessary for the cheese to develop the full characteristic flavour.

Milk for cheesemaking. Traditionally, Swiss cheese was made from raw milk, and the bacteriological condition of the milk has always been an important factor in producing cheese of good quality. Coliform bacteria in large numbers may interfere with acid production by the starter bacteria and reduce the quality of the cheese (*Enterobacter aerogenes* inhibiting lactobacilli and *E. coli* inhibiting streptococci).

Some micrococci promote, and some inhibit, acid production by the starter bacteria, and certainly fewer numbers of micrococci in the cheese milk reduce defects due to inadequate acid production (Langsrud and Reinbold, 1973). Anaerobic spore-formers (particularly *Cl. tyrobutyricum*) may cause late blowing of the cheese, and in bulk raw milk stored for several days at refrigerated temperatures, psychrotrophic bacteria are able to multiply and, at high levels, produce lipolytic changes in the milk.

Clarification of the milk improves the quality of Emmenthal, although it may reduce the moisture content of the cheese, and there may be loss of fat into the whey. The treatment also improves multiplication of the starter organisms, and while it reduces the number of eyes in the cheese, it increases their size. In addition, the ripening process is somewhat slower and the cheeses are firmer. Any detrimental effects of clarification can be reduced by adjustments to the temperature and flow rate of the operation (Langsrud and Reinbold, 1973).

Bactofugation is carried out at higher temperatures (60 °C), and has been used in Sweden to remove clostridial spores from milk used for making Grevé cheese, a round-eyed variety similar to Emmenthal. The treatment has proved beneficial for Emmenthal cheese, but the efficiency depends on the spore content of the original milk; this must be no more than one log cycle above the safety level of $200\,ml^{-1}$ (Mocquot, 1979). In large Emmenthal-producing plants, where milk is collected from distant farms and may have to be stored before use, the milk is always heat treated before cheesemaking, usually at 62 °C for 30 s. Pasteurisation of the milk reduces the numbers of unwanted bacteria and cheese of good quality can be made with milk heat treated within the range 68–74 °C with 15 s holding time (Langsrud and Reinbold, 1973); optimal conditions were 72 °C for 15 s. Excessive numbers of bacteria survived temperatures below 68 °C, and heating above 74 °C caused changes in the casein which resulted in body and eye defects, and splits in the cheese. Thermisation is used in some countries where pasteurisation is considered to be too severe.

Starters. Thermophilic lactic acid bacteria (Table I) play a principal part in the making and ripening of Swiss-type cheese. They must be heat-resistant and able to grow at relatively high temperatures to survive the cooking process, and continue to grow during processing. The rate of acid development in the making vat, and in the press, controls the pH, acidity, moisture content, body and texture of the cheese and determines its quality. The organisms most widely used are *Str.*

thermophilus and *Lac. helveticus*, although *Lac. bulgaricus* is generally used in the USA.

Propionibacteria are essential for development of the characteristic flavour and eye formation in Swiss cheese. They ferment lactic acid, carbohydrates and polyhydroxy alcohols to propionic and acetic acids and CO_2. Their growth is slow and preferably anaerobic; their optimum pH is 6·0–7·0 (maximum 8·5 and minimum 4·6); and their optimum growth temperature is 30 °C, but they remain viable for up to 8 weeks at storage temperatures of 5 °C. They have a low tolerance to salt, and while there may be strain differences in reaction to antibiotics, penicillin and streptomycin inhibit the normal ripening of Swiss cheese. Growth of propionibacteria is stimulated in association with *Lac. bulgaricus* and *Lac. helveticus*, and more propionic acid and CO_2 are produced, with improved flavour, in the cheese.

The role of the microflora in processing and ripening. During curd-making, the numbers of *Str. thermophilus* increase slowly and steadily, but more rapidly at the higher temperature towards the end of cooking; *Lac. bulgaricus* may decline in numbers. Low cooking temperatures may allow the survival of an unwanted flora, thus leading to defects in the cheese. Draining and pressing control the moisture content of the curd and promote acid production which assists the curd particles to mat together. The process of draining and pressing takes place at an ambient temperature of 20 °C, and lasts for 20–24 h. During this period the temperature of the cheese falls from about 50 °C, and growth of the starter bacteria and acid production increase with the falling temperature. Cooling is more rapid at the surface of the cheese, and the thermophilic lactic acid bacteria begin to multiply there during the first few hours of pressing. *Streptococcus thermophilus* multiplies first, producing lactic acid at the periphery of the cheese within 2–3 h, and reaching maximum numbers of $10^8 \, g^{-1}$ in the centre of the cheese and $10^9 \, g^{-1}$ at the periphery about 3 h later. The lactobacilli begin to multiply later, *Lac. bulgaricus* after 4–5 h and *Lac. helveticus* after 6–8 h, reaching maximum numbers in 20 h; the count again being about 10 times less at the centre of the cheese (see Fig. 6). Lactic acid formation is most rapid at the periphery, and hence lactose from the centre diffuses towards the periphery where it is utilised, so that less lactose is available in the centre when the temperature is eventually suitable for the lactic acid bacteria to metabolise. When the cheeses are removed from the press (within 24 h), no lactose remains, and there is more lactate at the periphery than in the centre. These biochemical

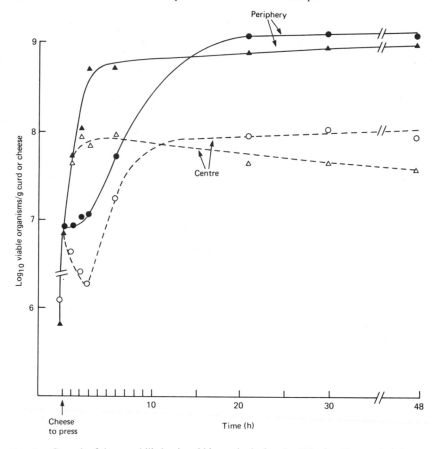

Fig. 6.   Growth of thermophilic lactic acid bacteria during the 48 h after Emmenthal cheese is put into press. Cheese periphery: ●——●, *Lac. helveticus* colony count; ▲——▲, *Str. thermophilus* colony count. Cheese centre: ○----○, *Lac. helveticus* colony count; △----△, *Str. thermophilus* colony count. (From: Accolas *et al.*, 1978.)

changes may affect the development of the propionibacteria, and the formation of CO_2 and volatile fatty acids (Mocquot, 1979). The pH of the cheeses from the press (24 h old) should be 5·2–5·3, and if higher there may be defects in eye formation and flavour, while if the starter bacteria have been inhibited by antibiotic residues in the milk, early blowing may occur.

Salting the cheese from the outside by immersion in brine, or by rubbing salt on the surface, concentrates the salt at the periphery of the cheese, and although it diffuses slowly throughout the cheese mass during the early months of ripening—eventually reaching an average salt concentration of 0·40–0·80%—a salt gradient exists within the cheese, with a maximum difference of 4–5 fold between periphery (highest salt content) and centre (Mocquot, 1979); this variation has an important effect on the microbial flora. Thus, propionibacteria are inhibited by relatively low salt concentrations, and must develop whilst the salt is still largely confined to the periphery of the cheese. Excessive eye formation, however, can be indicative of low salt levels in the cheese.

When the 7–10 day old cheeses are moved to the warm room, propionibacteria develop rapidly, usually reaching numbers of $10^9 g^{-1}$ at 4–8 weeks and remaining the major flora for the rest of the ripening period.

The maturation of Swiss-type cheese consists of a slow proteolytic and lipolytic breakdown of the curd leading to flavour production, as in other hard cheeses, and the fermentation of lactate by propionibacteria.

Defects

Lipolytic off-flavours. Storage of milk at refrigeration temperature for several days may create problems arising from the growth of lipolytic psychrotrophic bacteria.

Early blowing. Inhibition of starter bacteria by antibiotics in the milk, too low a heat treatment temperature, and/or low cooking temperatures can lead to the growth of unwanted bacteria (e.g. *Aerobacter aerogenes*) causing gas production in the curd whilst the cheese is in the press.

Secondary fermentation. Following the desired eye formation by propionibacteria, a subsequent, secondary fermentation may occur involving further eye production and the development of cracks in the cheese, because the texture of the more mature cheese is no longer suitable for the normal eye formation. This late formation of CO_2 has been ascribed to amino acid decarboxylation by enterococci (*Str. faecalis, Str. durans* and *Str. faecium*), or to stimulation of the propionibacteria by free amino acids (e.g. glutamic acid) formed by the enterococci or by the proteolytic activity of some lactic acid bacteria (e.g. *Lac. lactis*). *Lac. helveticus* is considered to inhibit late fermentation (Mocquot, 1979).

Late blowing. Abnormal production of gas (CO_2) during ripening is usually caused by lactate-fermenting species of clostridia (e.g. *Cl.*

tyrobutyricum). This group of anaerobic spore-forming bacteria, the butyric acid bacteria, produce large amounts of butyric acid and CO_2, and can cause severe spoilage of Swiss cheese. They can be present in large numbers in milk from silage-fed cows, and numbers in excess of $200 \, \text{litre}^{-1}$ are considered enough to put the cheese at risk (Mocquot, 1979).

Excessive eye formation or 'oversetting'. Large numbers of small eyes can result from excessive growth of propionibacteria, and this fault usually indicates a lack of salt or acid in the cheese.

Lack of eye formation. Propionibacteria are very sensitive to salt and acid, and high levels of these can inhibit their growth. Such high levels of salt and acid also cause loss of elasticity in the curd, which then cracks instead of stretching when 'eyes' are being formed. This is known as the glass or gläsler defect.

Various off-flavours. These may develop in the cheese if unwanted micro-organisms are able to grow. All the gas producing organisms form products which cause off-flavours, such as 'yeasty', 'fermented', 'unclean' and 'putrid'.

Rusty spot. These small rust-coloured spots occasionally appear in Swiss-type cheeses, and as in other hard cheeses, they are usually caused by *P. rubrum*.

Gruyère

This variety of cheese takes its name from the town of Gruyère in the canton of Fribourg in Switzerland. It is very similar to Emmenthal in appearance, nature and method of manufacture. The main differences are in the eye formation, in the size of the cheese wheels, and in the rind. Gruyère has fewer and smaller holes, is a smaller wheel, and the rind is a darker brown and rougher, with a slightly greasy, polished appearance (Davis, 1976*a*). A certain amount of surface ripening occurs (*Bre. linens*) giving Gruyère its particular flavour.

Semi-Hard Cheese

The cheeses in this group vary considerably, and demonstrate the diverse roles of the lactic acid bacteria in cheesemaking, ripening and flavour development. The varieties range from the fresh, acid curd of Caerphilly to the crumbly, flavourful, mature Lancashire and the close-textured, mild, washed curd cheeses, Edam and Gouda. All varieties are prepared from whole milk or low-fat milk with starter cultures of mesophilic lactic acid streptococci (Table I). The rennet curds are coarsely cut and lightly scalded, thus retaining a high level of moisture and lactose. Specific methods are used subsequently to control the growth of the starter bacteria and the development of acidity in the different curds. These differences govern the pH of the young cheeses, and account for the widely different varieties of cheese within this group. For certain varieties, e.g. Caerphilly and Lancashire, growth of the starter bacteria is encouraged during curd-making, and the pH of the young cheese is low (5·2–5·0); acid curds are produced and these varieties have a crumbly texture. In other varieties, e.g. Edam and Gouda, acid production is restricted by removing a portion of whey during stirring and replacing it with water. This reduces the lactose content of the young cheese, the pH is high (5·4–5·3), the body is plastic and the texture is close.

The equipment used for making Caerphilly and Lancashire is similar to that required for making any of the British hard cheeses. Cheese moulds for Caerphilly are fitted with special collars, and brine tanks are required for salting the cheese.

Caerphilly

This cheese is of Welsh origin, taking its name from the village in Glamorgan where much of it was made. It is a quick-ripening cheese made from full-cream milk, and has a clean, mild, lactic flavour which becomes sharper with age.

Pasteurised milk, cooked to 31–32 °C, is inoculated with a starter culture of mesophilic lactic streptococci (1–2 %) and ripened for 1·0–1·5 h. Rennet (20 ml per 100 litres milk) is added, and 35–40 min later the coagulum is cut into 6 mm cubes. The curds and whey are lightly scalded (33·5–34·5 °C), and after separation from the whey, the curd is drained, salted (1·0 % NaCl) and lightly pressed. The cheeses are floated in brine (18–20 %) for 24 h and then dried, film-wrapped and stored at 13–16 °C for 1–6 weeks. They are subject to spoilage by mould growth on the surface, unless suitably protected.

In the lightly scalded and lightly pressed curd, the starter bacteria produce lactic acid steadily during manufacture and pressing. Caerphilly is a fresh, acid, moist cheese which is not matured and is consumed in 10–30 days. The clean, mild, lactic flavour depends on the strains of lactic streptococci selected for the starter culture.

Lancashire

This cheese takes its name from the English county where it originated. It is characterised by a soft, crumbly texture derived from mixing curds of different ages and acidities. The flavour is mild and mellow at 3 months old, and becomes stronger and fuller if matured further.

Full-cream pasteurised milk, cooled to 30 °C, is inoculated with a starter culture of mesophilic lactic acid streptococci (0·1–0·75 %) and ripened for 1 h. Rennet (22 ml per 100 litres of milk) is added, and 40–60 min later the soft coagulum is cut into 7·5 mm cubes and stirred for 1 h. The starter bacteria multiply slowly, and the curd becomes slightly firmer. After removal of whey, the curd is lightly pressed, then cut, broken and lightly repressed at 15 min intervals for a further 1 h to encourage further drainage of the whey. During this time the acidity is developing, and when it reaches 0·20 % lactic acid, some of the curd is placed in containers and kept at 15–18 °C for 24–48 h to mellow and develop a high acidity. The following day a proportion of the high acid curd (1·5 % lactic acid) is mixed with a specific amount of fresh, new curd (0·18–0·20 % lactic acid—only 2–3 h old) to provide a cheese curd with an average acidity of 0·9 % lactic acid. The two curds are then milled in a fine, Cheshire-type mill, salt (2 %) is added and, after thorough mixing, the curd is filled into moulds, held overnight at 21–22 °C, and then pressed. The cheeses are waxed, or film-wrapped, to prevent excessive loss of moisture, and are stored at 13–15 °C for 3–4 months. Long-keeping cheeses are stored at 5–10 °C for 6–12 months.

The ripening process, like that of other varieties of hard-pressed cheese, is brought about by the combined action of proteinase, produced by the starter bacteria, and the rennet enzymes. Lancashire cheese is subject to the same faults (defects) as Cheddar and related varieties and, unless protected, the surface suffers spoilage by moulds.

Edam and Gouda

These varieties originated in Holland where they are the most important types of cheese made, and their manufacture is well mechanised. Edam is made in the form of balls, weighing about 2 kg, which are red waxed for sale. Gouda is made in the form of small wheels weighing 3·5 to 25 kg (Table II). Both varieties are sweet curd, renneted cheeses made by the system of washing the curds with water. The high pH of both curds (5·3–5·4), and the absence of lactic acid to suppress spoilage bacteria, emphasise the importance of using milk of good bacteriological quality for making these cheeses. Bactofugation of the milk for making Gouda improves the cheese quality by removing a considerable number of clostridial spores present.

Cheesemaking equipment varies with the size of the plant and the extent of mechanisation, but consists of suitable making vessels, ancillary tools, cheese moulds and presses, brine tanks and cheese stores.

The process. Milk is standardised, pasteurised and cooled to 30–31 °C. Calcium chloride (up to 0·02%), sodium nitrate (up to 0·02%) and colouring may be added before the milk is inoculated with 0·5% of a starter culture of mesophilic lactic acid streptococci (Table I). Rennet (30 ml per 100 litres of milk) is added, and 30 min later the coagulum is cut into large particles (10 mm cubes). These are lightly scalded by removing some of the whey and replacing it with water at 50–60 °C; this raises the vat temperature to 36–37 °C in around 30 min. The amount of water added is equivalent to 25% of the original milk, and this dilution reduces the lactose content of the curd and controls the growth of the starter bacteria. After the whey has been drained off, the high moisture, high pH (5·3–5·4) curds are dipped (filled) into moulds and pressed lightly. The cheeses are then salted in brine (21–36% NaCl) for 24–48 h, dried to form a coat, and ripened for about 6 weeks (12–15 °C and 85–90% RH). They are waxed or film-wrapped before sale.

The changes which take place during ripening are similar to those in other hard-pressed cheeses, namely the breakdown of casein by enzymes from the rennet, and by proteinases from the starter bacteria which further degrade the intermediate products of casein hydrolysis (proteoses and peptones). These are restricted, somewhat, by the high pH of the young curd (5·3 at 24 h old). The total count of the cheese increases rapidly at the start of ripening and then decreases slowly; *Str. cremoris* is present in large numbers during the first 10 days, and then quickly disappears; *L. dextranicum* grows quickly in the first week and then decreases slowly. As the starter bacteria grow and the pH of the cheese decreases (5·1 at 3 days), casein hydrolysis intensifies. Particular strains of lactic acid bacteria vary in their ability to break down the products of casein hydrolysis (Stadhouders, 1962) and in their effect on the quality and microflora of Edam cheese (Habaj *et al.*, 1966).

Gouda, with its slightly lower moisture content, can be matured for longer than Edam. This allows more protein breakdown and rather more flavour in the variety. A few small eyes, produced by propionic acid bacteria, are permissible in Gouda but are regarded as a defect in Edam.

Defects. Edam and Gouda may suffer any of the defects which spoil hard cheese, such as mould growth (mainly *Penicillium* and

Aspergillus species), gassiness, and the development of off-flavours and bitterness.

BACTERIAL SURFACE-RIPENED CHEESE

The Semi-Soft Cheeses
This group of cheeses contains a number of varieties which are ripened largely from the outside by the action of a surface flora. The group includes Limburger, Port du Salut, Münster (Table II). All the varieties are made from sweet, rennet curds and are ripened, to some extent, by proteolysis from the starter bacteria and the rennet, but the characteristic brownish-red surface growth of *Bre. linens* makes a major contribution to the ripening of some varieties, and affects the flavour of all of them.

Equipment required for making this type of cheese includes making vats, cutting tools, pressing plates, cheese moulds (cylindrical or rectangular), cheese followers and light weights to fit the moulds; brine tanks and ripening rooms (12–15 °C and 85–95 % RH).

The Process
The general process of manufacture is similar for all varieties. They are made from pasteurised milk with small amounts of a starter culture of lactic acid streptococci. This may be a culture of *Str. lactis* alone, or *Str. lactis* and *Str. thermophilus*, and the amount will vary from 0·1–0·2 % for Limburger to 0·8 % for Port du Salut. Rennet (22 ml per 100 litres) is added at 30–32 °C, the coagulum is cut in 30–40 min, and the curd particles are heated in the whey. The moisture content of the different varieties is controlled by the particle size, and the temperature of the scald, 35 °C for Limburger and up to 45 °C for Brick. After removal of whey, the curd is filled into suitable cheese moulds; followers and weights are placed on top of the curd to consolidate it, and the cheeses are turned frequently for several days. They are then immersed in brine (23 % sodium chloride) at 10 °C for 24 h, or dry salted on the surface for 1–2 days and wiped with cloths soaked in brine. Finally the cheeses are placed on shelves in the ripening cellar. Within a few days the characteristic brownish-red slime develops on the cheese surface, and this is smeared over the whole surface to distribute the bacteria uniformly. These varieties of cheese vary in size (Table II), and the relationship between depth of cheese and

surface diameter is important, as too large a diameter or too great a depth can accelerate or delay ripening, and the cheese could be over-ripe and bitter near the surface long before the interior is ripe.

Role of the Microflora in Processing and Ripening

Starter bacteria. During manufacture the lactic acid streptococci grow rapidly, and development of acidity continues whilst the curd is draining in the moulds. After 24 h the pH may be about 5·0, but this varies with different varieties. The rate of acid development governs the rate of whey drainage from the curd, and thus the final moisture content of the cheese. It is affected by the numbers, and the activity, of the acid producing bacteria, and by the room temperature. Rapid acid production, leading to a dry, over-firm curd, or slow, inadequate drainage, leading to a wet, sour curd, result in cheeses which do not ripen satisfactorily.

The surface flora. Within 2–3 days of manufacture, a whitish growth of aerobic, salt-tolerant yeasts and *Geotrichum candidum* appears on the surface of the cheese, and some days later, when the pH has been increased by their metabolism, the brownish-red growth of *Bre. linens* appears and the surface becomes slimy. *Brevibacterium linens* is very actively proteolytic, and the extent to which its growth is allowed to develop depends on the variety of the cheese. It is greatest in Limburger, where the cheese may resemble the consistency of warm butter.

Ripening. Within these cheeses, as the starter bacteria die out, the curd undergoes varying degrees of proteolytic and lipolytic changes associated with enzymes from the starter bacteria and the rennet. The function of the surface organisms, especially *Bre. linens*, is to contribute, in a greater or lesser degree, to the flavour of the particular variety. Their growth and contribution to the cheese flavour is governed by: (1) the moisture content of the curd, drier varieties, such as Bel Paesa and Monterey, show little surface ripening and have a low flavour intensity; (2) the area of cheese exposed during ripening, cheeses may be piled on top of one another to prevent excessive growth of *Bre. linens*; (3) the temperature and duration of ripening; and (4) removal of the surface growth. The flavour of these varieties ranges from mild, in the case of Bel Paesa and Monterey, to the medium flavour of Münster, Port du Salut and Brick, to the strong flavour of Limburger.

In Limburger the surface growth is responsible for many of the changes in ripening. Thus, *G. candidum* utilises the lactate formed and increases the pH at the cheese surface to enable *Bre. linens* to grow. The difference in pH from the surface (6·0–7·0) to the centre (5·4–5·7) suggests that neutralisation of the acid is associated with products of the surface flora rather than with products of casein hydrolysis within the cheese (Kelly and Marquardt, 1939).

Defects

Moulds. One of the most serious defects arising in bacterial surface-ripened cheese is failure of the smear to develop. This can occur if the surface of the cheese is too dry, or the humidity of the curing room is too low. Slow development of the smear allows unwanted moulds to grow. These may include species of *Cladosporium, Penicillium, Aspergillus* and others.

Early blowing. Unless good quality pasteurised milk is used, all varieties in this group may suffer early gas formation by coliform organisms during draining and salting. This defect can vary in intensity from a few, small pin-holes to a spongy condition.

Late blowing. If the curd, after manufacture, does not have a low enough pH (below 5·3), or high enough salt content, the cheeses are subject to late gas formation by the growth of clostridia.

Over-acidity. Excessive development of acidity by the starter bacteria during manufacture may cause the pH of the curd to fall too rapidly or too far (4·8–4·7). This is a common defect in Brick.

INTERNAL MOULD-RIPENED CHEESE

Blue-Veined Cheese

This group of cheeses, exemplified by Gorgonzola, Roquefort and Stilton, are made from high acid, semi-soft curd, and the process involves slow acid development during a long draining period. The cheeses are not pressed, instead the curd is allowed to consolidate under its own weight. Salt may be added to the drained, acid curd before it is placed in the cheese moulds (Stilton), or it may be rubbed into the surface of those cheeses where acid development takes place in the moulds (Gorgonzola and Roquefort). Initially blue-veined cheeses may undergo some proteolysis from the starter bacteria and rennet enzymes, but when they are pierced to admit air, the blue-green moulds, *Penicillium roqueforti* or *P. glaucum* spread through the cheese and bring about the proteolytic and lipolytic changes responsible for the ripening and distinctive flavour of these varieties.

Equipment required for making blue-veined cheese includes making vats; suitable cutting devices (curd knives); curd drainers; cheese moulds or hoops (perforated cylinders, 15–30 cm in diameter and 10–30 cm high); draining boards and mats. Rooms for draining and salting, for drying, and for ripening the cheese are necessary, and these should be provided with controls for temperature and humidity.

Cheesemaking

Milk. For the manufacture of blue-veined cheese, high-fat milk is desirable as lipolysis plays an important part in the flavour development. Sheep's milk is used extensively in France, and is mandatory for making Roquefort. Stilton and Gorgonzola are made from cow's milk, but cream may be added if the fat content is low.

Much of the milk for making blue-veined cheese is used raw, as milk lipase aids fat hydrolysis in ripening, but there is a move towards the pasteurisation of the milk to overcome defects arising from raw milk of poor bacteriological quality, and to contribute to a more standard quality cheese. Pasteurised milk is used for making Stilton, and it is not uncommon for the milk, or cream from the milk, to be homogenised. This process helps to form a smooth curd with maximum retention of moisture, and it increases the surface area of the fat, thus promoting the lipolytic action of the mould (*P. roqueforti*) and accelerating typical flavour production.

Starters. Mesophilic lactic acid bacteria (Table I) are used as starters, and their growth and acid production are essential for adequate draining of the curd. The starters in general use are strains of *Str. lactis* and *Str. lactis* sub-sp. *diacetylactis* and *Leuconostoc* species.

Moulds. The blue-veined cheeses are characterised in appearance and flavour by the growth and development of the blue mould *P. roqueforti.* Cultures of *P. roqueforti* in liquid suspension are generally introduced deliberately, either to the milk before rennet is added, or by spraying onto the curd with the salt. The cheeses are pierced, several times, during ripening to admit air to enable the mould to grow and metabolise. It is the products of this proteolytic and lipolytic activity within the cheese which give these blue-veined cheeses their characteristic piquant flavour.

The process. For various reasons the blue-veined cheeses have become closely associated with particular geographical localities. Stilton cheese was made originally in farmhouses in the Vale of Belvoir in Leicestershire, and the name Stilton can only be used to describe the cheese made in this county and the two neighbouring counties of Derbyshire and Nottinghamshire. Gorgonzola originates from a town of that name, near Milan, in northern Italy. Its history can be traced back to about 1000 AD. Originally made from ewe's milk by a two-curd system, it is now made from cow's milk by a single-curd process. Roquefort is the most famous of the cheeses made from ewe's milk. Its origin can be traced back into the 8th Century, and French legislation protects the designation by restricting the name to cheese made from ewe's milk produced in the province of Aveyron and six surrounding provinces, in Corsica and in the Pyrenees. These areas share similar farming systems, breeds of sheep, grassland and climate. The cheese may be made in local dairies, but it must be transported to ripen in the vast system of caves in the Causée du Combalou, a vast mountainous limestone ridge which overlooks the village of Roquefort (Mocquot and Béjambes, 1960).

Many varieties of blue-veined cheese are made in France from the milk of sheep, goats and cows, and named after the place of manufacture (e.g. Bleu d'Auvergne). Blue-veined cheeses are also made in the USA, Australia, Denmark and Ireland, and generally on the same basic principle of high acid development over a long slow drainage period; a pH of 4·5–4·7 is reached in about 24 h. The slow acid development controls the expulsion of moisture from the rennet curds, and produces a cheese with a soft, velvet-like body, a flaky texture, a low pH and a moisture content of 48–50%. Starter and rennet are added to the milk, and after cutting, syneresis of the curd is brought about by acid development only. The curds are not scalded during processing and the cheeses are not pressed mechanically. After the curds are filled into the moulds (hoops), frequent turning ensures adequate drainage and consolidation of the curd by its own weight.

Changes in manufacturing techniques to accelerate the making and ripening of blue cheese can cause considerable differences in flavour. In Danish Blue, the milk is homogenised, the curd is highly salted, and mould spores are incorporated into the curd; intensive stabbing encourages their rapid growth, and contributes to the strongly lipolytic, salty flavour of this cheese.

The role of the microflora in processing and ripening. The slow production of acid by the lactic acid bacteria makes it essential to use milk of good bacteriological quality. Thus the balance between acid production and moisture expulsion is vital, and any acceleration or hindrance of acid development affects the moisture level in the curd, which, in turn, will interfere with proper development of the mould.

Salting, either of the curd before moulding (hooping) or of the formed cheese, reduces the moisture content and, if applied to the cheese, helps to firm and harden the surface. At this stage, when the cheeses are being salted, the numbers of starter bacteria are high (10^9 cfu g^{-1}), but the low pH and the increasing concentration of salt in the moisture of the cheese inhibit these organisms, and after 2–3 weeks, few viable cells are found; as a rule no other bacteria are found in significant numbers during ripening. The extent to which the starter bacteria contribute to the ripening changes is not known, but it is possible that intracellular enzymes are released which, together with enzymes from the rennet, help to bring about some proteolysis of the casein in the early stages of ripening, and before the mould (*P. roqueforti*) begins to grow.

The characteristic process in the manufacture of blue-veined cheese is the pricking, piercing or stabbing of the cheese to admit air, and possibly mould spores, to the interior. Roquefort and Gorgonzola are pierced at 2–3 weeks old, and Stilton at 5–6 weeks old. At the temperature and humidity of the curing room (10–13 °C and 96 % RH), *P. roqueforti* generally appears within 8–10 days. The mould ramifies through fissures in the cheese and along the lines of the stabbings reaching maximum growth in 30–90 days. Maximum growth generally appears at the centre of the cheese, where initially the concentration of salt is lower. The cheeses are then wrapped in foil, or moved to a cooler curing room, to slow down the ripening during the remainder of the maturation process.

The establishment of *P. roqueforti* as the ripening agent in blue-veined cheese is accounted for mainly by the fact that this species of mould is more tolerant of salt and low concentrations of oxygen than most other species (Thom and Currie, 1913), and that it is added in large numbers to the cheese (Mocquot and Béjambe, 1960). Furthermore, as the cheeses ripen, the oxygen content is reduced and the CO_2 content is increased. These are adverse conditions for all moulds, but *P. roqueforti* appears to withstand them better than other species.

As the mould develops, casein is hydrolysed and the body of the cheese becomes softer. The proteinase formed by the mould causes this breakdown, and there is a considerable increase in the amount of amino nitrogen. This is most marked during the first months of the ripening period, but continues, slowly, throughout the life of the cheese. The proteinase is active over a pH range of 5·3 to 7·0, and its optimum is about pH 6·0. During ripening, the pH of the cheese rises from its initial minimum of 4·7–4·5 at 24 h to a maximum of 6·0–6·24 at 2–3 months, probably as a result of utilisation of lactic acid by *P. roqueforti*. There may be a subsequent fall in pH when butyric, caproic, caprillic and capric acids, as well as some higher fatty acids, are liberated from the fat by lipases produced by the mould. These lipases are active at the pH, temperature and salt concentration found in the ripening cheese (Morris and Jezeski, 1953). According to Hammer and Bryant (1937), part of the products of fat hydrolysis may be metabolised by the mould, the volatile acids increase and the characteristic flavour is intensified. The peppery flavour of blue-veined cheese can be attributed to methyl ketones derived from caproic, caprillic and capric acids.

The responsibility of *P. roqueforti* for ripening the blue-veined cheese is evidenced by the absence of characteristic flavour and proteolytic changes in cheese made without the mould (e.g. White Stilton).

Attempts to accelerate flavour development by adding animal lipase have increased the amount of volatile fatty acids in blue cheese and hastened their appearance, and the typical Roquefort flavour developed sooner than in cheese made with the mould alone, but a bitter flavour was present (Coulter and Combs, 1939). Parmelee and Nelson (1949) reported the use of *Candida lipolytica* cultures to accelerate ripening in blue cheese made in the USA from pasteurised, homogenised milk, but this does not appear to have been adopted by the industry.

Coating. Throughout the making process, and until a satisfactory rind or coat has formed, blue-veined cheeses are vulnerable to infection from air-borne moulds and flies. Formation of a relatively smooth, impervious coat is necessary to give protection to the cheese during ripening; the type of coat varies considerably with the variety of cheese. The surface flora is affected by such factors as the pH, the moisture content and the concentration of NaCl. Salt-tolerant yeasts usually develop first, increasing the pH so that other micro-organisms can grow. The reddish-brown growth of *Bre. linens* may then appear, and the cheese acquires a slimy type of surface (e.g. American Blue). Where the cheeses have a drier surface (e.g. Stilton), yeasts and lactobacilli are found regularly, with aerobic spore-formers, pigmented cocci and coliform organisms occurring irregularly and in variable numbers

(Brindley, 1954). The type of coat is often characteristic of particular curing rooms.

Defects

Gassiness. The low pH and high salt concentration in the young blue-veined cheese, combined with the fairly low curing temperature, provide unsuitable conditions for growth of spoilage bacteria, and gassiness is seldom troublesome. Volatile fatty acids, produced quite early during ripening, especially if the milk for manufacture has been homogenised, are inhibitory to many bacteria and may assist in the control of bacterial spoilage.

Failure to blue. Failure of *P. roqueforti* to develop properly, and the consequent low levels of proteolysis and lipolysis, produces a cheese which is too firm and lacks flavour. The poor mould growth may be the result of excessive development of acidity in the curd leading to too close a texture in the cheese, or of too rapid a development of acidity leading to a dry curd. Low humidity in the making rooms will also produce a dry curd and affect mould development. Poor colour of the mould may arise from the species or strain of *Penicillium* used, or, possibly from a lack of iron in the milk.

Over-bluing. Excessive mould growth may occur if the cheeses are pricked heavily, and the proportion of oxygen to carbon dioxide is upset. A good 'blue' is achieved, but the volatile acids are used for the moulds metabolism and the subtle flavour is not produced. At worst, a musty, unclean flavour is developed.

Surface defects. These can cause considerable losses. A black mould, usually a species of *Cladosporium*, may grow on the surface and in the stab-holes causing a musty flavour. The application of paraffin wax or Cryovac film or Al-foil after satisfactory mould growth is sometimes used to protect the cheese (Bakalor, 1962).

Browning of the mould is a common defect which Czulak (1960) has associated with high pH (8·0–8·5) in the cheese.

Slip-coat in Stilton cheese and soft edges in Roquefort are defects associated with the collection of moisture under the rind of the cheese. Strongly proteolytic organisms, *Proteus vulgaris* and *Str. liquefaciens* were found to be associated with slip-coat (Brindley, 1954), but they were not the only factors involved.

Gammelöst

This semi-hard blue-veined cheese has been made in Norway for centuries. Its manufacture is very different from that of most other varieties of cheese in that the milk is not coagulated with rennet; the curd is heated to boiling point and the cheese is not salted. There are two methods of making this cheese. In the Hardanger method, a *Str. lactis*-type starter (2%) is added to skim-milk at 20°C, and forms an acid curd in about 3h. This curd is then broken and heated to 63°C. The whey is drained off, and the curd is packed in perforated wooden moulds and held for about 3h in boiling whey. The cheeses are dried in a warm room and pierced with steel wires coated with spores of *Penicillium* spp. In the Sogn method, 2% of the *Str. lactis*-type starter is added to skim-milk at 22–24°C, and the resultant acid curd is heated slowly to 100°C. The whey is drained off and the curd cooled, inoculated with spores of *Penicillium* spp., and packed into the moulds.

In both methods of make the high temperature kills all vegetative bacteria, and a similar ripening procedure is followed. The surfaces of all the young cheeses become infected with species of *Mucor* from older cheeses in the ripening room. The cheeses are held at 10–12°C and 90% RH for several weeks, during which time *Penicillium* ramifies throughout the interior and *Mucor* covers the surface. The pH rises rapidly reaching 6·0–7·5 in the ripe cheese, depending on the extent of proteolysis. Most of the protein is converted into water-soluble derivatives, and much of this fraction is present as ammonia and amino nitrogen. The changes during ripening are brought about by the moulds, and Funder (1946) reported finding *P. roqueforti* and *P. frequentans* inside the cheese, and various species of *Mucor* and *Rhizopus* on the surface.

SOFT CHEESE

Soft cheeses constitute a large group of varieties (Table II) characterised by high moisture content (55–80%) and consequent perishability. They are made from cream, whole milk, or whole milk standardised with cream, or skim-milk. The curds may be formed by rennet and/or acid, and if starters are used, they are mesophilic lactic streptococci. The curds receive little or no cutting and little or no scalding, and the cheeses are not pressed; any necessary drainage is done by gravity. High acidity may be developed in certain varieties. The cheeses are usually made in the shape of small cylinders, rectangles or logs. Some varieties are consumed fresh, i.e. within a few days of make, and these have a clean lactic flavour and there is little breakdown of protein or fat. Other varieties are ripened by surface moulds (e.g. Brie and Camembert) or surface bacteria (e.g. Romadour), and undergo varying degrees of proteolysis during a short ripening period of 1–6 weeks.

Surface Mould-Ripened Soft Cheese

This group includes many varieties of soft cheese made in France, the most notable of which are Camembert and Brie. These are full fat soft cheeses containing 45–50% FDM and a maximum of 55% moisture (Table II). They are generally made from cow's milk, but ewe's or goat's milks are used for some varieties (e.g. Chabris). The cheeses are prepared from whole milk with the addition of starter and rennet to form a medium acid, soft curd. This is transferred, without cooking, to suitable perforated moulds and drainers. Frequent turning, in the moulds, helps to achieve adequate drainage. This is important for subsequent ripening, and initially gives the cheese a sufficiently dry surface for yeasts and moulds to become established. Strong proteolytic enzymes, produced by the mould (*P. camemberti*), diffuse into the curd converting it to the characteristic soft, smooth condition during a relatively short ripening period. Surface mould-ripened soft cheeses are made in small sizes to provide the maximum surface for mould growth, and the optimum distance for diffusion of their enzymes. Mature cheeses deteriorate rapidly and, in the absence of refrigerated storage, they must be marketed quickly.

Camembert

This is the most important of the French soft cheeses. It originated in Normandy, and is made extensively in small and large dairies in that province. The making room should be kept warm (22–25 °C), and should be equipped with suitable vats for curd-making, cylindrical cheese moulds, and draining tables.

The cheese is made from fresh milk, sometimes used raw, but it is generally heat-treated (72 °C for 16 s), cooled (33–34 °C), and delivered to the making vats. A culture (1·5–2·5%) of lactic acid streptococci is added and allowed to ripen the milk for about 1 h before rennet (16–17 ml per 100 litres) is added. Coagulation begins in about 25 min, and the curd should be firm enough for distribution in 70–75 min from renneting. The curd is ladled into perforated, cylindrical moulds, standing on slats on a draining table. Drainage proceeds for about 6 h, and in this time, the curd should have shrunk to about half of its original height, and the acidity of the whey should be about 0·60–0·70% lactic acid. The cheeses are then turned and left to drain overnight (minimum temperature 22 °C).

Changes during processing and ripening. When the moulds are removed from the young cheese the following morning, the moisture content of the curd should be about 60%. Control of temperature and humidity in the making room helps to achieve the balance between increasing acidity and decreasing moisture content which is essential for the development of the surface flora. Low temperatures in the making room impair acid development and drainage, whilst high temperatures accelerate acid development and cause excessive drainage. At 24 h old the cheeses are dry-salted, and held at 18–20 °C for a further 24 h. This helps to dry the surface of the cheese further, and to control the type of organisms which are able to grow on it. The cheeses are then sprayed with a culture of *P. camemberti*, and 24 h later, they are transferred to the ripening room. During this time, the starter bacteria have multiplied and risen to very high numbers in the curd. Within a few days, however, at the start of the ripening period, their numbers begin to decline, and the combination of low pH (4·7–4·9), shortage of lactose and the rising salt content of the curd prevents further growth of these organisms. Species of lactobacilli may be found in the cheese during ripening, but their effect is considered to be insignificant (Eigel, 1948).

For the first 10–14 days of maturation, the cheeses are held at a temperature of 11–14 °C and relative humidity of 85–90%, and within 2–3 days, species of film yeasts (*Mycoderma*) and of *Geotrichum* appear on the surface. These organisms are able to withstand the high acid and high salt content of the cheese surface. A few days later, when the surface has become drier and the pH more favourable to its growth, *P. camemberti* begins to develop and spreads over the whole surface. Maximum development is reached in 10–12 days then, when the acidity at the surface has been further reduced by the yeasts and moulds, the red and brown spots of *Bre. linens* appear.

A ripening Camembert undergoes a complex series of microbiological changes. The yeasts which grow first may ferment residual lactose at the surface of the cheese and reduce the acidity, but their growth should not be excessive, otherwise the spores of *P. camemberti* may have difficulty in penetrating the yeast mat and establishing themselves. The yeasts are restricted to a certain extent by the salt content of the cheese surface, and their development can be controlled by regulating the temperature and humidity so that the surface is fairly dry. Opinions vary regarding the purpose served by these yeasts. Some workers consider them to be unnecessary, believing *P. camemberti* to be capable of producing all the changes involved in ripening. Others regard their growth as necessary to soften the cheese initially, so contributing to the flavour. It has been shown that spores of *P. camemberti* do not grow well at the pH and salt content pertaining at the surface of fresh Camembert (Kundrat, 1952).

The luxuriant snow-white growth of *P. camemberti* spreads over the whole surface of the cheese, and is responsible for the subsequent changes in body and flavour of the cheese. Its proper development is therefore essential, and can only be ensured by efficient control of the atmospheric conditions in the ripening room. If the temperature is too low, mould growth is slow and the ripening sequences are delayed. High temperatures, on the other hand, can lead to excessive mould growth and strong flavours in the cheese. If humidity is too low, the cheese surface may become too dry, particularly if the movement of air is too great. This prevents normal yeasts and moulds from growing, and wild species, such as the green-spored penicillia, will develop. Alternatively, should the humidity be too high, and the movement of air inadequate, evaporation of moisture from the cheese surface is insufficient. This allows excessive growth of film yeasts, *Geotrichum* and *Mucor* species, and *P. camemberti* may be inhibited.

About 10–14 days after ripening begins, the edges of the cheese begin to soften, and it is at this stage that the cheeses are generally wrapped and boxed. By the time they reach the consumer, 2–3 weeks later, this softening will have extended to the centre of the cheese; the fine flavour will have developed and ripening will be complete.

As the powerful proteolytic and lipolytic enzymes produced by the mould diffuse inwards towards the centre of the cheese, the firm lactic curd becomes softer and smoother and more wax-like. Proteolysis proceeds slowly for about 10 days, then increases steadily and may reach 80 % in the fully ripened cheese. There is a concurrent decrease in acidity, both within the cheese, as alkaline products are formed, and on the surface, as the lactate is utilised by yeasts and moulds.

A process for making Camembert, using skim-milk concentrated by membrane filtration (by a factor of five) and standardised with cream, is being used successfully to make a 'pre-cheese', which is then shaped in moulds but requires no further draining (Maubois and Mocquot, 1971). This has led to the development of a continuous process for the manufacture of Camembert.

Brie
This is another of the great cheeses of France. Basically the same as Camembert, it is a thin, cylindrical, surface mould-ripened soft cheese, but it is made in several diameters, all of which are larger than Camembert. These cheeses are made from milk of different fat contents, so that they may differ in the extent of ripeness when consumed. The fresh curd, and the ripened cheese, has a higher moisture content than Camembert (Table II), so the sequence of ripening takes place rather more quickly. The mould often becomes reddish, as *Bre. linens* develops to a greater extent, and the curd becomes creamy with the characteristic soft, smooth body and fine flavour.

Defects
Many defects in surface mould-ripened cheese occur as the result of incorrect curd composition, or incorrect atmospheric conditions during ripening. Cheeses with a high initial moisture content, or exposed to high temperatures in the ripening room, develop excessive proteolysis and a strong flavour. Cheeses with a dry curd, or dry surfaces owing to low humidity, will not support normal mould development; over-salting or under-salting can also interfere with proper surface growth. Cheeses with wet surfaces are at risk from excessive growth of *Bre. linens*, and the consequent development of a surface smear rather than a surface mould.

Early gas. This is sometimes formed during curd draining, especially if raw milk is used, but later gas formation is uncommon as the high acid and salt content of the curd prevents the growth of clostridia. Wild moulds, such as *P. glaucum, P. roqueforti* and *P. bruneo-violaceum*, can be a problem if the cheeses become too dry for the *P. camemberti* to grow. Careful hygienic precautions are essential in the factory to prevent contamination by these moulds.

Prolonged ripening or failure to hold ripened cheese at low temperatures can lead to rapid deterioration.

Surface Smear-Ripened Soft Cheese
Romadour
Made in Austria to fat standards of 25 and 45% FDM, Romadour is a typical variety of this type of soft cheese. It is made in the shape of small logs, and the making process resembles that of Limburger. The cheeses develop the same reddish-brown smear, produced by *Bre.*

linens, during a short ripening period (about 4 weeks). Proteolytic and lipolytic changes produce ammonia, fatty acids and other volatile organic compounds responsible for the characteristic odour and flavour of this cheese.

Unripened Soft Cheese

This group contains many varieties of soft cheese produced in various countries. They may be made from cream, milk or skim-milk, which allows a wide range of fat and moisture contents (Table II); the high fat varieties generally have more flavour and are softer and smoother. All are characterised by a high moisture content which contributes to their soft body and to their perishability. Refrigerated storage and transport can prolong their shelf-life.

All the varieties are made according to the same, simple, basic principles, whereby milk with a starter culture of lactic streptococci is used to make a soft, rennet curd from which some whey is strained. Salt is mixed into the curd, and various flavouring materials may be added (e.g. herbs, nuts, caraway seeds). The curd is packed in paper cartons or plastics pots, and is consumed within a few days. Some varieties of soft cheese are homogenised to increase the smoothness of the product.

The soft cheesemaking process has been well mechanised, and concentration of the milk by membrane filtration has led to the development of a continuous manufacturing process.

Microbiological changes during the manufacture of unripened soft cheese are essentially those concerned with the production of acid by suitable lactic streptococci. Fast production of acid is required to produce the clean, acid flavour in the drained curd, and starters containing cultures of *Str. lactis* and *Str. cremoris* and aroma-forming bacteria, such as the leuconostocs, are suitable. *Str. lactis* sub-sp. *diacetylactis* is often included in the starter, but it can produce a 'yoghurt flavour' defect due to high acetaldehyde production at certain times of the year (Galesloot, 1968).

Pasteurisation of the milk for cheesemaking destroys most of the vegetative forms of micro-organisms, but the cheese milk, the curd and the cheese are at risk from post-pasteurisation contaminants, and defects in these unripened soft cheeses arise mainly from contamination by yeasts and moulds. UK Preservatives in Food Regulations (1974) allow up to 1000 ppm sorbic acid in all soft cheeses.

PASTA FILATA

This is a term used to group certain varieties of Italian cheese whose manufacture includes a drawing-out, or pulling, series of manipulations to produce cheeses of characteristic plasticity. The group includes hard cheeses, such as Cacciocavallo and Provolone, and the soft, uncured variety, Mozzarella. The plasticity depends on the fact that dicalcium paracasein is converted to monocalcium paracasein when a high acid curd (pH 5·3) is softened in water at 85 °C, and then kneaded and stretched until it is smooth and elastic (see Fig. 7). The cheeses are moulded by hand into shapes, characteristic of the variety, and hardened in cold water; they are not pressed. Mozzarella cheese is brined for 2–3 days (8–10 °C and 16–20 % NaCl), and is then ready for consumption. Cacciocavallo and Provolone are brined for 3–4 days at 10 °C, and then cleaned and oiled regularly during curing (2–12 months).

FIG. 7. The typical elastic curd of a 'pasta filata'-style Italian cheese. Reproduced by courtesy of Miles Laboratories Inc., Elkhart, Indiana, USA.

Mozzarella

Mozzarella cheese is made with a starter culture of *Str. lactis, Str. thermophilus* and *Lac. bulgaricus*. Rapid acid production is required during manufacture, and the cheese is eaten fresh. The cheese has a soft, waxy body and a mild, slightly acid flavour. Systems for cooling Mozzarella and for the direct injection of salt into the cheese are described by Olson (1979), and as the demand for Mozzarella increases, modifications to the process, to accelerate acid development, can be considered (Nilson *et al.*, 1979).

Cacciocavallo and Provolone

Cacciocavallo and Provolone cheeses are made with starter cultures of the thermophilic *Lac. bulgaricus* (Table I), which withstands the high cooking temperature (47·8 °C) employed in their manufacture. This high temperature and the high acid developed in the curd (pH 5·1–5·2), followed by the kneading and stretching process in hot water (85 °C), give both these varieties of cheese a very smooth, close texture, and prevent the development of unwanted bacteria.

During ripening, proteolytic and lipolytic enzymes derived from the starter bacteria and the rennet paste give these cheeses a mild, not unpleasant, slightly rancid flavour. It has been described as 'piquant' and, according to Kosikowski (1966), is attributed largely to 'a desirable ratio of free butyric and glutamic acids'.

There is very little information regarding the microbiology of these cheeses. Maskell *et al.* (1951) report a decrease in the number of bacteria in Provolone cheese from 10^9 cfu g^{-1} at 1 month to less than 10^6 cfu g^{-1} at 12 months. They found *Lac. lactis* and *Str. faecalis* in most samples of ripened cheese made from pasteurised milk.

Both varieties have clean rinds and, if not waxed, the cheeses are oiled to keep them free from any surface growth of micro-organisms.

MISCELLANEOUS

Cottage Cheese

This is a low-acid, soft curd product which differs from similar ones, such as lactic cheese, Bakers' cheese and Quarg, in that the curd particles are left separate and retain their identity. Only one type, creamed Cottage cheese, is made in Britain, but several variants are produced in the USA. Cottage cheese is prepared from acid coagulated skim-milk, cooked to achieve a moisture content of 80% or less, and washed to cool the curd and remove whey from between the curd particles. Stabilisers, such as alginic acid, calcium and sodium alginates, carrageen, edible gums and starches, and the colouring agents, carotene and annatto, are permitted additions, whilst up to 1000 ppm sorbic acid may be used as a preservative. Flavourings, such as chives or pineapple, may be incorporated into the curd. A creaming mix or dressing, is generally added to give a fat content of 4% to the finished product.

Manufacturing Methods

Equipment required for the manufacture of creamed Cottage cheese should include: a pasteurising plant; steam-jacketed making vats; curd-knives, agitators, shovels; a supply of filtered, chlorinated chilled water; a cream tank for preparing the dressing; a homogeniser or suitable blending equipment; cartons and a filling machine.

Good quality skim-milk, free of antibiotics, is fortified, if necessary, with skim-milk powder to provide a total solids content of 9·5–10·5%. The milk is heat treated, usually at temperatures slightly above the standard 72 °C for 15 s, and delivered to the vats, where it is coagulated by acid produced by the starter bacteria (Table I) employing either the long-set or short-set making process. The long-set method uses a 1% inoculation of starter into milk held at 21–22 °C to produce coagulation in 12–13 h, and the short-set method uses an inoculation of 4–5% starter into milk held at 31–32 °C to bring about coagulation in 4–5 h. The latter is the generally adopted system. A little rennet is added (0·1–0·2 ml per 100 litres for the long-set, and 0·2–0·3 ml per 100 litres for the short-set process), to firm the coagulum at cutting, and to assist matting of the curd during cooling. When a pH of 4·65–4·75 is obtained, the coagulum is cut into cubes and cooked, with the temperature being raised gradually over a period of 1 h to 53–57 °C. Acid production ceases at about 39 °C, and the rate of heating can then be increased to 1·5–2·0 °C every 5 min. The final cook temperature should be at least 53 °C to control spoilage organisms (Collins, 1961), and may be as high as 65 °C if a very dry, firm curd is required. The whey is drained off to curd level, and the curd is washed three times at approximately 20 min intervals; first with warm water (22–24 °C), as sudden contact with cold water is very damaging, then with water at 10 °C and finally with chilled water at 3–4 °C. The curd is then rested for about 1 hour at 1 °C. The cream dressing may be mixed into the curd in the vat, or the curd may be transferred to special creaming and blending equipment. Finally the creamed Cottage cheese is packaged in small cartons or bulk containers.

The cream dressing is made separately, from pasteurised cream ripened with *Str. lactis* sub-sp. *diacetylactis* and *L. cremoris*. The dressing may also contain skim-milk powder, to increase viscosity and ensure a minimum total solids content in the product, stabilisers (0·1–0·3 %), and salt (1 % of the final product). The dressing is pasteurised (80 °C for 15 s), homogenised (2500–3000 lb in^{-2}), and cooled before amalgamation with the curd. The proportion of dressing to curd varies from 25–40 % of the final product, and depends largely on its fat content.

Microbiological Changes During Manufacture

The microbiological changes which occur in making Cottage cheese are essentially those concerned with the production of lactic acid by the starter streptococci during the setting period at the start of manufacture, and undesirable contamination of the curd from various sources at all stages of the process.

The skim-milk should be used very shortly after separation, or held for a maximum of 24 h at 5 °C; excessive growth of psychrotrophic bacteria during long-term storage of the original milk may give rise to undesirable flavours in the cheese due to production of lipases and proteinases (Law *et al.*, 1976).

The starter for Cottage cheese usually contains a blend of several strains of *Str. cremoris* with a single strain of *Str. lactis*, selected on the basis of rapid acid production and freedom from agglutination sensitivity. The starters must not give rise to bitter flavour development in the curd, as sometimes occurs with rapid acid developing strains, and agglutinin-sensitive strains lead to clumping of the streptococcal cells. These clumps fall to the bottom of the vat and fail to produce sufficient acidity in the main bulk of the cheese milk. If *Str. lactis* sub-sp. *diacetylactis* is included in the starter, it may produce CO_2 from the citrate in the milk which may cause pinholes to appear in the curd and, in extreme cases, can lead to a floating curd and poor textured cheese (Sandine *et al.*, 1959). For this reason *Str. lactis* sub-sp. *diacetylactis* and *L. cremoris* are used to ripen the cream for the dressing. Both produce the diacetyl that imparts the desired flavour to the Cottage cheese, but as *L. cremoris* only produces diacetyl in milk or cream under acid conditions, it is best used with *Str. lactis* sub-sp. *diacetylactis* to initiate acid production. In addition, the *Leuconostoc* sp. can convert acetaldehyde produced by *Str. lactis* sub-sp. *diacetylactis* to ethanol, and thus mitigate the off-flavour that might otherwise arise.

During manufacture, extreme care should be exercised to ensure that all cheesemaking equipment has been thoroughly cleaned and sanitised before use. Microbial contamination from vats, knives, agitators, etc., can give rise to spoilage defects. The walls and floor of the making room should be clean and, if necessary, the air should be filtered to overcome the problem of air-borne yeasts and moulds.

Post-pasteurisation contamination of the milk during the setting period, especially with species of *Aerobacter* and lactose-fermenting yeasts, may interfere with normal starter development. In large numbers these contaminants may give rise to gas formation in the curd, or to unclean or yeasty flavours in the cheese. Gas formation may not become evident until after the dressing has been added to the curd and the cheese is in store.

After the curd has been cut, the high temperatures of the cooking operation destroy coliform bacteria and psychrotrophs, but the chilled water used for washing the curd can be a major source of contamination with spoilage organisms, particularly psychrotrophs. Unless it is of potable grade, it is generally advisable to treat the wash water, after filtration, either by irradiation with UV light or by chlorination with 5–10 ppm available chlorine.

The cream dressing, the equipment used for filling the cheese into containers, the containers themselves and the air of the packaging room may all be further sources of contamination. Sprays or UV lighting may be used to destroy air-borne mould spores during the final stages of production.

Defects

Undesirable off-flavours can result from using milk or skim-milk which is highly contaminated with lipolytic or proteolytic pseudomonads or coliform bacteria. An over-acid curd at cutting may be caused by an over-active starter, and this acidity can lead to the development of a sour flavour. Sourness can also develop if the curd is not washed sufficiently, or is stored at temperatures (above 10 °C) which allow the starter bacteria to multiply; over-acidity can also lead to bitter, and sometimes yeasty flavours. Insufficient acid, on the other hand, can be responsible for the curd having a flat flavour, and for the growth of any spoilage bacteria present. Lack of aroma in the curd may arise simply from poor growth of the *Leuconostoc* spp., but Parker and Elliker (1952) have shown that it can arise from the growth of *Pseudomonas fragi* and *Ps. viscosum* which convert the diacetyl in the cheese to acetylmethyl carbinol during storage. Coliform bacteria, especially *Aerobacter*, and lactose-fermenting yeasts if present in the milk will produce gassiness in the curd.

Cottage cheese curd is an ideal substrate for film yeasts and moulds to develop. *Geotrichum candidum* and various species of *Penicillium*, *Mucor* and *Alternaria* have all been found in Cottage cheese during storage.

Another troublesome defect which may occur during storage is the development of a slimy film on the curd particles, due to the growth of the psychrotrophs *Alcaligenes*, *Pseudomonas*, *Proteus*, *Aerobacter*, *Aeromonas* and *Achromobacter*. This slime may be thin and watery, or quite thick and even ropy. It varies in colour from white to yellow, to brown or fluorescent brown, and it may have a rancid, putrid or fruity odour (Davis and Babel, 1954).

White Brined Cheese

This type of cheese, made in many countries of eastern Europe, has different names in the different countries, but the simple, basic principles of manufacture are the same. Traditionally the milk used to make these varieties was produced by goats and sheep, but the use of cow's milk is increasing. An acid coagulum is formed with cultures of *Str. lactis* and *Str. cremoris* (2:1) or *Str. thermophilus* and *Lac. bulgaricus*. This coagulum is ladled into bags or moulds and drained for about 6 h. Drainage is spontaneous at first, but later the curd may be pressed (1 kg weight per kg curd). The curd is then cut into 10 cm cubes, and immersed in brine at 12–14 °C for 12–14 h. The salted curd is packed into cans or cartons, covered with a special brine made from sour whey (0·36% lactic acid) with 8–12% salt, and ripened for 30 days.

Feta is a white brined cheese of Greek origin. It is made from an acid-coagulated curd which is drained in moulds and cut into blocks. These blocks of curd for Feta cheese are dry salted, with very coarse salt, for 24–48 h depending on atmospheric temperature. If the temperature is hot, the salting time is short. The young cheeses are ripened for 8–15 days, during which time the typical microflora derived from the starter should have developed. The cheeses are then packed in barrels, cans or bags which are filled with brine and stored at 2·5 °C.

The cheeses are soft, somewhat pasty, and salty, and the texture is close but full of mechanical openings as the curd is not pressed. Where the milk for manufacture is pasteurised and cultures of *Str. lactis* and *Str. cremoris* are used, these should dominate the flora of the curd. The salt content of the cheese is 6% which helps to control development of undesirable bacteria, but common defects in white brined cheese are early gas production caused by coliform bacteria, and rusty spot caused by *Lac. plantarum* sub-sp. *rudensis* (Chomakov, 1967).

Bakers' Cheese

Like the Italian Impastata, Bakers' cheese is a lactic curd produced especially for bakers and confectioners. It is made from skim-milk with a good active starter culture of mesophilic lactic streptococci and a very small amount of rennet. The starter is allowed to develop acidity in the milk until the pH reaches 4·5, at which point, refrigeration is essential to prevent further development of acidity. The curd is then drained in bags or by the Berge system, but this operation can be mechanised by introducing a desludging type of separator. Bakers' cheese should have a dry, but soft, pliable curd which can be kneaded or worked easily. If the starter is slow, or insufficient acid is produced, the curd will be heavy and grainy, and its absorption properties will be impaired.

Lactic Cheese

These are the most simple of all cheeses. They are made from soured milk of any fat content up to 25%, but generally they are made from skim-milk with the addition of starter bacteria. They are sometimes called Starter cheese, because their manufacture can be a means of utilising surplus starter cultures. The acid coagulum may be cut, to assist whey drainage, and is then transferred to drainers lined with cloths. The curd is left undisturbed for 1 h and is then suspended, in the cloth, and the whey allowed to drain from it. When sufficiently dry, the curd is salted to taste and packaged, usually in 112 g portions. Herbs or other foodstuffs can be incorporated into the curd at the time of salting. The curd can be put through a curd homogeniser to give it a smooth finish.

Lactic cheeses have a typical, sharp, lactic flavour. They are best eaten fresh, but cold storage can extend their shelf-life. The moist, high-acid curd is readily spoilt by the growth of yeasts and moulds.

Quarg

In essence Quarg is similar to lactic cheese, but it is always produced from skim-milk and the manufacturing process is very well mechanised and highly automated. Pasteurised skim-milk is cooled to 28–30 °C and delivered to the making vats. A starter culture (2 %) of *Str. lactis* and *Str. cremoris* (0·7 to 0·8 % acidity) is mixed into the milk and rennet (2·0–4·0 ml per 100 litres) is added. The vat is covered and left for 3–6 h (short-set) or 16 h (long-set). The coagulum is cut once only, with wire knives, then stirred for 5 min before the curds and whey are run off into a drainer covered with a nylon cloth. (Special Quarg separators are now available to remove the whey from the curd at this stage.) When the curd reaches a satisfactory moisture content, it is mixed in a bakery-type mixer, and flavours, and up to 1 % salt, may be added. Finally, the curd is pumped through a cooler (chilled water type), filled into suitable containers and held in cold store. Fresh Quarg has a typical sharp, lactic flavour, and like lactic cheese, the moist, high acid curd is readily spoilt by yeasts and moulds.

Whey Cheese

This type of cheese is made in a number of European countries (Table II), generally using whey produced during the manufacture of a hard type of cheese. The whey is not separated, but may be partly neutralised if over-acid, and 5–10 % skim-milk or whole milk is added to improve the yield. The whey is heated and concentrated in an open (or closed) vessel for several hours at high temperatures (85–90 °C), with continual stirring. This coagulates and precipitates the proteins, and when the temperature reaches 40–45 °C, up to 0·1 % salt may be added. The precipitated protein becomes enmeshed with any fat, residual or added, and rises to the surface where it forms a layer of curd when stirring ceases. The curd is removed, with perforated skimmers, filled into perforated moulds and allowed to drain and cool for several hours, preferably in a chilled atmosphere. The drained cheeses are solid, but soft, and have a short shelf-life; firmer cheeses are obtained by applying pressure during draining. The lower moisture content of these latter cheeses improves their keeping quality.

One type of Ricotta is a mature cheese and becomes dry enough to be used as a grating cheese (Davis, 1976b), but fresh Ricotta is a white, soft, unripened grainy curd with a bland flavour. Mysöst is a brown and sticky cheese, with a cooked flavour. During the manufacture of whey cheese, microbiological changes are obviated by the high temperatures used in the process, and the highly concentrated nature of the product. The cheeses do not ripen like natural cheese as they have no lactic acid bacteria present, so any microbiological changes during storage are concerned with contamination and spoilage.

Processed Cheese

This is made from a single variety, or a mixed variety, of natural cheese(s). A cheese mix is prepared to give the required properties and flavour, and to this, emulsifying salts are added and, if desired, skim-milk powder or whey powder, flavouring, such as chopped ham, gherkins, etc., and water. The amount of water is controlled by the final composition of the product, i.e. whether it is required for spreading or for cutting. The whole mixture is heated to 80–85 °C for 4–8 min (processed cheese) or 85–98 °C for 8–15 min (cheese spread), and filled directly into lacquered cans, tubes, or moulds lined with aluminium foil. The cheeses are packed in suitable outer cases and cooled before going into store. No ripening is required (Meyer, 1973).

The low moisture and high salt content combine to give processed cheese its good keeping qualities; only microbial spores survive the high temperature of cooking. They are usually species of clostridia which come from the original cheese or from the added ingredients, and their germination is stimulated by the heat treatment. Gas production, rancidity and putrefaction result from their growth. At one time blowing caused by clostridia was a common defect, but this can be controlled by incorporating nisin into the mix (Meyer, 1973).

Accelerated Ripening of Cheese

The manufacture of most types of cheese is now a capital-intensive industry, and one where a high turnover rate is required. The storage of cheese, particularly hard varieties, until it has acquired an acceptable flavour and consistency adds significantly to operating costs, and a controlled method of accelerated cheese ripening, achieved without upsetting the flavour balance, would give large cost savings; although other problems might arise, e.g. fast ripened cheese would be likely to become over-ripe more rapidly.

Different methods used to accelerate ripening have been reviewed in detail by Law (1980). Potentially useful methods are mainly concerned with increasing the concentrations of those enzymes which are considered to contribute to cheese flavour; this can be done by adding the enzymes

TABLE IV
EXAMPLES OF THE USE OF ADDED ENZYMES TO ACCELERATE CHEESE RIPENING

Type of cheese	Type	Enzymes added Source	Stage of addition
Cheddar	Acid and neutral proteinases, peptidases, lipases, decarboxylases	Various commercial and animal enzymes	Curd
Cheddar, Romano, Parmesan	Lipase Proteinase	Lamb gastric tissues	Milk and curd
Gouda	Proteinase	*Aspergillus oryzae*	Milk or curd
Edam Cheddar[a]	Proteinases and peptidases	*Pseudomonas fluorescens*	Milk
Rossiiskii	Proteinase	Pancreatin	Milk
Mozzarella	Lipases, esterases (micro-encapsulated)	Calf pregastric secretions	Milk
Blue	Lipase	*Aspergillus* sp.	Curd

[a] Data unpublished.
From Law (1980).

directly, or by the use of modified starters to increase the level of starter enzymes present.

Enzymes added exogenously to the cheese milk are usually lipases, proteinases and/or peptidases (Table IV). In general, lipase addition is useful in strong flavoured cheese, such as a blue-veined type. With Cheddar, lipases tend to produce off-flavours, usually rancidity, and proteinases produce bitterness, but proteinases together with peptidases have shown promising results in increasing the rate of flavour production without excessive bitterness. Pretreatment of cheese milk with commercially prepared β-galactosidase (lactase), which hydrolyses the milk lactose to glucose and galactose is claimed to increase Cheddar flavour production. This is attributed either to stimulation of the starter, which leads to increased populations and higher proteinase content, or else to contaminating proteinases in the lactase.

To increase starter enzyme concentrations in cheese, the normal starter is supplemented by starter preparations which have been treated so as to prevent them from metabolising and producing acid during cheesemaking, and yet to retain intact their degradative enzymes. The use of lysozyme-treated mesophilic starter streptococci, which had increased proteinase and peptidase activity equivalent to 10^{10} cells g^{-1} cheese, had little influence on Cheddar flavour. However, sub-lethally heat shocked *Lac. helveticus* starter cells increased flavour scores in Swedish semi-hard cheese, and a similarly treated *Str. lactis* starter has shown some promise in Cheddar cheese. The use of *lac*$^-$ (not producing acid from lactose) mutants of *Str. lactis*, or of X-ray mutants of *Lactobacillus* starters with increased selective proteolytic activity, have been described but require further work to determine their effects.

Adjustments to making conditions to increase starter populations, by the use of a large starter inoculum and a prolonged ripening period, followed by in-vat neutralisation of excess acid by NaOH has also led to increased flavour in Swedish cheese.

In a different approach, ripening of Cheddar cheese has been accelerated by the use of cheese slurries. Cheddar cheese curd with increased moisture and incubated at 30 °C develops a strong cheese flavour in about 7 days. Such slurries have been added to curd and resulted in more rapid flavour development, often accompanied, however, by off-flavours.

Raising the storage temperature of the cheese is unacceptable, as not only is flavour formation accelerated, but also that of off-flavours, whilst the growth of microbial contaminants is encouraged.

All these methods require further experimental work before they could be considered for large-scale cheesemaking. It must be remembered also that methods suitable for one type of cheese may not be useful for another type, and that the body and texture of the cheese may be adversely affected.

Moulds on Cheese

Moulds are a common form of spoilage of dairy products, particularly of soft unripened cheeses such as Coulommier, Cottage cheese and cream cheese. They also occur very frequently on the surface of hard cheeses during ripening and curing. Some moulds are able to grow at very low temperatures, such as 4–10 °C, *Aspergillus* and *Penicillium* spp. being very common. Their growth may result in a musty off-flavour, their appearance is commercially undesirable, often resulting in downgrading of the cheese, and there is the potential that mycotoxins might be produced (see below).

Pimaricin

This non-toxic antibiotic, a polyene produced by *Streptomyces* which suppresses yeasts or moulds but not bacteria, is used in some countries to combat mould growth on cheese (Morris and Hart, 1978). Applications of pimaricin to the surface of a hard cheese at concentrations of about 1000 ppm, resulting in 5 ppm on the cheese surface, are recommended; this inhibitor can also be used in Cottage cheese where it can be added to the curd. Objections to the use of pimaricin as a cheese preservative lie in its clinical use as a human antifungal agent, although problems due to development of resistant strains are unlikely to arise in this context.

Mycotoxins

Mycotoxins are toxic metabolites of moulds, and can be produced during growth on substrates including cereals, ground nuts and oil seeds, which may be used as feed for dairy cattle. Of all known mycotoxins, the aflatoxins, which are normal metabolites of *Aspergillus flavus* and *A. parasiticus*, are of the most concern, being potent liver carcinogens, particularly aflatoxin B_1. If lactating cows ingest feed contaminated with aflatoxin B_1 (the most commonly occurring in feedstuffs), this toxin is converted into a carcinogenic derivative M_1 which is transferred to the milk (Stoloff, 1980). If such milk is used for making cheese, aflatoxin M_1 will be present; in the USA, the permitted level of M_1 in milk is $0.5 \, \text{ng ml}^{-1}$.

Investigations using either milk from cows fed aflatoxin B_1 and thus excreting M_1 in their milk, or milk with M_1 added directly to it, have shown that pasteurisation or sterilisation of such milk has little effect on the aflatoxin. Cheesemaking experiments have shown that, with Cheddar cheese, 47% of the toxin present in the milk was detected in the cheese, while with Camembert, about 50% was recovered in the cheese and 45% in the whey. The level in the cheese was not reduced after 40 days storage. In Gouda cheese there was no loss in toxin level after 6 months ripening. Thus aflatoxin is stable in cheese and does not deteriorate on storage. Aflatoxin M_1 has been detected in 17.5% of a variety of cheeses at levels of $0.02–1.3 \, \text{ng g}^{-1}$ (Stoloff, 1980). In general, cheese can be expected to contain 3.5–5 times the aflatoxin M_1 level present in the milk from which it was made.

The other important aspect of mycotoxins is that the growth of moulds in or on cheese may give rise directly to toxin production. These moulds include not only *A. flavus* or *A. parasiticus* but also *Penicillium* spp. whose known mycotoxins include penicillic acid, patulin, cyclopiazonic acid, roquefortins or PR toxins. In one study, 20% of mould isolates, mostly *Penicillium* spp., from Cheddar cheese were found to produce mycotoxins, and 32% of those from Swiss cheese (Bullerman, 1979).

Aflatoxins can be produced by *A. flavus* and *A. parasiticus* on Cheddar cheese at room temperature, but not at 4·4 or 7°C. They may also be produced on Tilsit and Emmenthal cheese, but not on Camembert. In general, *A. flavus* and *A. parasiticus* will not produce aflatoxins at temperatures below 10°C (Marth, 1979). Patulin and penicillic acid-producing strains of *Penicillium* spp. can grow extensively on Cheddar, Swiss and Mozzarella cheese at 5, 12 and 25°C. However, neither toxin has been detected in any of the cheeses examined, except for traces of patulin in a Cheddar cheese kept at 25°C (Bullerman, 1979). In general, cheese is not well suited to the production of mycotoxins as it lacks the high level of carbohydrate necessary for toxin production, and also it is ripened at temperatures below the minimum necessary for toxin production (Marth, 1979). If necessary, and permitted by health regulations, pimaricin already mentioned as a mould suppressor, can be applied to the cheese surface.

There has been some concern about toxin production by strains of *Penicillium* spp. used for inoculation of mould-ripened cheese. Rigorous investigations of cultures of *P. camamberti* and *P. roqueforti* used for making blue cheese have not detected aflatoxin, penicillic acid or patulin production in the cheeses (Engel and Milezewski, 1977). When patulin, penicillic acid or PR toxin were isolated from *P. roqueforti* cultures, these toxins were only found when yeast sucrose laboratory media were used. Engel and Prokopek (1979) detected no PR toxin in 13 kinds of blue cheese made from *P. roqueforti*. However, traces of roquefortin and cyclopiazonic acid have been found in blue cheese by other workers (Le Bars, 1979).

The toxicity of some of these mycotoxins, other than aflatoxins, is not easy to assess, and doses ingested from cheese would be very low. Whilst conditions in cheeses are unfavourable for mycotoxin production, the screening and selection of non-toxin producing strains of *Penicillium* for blue cheese production is being practised in some countries.

Pathogens in Cheese

For a detailed review on this subject see International Dairy Federation Report 122 (IDF, 1980).

Many pathogenic organisms may be present in raw milk (Sharpe and Bramley, 1978). They may be derived from an infected (mastitic) udder, the faeces or other excreta of infected cows or symptomless (carrier) cows, human sources, a contaminated environment or dairy equipment. The group includes *Staph. aureus*, *Streptococcus* spp., *Salmonella* spp., *E. coli*, *Campylobacter* spp., *Yersinia enterocolitica*, *B. cereus*, *Cl. perfringens* and, in some countries, *Brucella* spp. and *Mycobacterium tuberculosis*. All of these, except the spore-formers and the enterococci, will be destroyed by full pasteurisation. However, in some instances cheese is made from raw milk, or post-pasteurisation contamination may occur. If pathogens are present in such milk and survive the cheesemaking process, this may result in incidents of food poisoning due to the ingestion of cheese contaminated with pathogenic organisms or with their enterotoxins, or to infections with pathogens.

Because of the frequent contamination of milk from dairy personnel, and the high incidence of staphylococcal mastitis in dairy herds, the most commonly occurring type of food poisoning is due to enterotoxin-producing strains of *Staph. aureus*. *Salmonella* spp. are infrequent causes of food poisoning due to cheese, but enteropathogenic strains of *E. coli* have caused outbreaks; *E. coli* is often present in milk, and post-pasteurisation contamination is not uncommon. Cases of Brucellosis due to eating unripened cheese made from raw milk, particularly goat's milk, in countries where *Brucella* sp. are still common, have been reported. The role of enterococci in producing toxic symptoms is likely to be due to production of high levels of pressor amines, which may interact with drugs, used in clinical treatment, for suppressing monoamine oxidase.

Lactic starters used in cheesemaking can effectively protect the cheese against pathogens by the production of lactic acid, acetic acid, H_2O_2 and antibiotics (e.g. nisin). A vigorous starter is essential to inhibit pathogens and enterotoxin. When the starter produces insufficient acid, due to phage, antibiotics, or the use of slow acid-producing strains, staphylococci and other pathogens multiply much more rapidly, and it is in such cheeses that the danger from food poisoning organisms and other pathogens chiefly lies. However, even under optimal conditions of acidification, the multiplication and death rate of pathogens may be influenced by the strain of starter used, some strains being more inhibitory than others.

Coagulase-Positive Staphylococci

With enteropathogenic strains of *Staph. aureus* in general, high levels of contamination result in enterotoxin production during cheesemaking, but whether certain conditions stimulate enterotoxin production is not yet clearly understood.

Conditions of cheesemaking. With Cheddar cheese, inhibition of the starter by phage resulted in a pH at milling of 6·6 instead of 4·95, and a 5–10 fold greater multiplication of *Staph. aureus* during cheesemaking than in normal cheese (Reiter *et al.*, 1964). In commercial cheesemaking with pasteurised milk heavily recontaminated with staphylococci, a high level of these organisms produced detectable enterotoxin only when a low acidity (<0·40) was recorded, but not with normal acidity. Variable rates of staphylococcal multiplication are recorded with different starters. At normal acidity, it is thought that at least $2·8 \times 10^7$ staphylococci g^{-1} of cheese must occur for enterotoxin to develop, but with phage-induced starter failure, enterotoxin has been detected when only 4×10^6 staphylococci g^{-1} are present (Tatini *et al.*, 1971). With Swiss cheese, high levels of *Staph. aureus* inoculated into cheese milk can multiply during cheesemaking of normal acidity, and enterotoxin was produced by two strains at levels of 10^7 and $10^8 g^{-1}$ cheese (Tatini *et al.*, 1973). However in Emmenthal cheese, *Staph. aureus* disappears within 24 h if an active starter is used. Gouda made from raw milk containing 10^2–10^3 indigenous staphylococci per ml may contain up to 10^4–$10^5 g^{-1}$ cheese after 24 h, even under normal acidification conditions. The death rate of cells in this cheese is influenced by the starter strain, and high levels of staphylococci of $10^8 g^{-1}$ of cheese are necessary before enterotoxin is detected.

Cheese ripening. In low-acid Cheddar cheese, staphylococci may continue to multiply during the first few weeks, and then die out only slowly, but in normal cheese, staphylococci may decrease 100 times in 16–24 weeks; in low-acid cheese, the numbers may not fall even after 18 months ripening (Reiter *et al.*, 1964). Numbers decrease more rapidly in cheese ripened at 10 or 13°C rather than at 7°C. In Swiss cheese of normal acidity, the survival rate of staphylococci is very low. In Camembert no staphylococci have been detected after 10 days, whereas in Cheshire cheese it took up to 22 days for their disappearance. If staphylococci have grown to levels of at least $10^6 g^{-1}$ in the milk or during manufacture, then detectable amounts of enterotoxin could have formed. This toxin remains potent in cheese for many months or even years (Tatini *et al.*, 1971).

Salmonellae

In Cheddar cheese with slow acid production, salmonellae can multiply rapidly during cheesemaking, $10^2\,ml^{-1}$ in milk giving rise to $10^4\,g^{-1}$ in the curd, and $10^5\,g^{-1}$ overnight (pH at pressing, 5·7). Even in normal cheese, salmonellae may develop during cheesemaking, although more slowly. In blue cheese the rate of destruction depends on the pH at the end of cheesemaking, there being little survival at pH 5·3. With soft cheeses, salmonellae are destroyed in normal cheesemaking (at pH 4·55), but multiply slowly in low-acid curd (pH 4·95). During the ripening period salmonellae slowly decline, but in most soft cheeses, they survive for periods up to or beyond the storage time of consumption.

Enteropathogenic E. coli (EEC)

In semi-hard, soft and fresh cheeses, *E. coli* can survive beyond the stage of ripeness for consumption, and with Camembert cheese, even a slight contamination of the milk with these organisms can result in an unacceptably high level in the cheese (Mourgues *et al.*, 1977).

When milk inoculated with EEC was used to make Camembert cheese, the EEC remained present in the cheese for up to 6 weeks, and in cheese made from milk containing $500\,EEC\,ml^{-1}$, $700-20\,000\,g^{-1}$ were present after 7 weeks ripening. When 106 retail samples of soft cheese were examined, 17 % contained $>10\,000\,g^{-1}$ faecal coliforms, but none contained enteropathogenic strains (Frank, 1978). In Cheddar cheese, *E. coli* declined to a low level during 3 months ripening, but no data is available on enteropathogenic strains.

Brucella

Brucella abortus and *Br. melitensis* survive ordinary cheesemaking conditions and remain viable for many months in hard cheese. *Brucella* may survive up to 90 days in Pecorino cheese, and in Cheddar cheese for 6 months, but in Camembert, using naturally infected milk, no organisms survived for more than 20 days.

Yersinia enterocolitica

Unlike most pathogens, this organism is a psychrotroph and able to multiply at 4 °C. In commercially made Cheddar and Italian (including Provalone and Mozzarella) cheeses manufactured from raw milks, 18 % of which were naturally contaminated with *Y. enterocolitica*, it was found in the curd of 9 % of the Cheddar cheeses, but in none of the Italian cheeses. All cheeses were negative for this organism when examined at 23 days (Cheddar) and 10 days (Italian) (Schiemann, 1978).

Mycobacterium tuberculosis

In countries where bovine tuberculosis is still endemic and raw milk is used for cheesemaking, *M. tuberculosis* can present a problem. Although killed by pasteurisation, it is resistant to acid, and therefore, when present is unlikely to be affected by the pH of the cheese. Its long survival in cheese is recognised, and although this is affected by the type of cheese, survival for 220 days (Cheddar), 300 days (Tilsit), 90 days (Camembert), and over 60 days (Edam) has been reported. However in Emmenthal and Gruyère cheese, the inhibitory products of the propionic acid bacteria have resulted in death of the tubercle bacillus in 20–40 days.

Viruses

Little is known about the survival of human pathogenic viruses in cheese. Experimental work done with milk infected with polio virus, certain enteric viruses and influenza virus suggests that pasteurisation greatly reduces or destroys the viruses present in the milk. With unheated milk, cheesemaking conditions result in a considerable decrease in virus level, but survivors may persist in the cheese for long periods (polio virus: 7 months in Cheddar cheese, 5–6 weeks in Cottage cheese).

REFERENCES

ACCOLAS, J. P., VEAUX, M., VASSAL, L. and MOCQUOT, G. (1978) *Lait*, **58**, 118.
BAKALOR, S. (1962) *Dairy Sci. Abstr.*, **24**, 11, 12; *Rev. Article*, 107.
BERRIDGE, N. J. (1956) *J. Dairy Res.*, **23**, 336.
BERRIDGE, N. J. (1972) *Biennial Reviews NIRD*, National Institute for Research in Dairying, Shinfield, UK, p. 39.
BILLE BRAHE, J. C. (1974) *A New View of International Cheese Production*. Chr. Hansen's Lab. A/S, Copenhagen, p. 4.
BOTTAZZI, V. (1959) *Ann. Microbiol., Milano*, **9**, 52.
BOTTAZZI, V. (1960) *Ann. Microbiol., Milano*, **10**, 57.
BRINDLEY, M. (1954) *J. Dairy Res.*, **21**, 83.
BRITISH STANDARDS INSTITUTION (1977) BS 5305. BSI, London.
BULLERMAN, L. B. (1979) *J. Fd Protection*, **42**, 65–86.
CHAPMAN, H. R. (1974) *Dairy Ind. Internat.*, **39**, 1.
CHAPMAN, H. R. and HARRISON, A. J. W. (1963) *J. Soc. Dairy Technol.*, **16**(3), 169.
CHAPMAN, H. R., MABBITT, L. A. and SHARPE, M. E. (1966) *XVIIth Internat. Dairy Congr.*, **D**, 55.
CHAPMAN, H. R., HOSKING, Z. D. and BINES, V. E. (1974) *XIXth Internat. Dairy Congr.*, **1E**, 683.
CHOMAKOV, C. (1967) *Milchwirtschaft*, **22**, 569.

COLLINS, E. B. (1961) *J. Dairy Sci.*, **44**, 1989.
COLUMELLA, L. J. M. (*c.* 50). '*De Re Rustica*', Bk. 7, Ch. 8, Trans. Curtis, M. C. (1945). Heinemann Press, London.
COULTER, S. T. and COMBS, W. B. (1939) *J. Dairy Sci.*, **22**, 521.
CZULAK, J. (1960) *Australian J. Dairy Technol.*, **15**, 118.
DAVIS, J. G. (1949) *Dictionary of Dairying*. Leonard Hill, London, p. 102.
DAVIS, J. G. (1965a) *Cheese*, Vol. 1. Churchill, London, p. 94.
DAVIS, J. G. (1965b) *Cheese*, Vol. 1. Churchill, London, p. 278.
DAVIS, J. G. (1976a) *Cheese*, Vol. 3, Churchill, London, p. 741.
DAVIS, J. G. (1976b) *Cheese*, Vol. 3. Churchill, London, p. 940.
DAVIS, P. A. and BABEL, F. T. (1954) *J. Dairy Sci.*, **37**, 176.
DAWSON, D. J. and FEAGAN, J. T. (1957) *J. Dairy Res.*, **24**, 210.
DOLBY, R. M. (1966) *XVIIth Internat. Dairy Congr.*, **D1**, 67.
DRIESSEN, F. M. and STADHOUDERS, J. (1971) *Netherlands Milk and Dairy J.*, **25**, 141.
DUMONT, J. P., DELESPAUL, G., MIGUOT, B. and ADDA, J. (1977) *Lait*, **57**, 619.
EIGEL, G. (1948) *Milchwissenschaft*, **3**, 46.
ENGEL, G. VON and MILEZEWSKI, K. E. VON (1977) *Milchwissenschaft*, **32**, 517–20.
ENGEL, G. VON and PROKOPEK, D. (1979) *Milchwissenschaft*, **34**, 272–4.
FEUILLAT, M., LE GUENNEC, S. and OLSSON, A. (1976) *Lait*, **56**, 521.
FLÜCKIGER, E., WAES, G. and WINTERER, H. (1980) *IDF Document No.* 120, International Dairy Federation, Bruxelles, Belgium.
FRANK, J. F. (1978) *Dissertation Abstr. Internat.*, **B38**, 5258.
FRYER, T. F. (1969) *Dairy Sci. Abstr.*, **31**, 471.
FRYER, T. F., REITER, B. and LAWRENCE, R. C. (1967) *J. Dairy Sci.*, **50**, 477–84.
FUNDER, S. (1946) *The Chief Moulds in Gammelöst and the Part Played by These in the Ripening Process*. A. W. Brøggers, Oslo.
GALESLOOT, TH. E. (1968) *XVIth Internat. Dairy Congr.*, **D**, 143.
GOUDKOV, A. V. and SHARPE, M. E. (1965) *J. Appl. Bacteriol.*, **28**, 63.
GREEN, M. L. (1972) *J. Dairy Res.*, **39**, 261.
GREEN, M. L. (1977) *J. Dairy Res.*, **44**, 159.
HABAJ, B., POZNÁNSKI, ST., RAPCZYŃSKI, T. and RYMASZEWSKI, J. (1966) *XVIIth Internat. Dairy Cong.*, **D2**, 471.
HAMMER, B. W. and BRYANT, H. W. (1937) *Iowa State Coll. J. Sci.*, **11**, 281.
IDF (1980) *Document No.* 122, International Dairy Federation, Bruxelles, Belgium.
KELLY, C. D. and MARQUARDT, J. C. (1939) *J. Dairy Sci.*, **22**, 309.
KEOGH, B. P. (1964) *Australian J. Dairy Technol.*, **19**(2), 86.
KOSIKOWSKI, F. V. (1966) *Cheese and Fermented Milk Foods*. Food Trade Press Ltd, London, p. 184.
KOSIKOWSKI, F. V. and O'SULLIVAN, A. C. (1966) *XVIIth Internat. Dairy Congr.*, **D1**, 25.
KUNDRAT, W. (1952) *Dairy Sci. Abstr.*, **14**, 618.
LANGSRUD, T. and REINBOLD, G. W. (1973) *J. Milk Fd Technol.*, **36**, 487.
LANGSRUD, T. and REINBOLD, G. W. (1974) *J. Milk Fd Technol.*, **37**, 26.
LAW, B. A. (1979) *J. Dairy Res.*, **46**, 573–88.
LAW, B. A. (1980) *Dairy Ind. Internat.*, **45**, 5, 15, 17, 19–20, 22.
LAW, B. A. and SHARPE, M. E. (1977) *Dairy Ind. Internat.*, **42**(12), 10.
LAW, B. A. and SHARPE, M. E. (1978a) *J. Dairy Res.*, **45**, 267.

LAW, B. A. and SHARPE, M. E. (1978b) *Streptococci in the Dairy Industry.* Academic Press, London.

LAW, B. A., SHARPE, M. E., MABBITT, L. A. and COLE, C. B. (1973) *Soc. Appl. Bacteriol. Tech. Ser.*, 7, 1, Eds. Board, R. G. and Lovelock, D. W. Academic Press, London.

LAW, B. A., SHARPE, M. E. and REITER, B. (1974). *J. Dairy Res.*, **41**, 137.

LAW, B. A., SHARPE, M. E. and CHAPMAN, H. R. (1976) *J. Dairy Res.*, **43**, 459.

LAW, B. A., HOSKING, Z. D. and CHAPMAN, H. R. (1979). *J. Soc. Dairy Technol.*, **32**(2), 87.

LE BARS, J. (1979) *Appl. Environ. Microbiol.*, **38**, 1052–5.

LOWRIE, R. J., LAWRENCE, R. C. and PEBERDY, M. F. (1974) *New Zealand J. Dairy Sci. Technol.*, **9**, 116.

MABBITT, L. A., CHAPMAN, H. R. and SHARPE, M. E. (1959). *J. Dairy Res.*, **26**, 105.

MANNING, D. J., CHAPMAN, H. R. and HOSKING, Z. D. (1976). *J. Dairy Res.*, **43**, 313.

MARTH, H. (1979) *Marschall Internat. Cheese Conf.*, 1979–6, 21.

MARTLEY, F. G. and LAWRENCE, R. C. (1972) *New Zealand J. Dairy Sci. Technol.*, 7, 38.

MASKELL, K. T., HARGROVE, R. E. and TITTSLER, R. P. (1951). *J. Dairy Sci.*, **34**, 476.

MAUBOIS, J. L. and MOCQUOT, G. (1971) *Lait*, **51**, 508.

MEYER, A. (1973) *Processed Cheese Manufacture.* Food Trade Press Ltd, London.

MIAH, A. H., REINBOLD, G. W., HARTLEY, J. C., VEDAMUTHU, E. R. and HAMMOND, E. G. (1974) *J. Milk Fd Technol.*, **37**, 47.

MOCQUOT, G. (1979) *J. Dairy Res.*, **46**, 133–60.

MOCQUOT, G. and BÉJAMBES, M. (1960) *Dairy Sci. Abstr.*, **22**, 1.

MORRIS, H. A. and HART, P. A. (1978) *Cuthisid Dairy Products J.*, **13**, 22–4.

MORRIS, H. A. and JEZESKI, J. J. (1953) *J. Dairy Sci.*, **36**, 1285.

MOURGUES, R., VASSAL, L., AUCLAIR, J., MOCQUOT, G. and VANDEWEGHE, J. (1977) *Lait*, **57**, 131.

NILSON, K. M., RADKE, W. O. and PARTRIDGE, J. A. (1979) *Marschall Internat. Cheese Conf.*, 1979–11, 63.

OLSON, N. F. (1979) *Marschall Internat. Cheese Conf.*, 1979–2, 3.

PARKER, R. B. and ELLIKER, P. R. (1952) *J. Dairy Sci.*, **35**, 482.

PARMELEE, C. E. and NELSON, F. E. (1949) *J. Dairy Sci.*, **32**, 993.

PINHEIRO, A. J. R., LISKA, B. J. and PARMELEE, C. E. (1965) *J. Dairy Sci.*, **48**, 983.

REITER, B. and SHARPE, M. E. (1971) *J. Appl. Bacteriol.*, **34**, 63.

REITER, B., FREWINS, B. G., FRYER, T. F. and SHARPE, M. E. (1964) *J. Dairy Res.*, **31**, 261.

REITER, B., FRYER, T. F., PICKERING, A., CHAPMAN, H. R., LAWRENCE, R. C. and SHARPE, M. E. (1967) *J. Dairy Res.*, **34**, 257.

RENWICK, R. S. (1978) *J. Soc. Dairy Technol.*, **31**, 18.

SANDINE, W. E., ANDERSON, A. W. and ELLIKER, P. R. (1957) *Fd Eng.*, **29**, 84.

SANDINE, W. E., ELLIKER, P. R. and ANDERSON, W. A. (1959) *J. Dairy Sci.*, **42**, 799.

SCHIEMANN, D. A. (1978) *Appl. Environ. Microbiol.*, **36**, 274–7.

SHARPE, M. E. (1962) *Dairy Sci. Abstr.*, **24**, 169.

SHARPE, M. E. (1972) *Proc. 3rd Nordic Aroma Symp., Finland.* Hameenlinna, Finland, p. 64.

SHARPE, M. E. (1978) *Streptococci in the Dairy Industry.* Academic Press, London.

SHARPE, M. E. (1979) *J. Soc. Dairy Technol.*, **32**, 9.
SHARPE, M. E. and BRAMLEY, J. (1978) *Dairy Ind. Internat.*, **42**, 24.
STADHOUDERS, J. (1962) *XVIth Internat. Dairy Congr.*, **B**, 353.
STOLOFF, L. (1980) *J. Fd Protection*, **43**, 226–31.
TATINI, S. R., JEZESKI, J. J., OLSON, J. C. and CASMAN, E. P. (1971) *J. Dairy Sci.*, **54**, 312.
TATINI, S. R., WESALA, W. D., JEZESKI, J. J. and MORRIS, H. A. (1973) *J. Dairy Sci.*, **56**, 815.
THOM, C. and CURRIE, J. N. (1913) *J. Biol. Chem.*, **15**, 249.

6

Microbiology of Fermented Milks

R. K. ROBINSON

Department of Food Science,
University of Reading, UK

and

A. Y. TAMIME

West of Scotland Agricultural College,
Auchincruive, UK

The origins of fermentations involving the production of lactic acid are lost in antiquity, but it is not difficult to imagine how nomadic communities gradually acquired the art of preserving their meagre supplies of milk by storing them in animal skins or crude earthenware pots. Initially the intention could well have been simply to keep the milk cool through the evaporation of whey from the porous surface, but the chance transformation of the raw milk into a refreshing, slightly viscous foodstuff would soon have been recognised as a desirable innovation. The introduction of refinements, such as concentrating the raw milk over an open fire to give a thicker coagulum, must also have evolved over a considerable period of time, but the end result was that many communities in the Middle East and along the Eastern seaboard of the Mediterranean gradually acquired the skills of making yoghurt and related products.

In some places, the basic character of the yoghurt was changed to give a product with improved keeping qualities. The derivation of condensed yoghurt, for example, in which some of the whey is expressed from the coagulum, or 'Kishk', a product made by sun-drying a mixture of yoghurt and a cereal flour (Robinson and Cadena, 1978), could well have been instigated by the desire for some method of storing yoghurt in the absence of refrigeration, and numerous local variations on these themes can still be found. Occasionally yeasts would figure in the fermentation as well, and mildly alcoholic beverages like Kefir and Kumiss made their appearance among the multitude of fermented milks.

Nevertheless, despite the popularity of these products in certain societies, communities in Northern Europe and elsewhere paid scant attention to their properties until the classic text by Metchnikoff (1910) stirred the imagination of those attracted by his hypothesis. Thus, his proposal that the apparent longevity of the hill tribesmen of Bulgaria was a direct result of their life-long consumption of yoghurt, inspired an interest in the nutritional characteristics of the product that has never abated. Even today the controversy smoulders on, and although modern commercial yoghurt bears little resemblance to its Balkan counterpart, there are still many consumers who believe that it is more than 'just another foodstuff' (Robinson, 1977*b*).

The validity, or otherwise, of Metchnikoff's views will be debated for many years to come, but one indisputable effect of his work was a marked increase in the popularity of yoghurt throughout Europe. The almost 'mystical' properties of natural yoghurt were not, however, enough to sustain the interest of the market for very long, and it was not until the introduction of fruit/flavoured varieties in the late 1950s that yoghurt became a major dairy product. Since that time, its popularity as an inexpensive and convenient dessert has increased almost annually, so that, at the present time, fruit yoghurts represent a major source of income to the dairy industry. This massive growth has, of course, been accompanied by considerable changes in the process plant employed for production, while, in a modern factory, control of the microbiological aspects of the fermentation have reduced the chances of unacceptable variation between batches. Nevertheless, the principles underlying the various aspects of manufacture have altered little with time, and an understanding of these basic tenets is essential for efficient control at plant level.

TYPES OF YOGHURT

The types of yoghurt that are produced world-wide can be divided into various categories, and the sub-divisions are usually erected on the basis of:

(1)   existing or proposed legal standards;
(2)   method of production;

(3)   flavours; and
(4)   post-incubation processing.

Legal Standards

Legal standards for yoghurt are mainly based on the chemical composition of the product, i.e. percentage of fat, solids-not-fat (SNF) or total solids (TS). A minimum specification for SNF or TS is included by some countries (Robinson and Tamime, 1976), but the main divisions are on the basis of fat content, and, according to FAO/WHO (1973), yoghurt may be designated 'full' (above 3·0%), 'medium' (3·0–0·5%) or 'low' (0·5% or below) fat.

Although these percentages may vary slightly, such compositional standards aim to facilitate standardisation of the product, and to protect the consumer. In some countries, e.g. The Netherlands, West Germany and the Soviet Union, a fourth type of yoghurt, designated as 'Balkan Yoghurt', may be produced, with a fat content that ranges from as little as 4·5% to as high as 10%.

Methods of Production

The industry recognises two main types of yoghurt, set and stirred, and this classification is based on the system of manufacture, and the physical structure of the coagulum. *Set yoghurt* is the product where incubation/fermentation of the milk takes place in the retail container, and thus the characteristic coagulum is a continuous, semi-solid mass. *Stirred yoghurt*, however, is the type of yoghurt produced after the fermentation has been carried out in bulk, and the coagulum has then been broken prior to final cooling and packaging. *Fluid yoghurt* can be considered as stirred yoghurt with low viscosity, e.g. 11% total solids or less (Rousseau, 1974).

Flavours

The incorporation of flavours into yoghurt has led to the appearance of three different types of product, namely: *Natural or plain yoghurt* which is the traditional type with a sharp, acidic taste; *fruit yoghurt* made by the addition of fruit and sweetening agents to natural yoghurt; and *flavoured yoghurt*, in which the fruit ingredient is replaced with synthetic flavouring and colouring compounds.

Post-Incubation Processing

Various types of 'modified' yoghurt can now be found on the market, and a few of the more notable developments in this field are mentioned

TABLE I

SCHEME OF CLASSIFICATION FOR
YOGHURT

Physical state	Product type
I Liquid/viscous	Yoghurt
II Semi-solid	Concentrated/condensed yoghurt
III Solid	Soft/hard frozen yoghurt
IV Powder	Dried yoghurt

below. *Pasteurised yoghurt* is processed by conventional methods of manufacture, but after fermentation the yoghurt is heat treated in order to extend the shelf-life. Incidentally, the designation of this type of product as 'yoghurt' is still disputed, and the objections against this practice have been well documented by Kroger (1978) and Speck (1977). *Frozen yoghurt* is prepared in a conventional manner, but is then deep-frozen to at least −20 °C. It may, in addition, require a higher than normal level of sugar and stabiliser in order to maintain the integrity of the coagulum during freezing. *Dietetic yoghurts* may include low calorie yoghurt, low lactose yoghurt or vitamin/protein fortified yoghurt. Finally, *concentrated and dried yoghurts* must be mentioned, where the former product has a total solids of around 24%, and the latter type between 90–94% (Robinson and Tamime, 1975; Robinson, 1977a; Tamime, 1978).

It can be observed from this brief review that it is difficult, if not impossible, to find a common definition for all those products at present termed 'yoghurt'. However, Tamime and Deeth (1980) have proposed a scheme of classification that does cover all types of yoghurt, with the groupings based primarily on the physical state of the product (Table I). If such a view was accepted, then it might be possible to envisage a separate definition emerging to cover each different 'type of yoghurt' mentioned in Table I. Whether this proposal would offer any real advantages remains a moot point, but in any event, colloquial usage of the word 'yoghurt' as a blanket term for all manner of products will continue for many years.

PRODUCTION OF YOGHURT

The steps involved in the manufacture of some different types of yoghurt are illustrated in Fig. 1, and it can be observed that the stages of

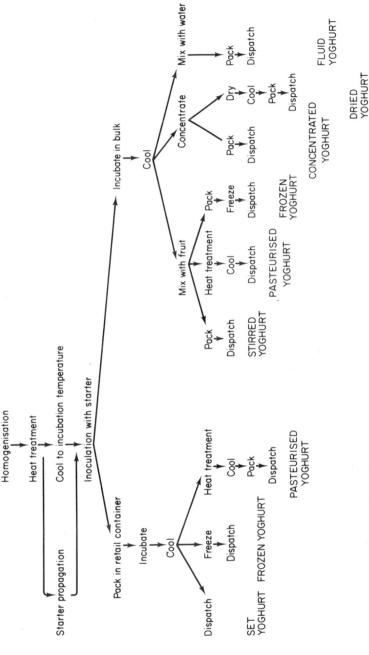

FIG. 1. An outline flow diagram indicating the processes involved in the production of various types of yoghurt. Note: (1) flavouring agents, sweeteners, stabilisers and colouring matter could be added to any of the above products; (2) fluid yoghurt can be produced directly from skim-milk of the required total solids.

production for set or stirred yoghurts are similar until after the addition of the starter culture.

Preliminary Treatment of Milk

The preparation of the basic mix involves the fortification and/or standardisation of milk, and fortification of the yoghurt milk means, in this context, increasing the level of solids in the mix in order to achieve the desired rheological properties of the manufactured product. The different methods for fortification/standardisation of the basic mix are summarised in Table II, which also indicates the possibilities that exist for either increasing or decreasing the various milk constituents. However, the choice of any particular system is primarily governed by the following factors:

(1) cost and availability of the raw materials;
(2) scale of production;
(3) capital investment in the processing equipment.

The composition of yoghurt (percentage total solids) varies considerably, and the available information is well documented in the reviews by

FIG. 2.   A section of a typical yoghurt line showing the tanks where pasteurised skim-milk is fortified by the addition of milk powder. The powder (and, if required, other dry ingredients) are fed into the recirculation line through the hopper. Reproduced by courtesy of the APV Co. Ltd, Crawley, UK.

TABLE II

SOME METHODS OF FORTIFICATION/STANDARDISATION OF THE PROCESS MILK EMPLOYED
FOR THE MANUFACTURE OF YOGHURT

Raw materials	Process/addition	Principal effect on process milk
Liquid whole milk	Evaporation	Increases all constituents
	Whole milk powder	Increases all constituents
	Skim-milk powder	Increases SNF
	Butter milk powder	Increases SNF and phospholipids
	Caseinate	Increases SNF, i.e. casein
	Whey powder (spray dried)	Increases SNF, i.e. whey proteins, lactose and minerals
	Evaporated whole milk	Increases all constituents
	Evaporated skim-milk	Increases SNF
	Ultrafiltered whole milk	Increases protein and fat
	Ultrafiltered skim-milk	Increases protein
	Concentrated milks (reverse osmosis)	Increases all constituents
Liquid skim-milk	(As above)	(As above)
Milk powders	Full cream milk powder or skim-milk powder can be reconstituted either to the desired total solids, or can be made equivalent to liquid milks and amended as above	

Robinson and Tamime (1975) and Tamime and Deeth (1980). However, commercial yoghurts tend to contain around 15% milk solids, and although the most popular method of fortification is the addition of milk powder to liquid milk (see Fig. 2), the use of ultrafiltration to concentrate the solids in skim-milk is being considered as a feasible alternative (Bundgaard *et al.* (1972)). It is also at this stage that stabilisers, such as pregelatinised starch (up to 1%) or plant gums (up to 0·5%) may be added to the process milk, for syneresis from the coagulum of stirred yoghurts is most easily avoided through the judicious use of hydrocolloids (Boyle, 1972).

The importance of total solids in this context stems from the improved consistency imparted to the yoghurt coagulum, an improvement that is carried further by the homogenisation stage that follows fortification. The effect of this treatment on the milk, and particularly on the lipid fraction, has been discussed by Banks *et al.* (1981), and these changes tend to impart a rather 'smooth' texture to the coagulum. It is also reported that homogenisation can reduce the incidence of 'pips' in yoghurt, i.e. the white flecks that appear in some fruit varieties (Davis, 1967; Nielsen, 1972; Robinson, 1980), and, although the origin of the 'pips' is not known for certain, diminishing their visual impact goes some way towards reducing consumer complaints.

Although homogenisation is widely practiced by the industry, its effects are not so apparent as those associated with the subsequent heat treatment. The optimum conditions for this treatment are 80–85 °C with a holding time of 30 min (Galesloot and Hassing, 1969), but, in many factories, time and temperature are dictated by the available plant (see below). However, although the impact of a given heating regime will vary with the conditions employed, the effects of heat treatment can be summarised as follows.

1. There is a denaturation of the whey proteins (albumins and globulins), and an aggregation of the casein molecules into a three dimensional network. This network traps the whey proteins, and the yoghurt coagulum produced subsequently is rendered more viscous (Jenness and Patton, 1959).

2. The bacterial load in the milk is reduced, and hence the starter culture has less competition from adventitious organisms.

3. There is a reduction in the amount of oxygen in the milk (Ling, 1963), and as the normal yoghurt cultures are micro-aerophilic, the lowered oxygen tension encourages their growth.

4. Some limited damage to the milk proteins may occur during heating, and the breakdown products can stimulate starter activity (Greene and Jezeski, 1957a).

Thus, the heating of milk can either inhibit or stimulate the activity of the lactic starter cultures, and the work of Greene and Jezeski (1957b) summarises the overall events as:

(1) stimulation of starter in milk heated between 62 °C for 30 min and 72 °C for 40 min;

(2) inhibition of starter in milk heated between 72 °C for 45 min and 82 °C for 10–120 min, as well as 90 °C for 1–45 min;

(3) stimulation of starter in milk heated to 90 °C for 60–180 min or autoclaved (120 °C for 15–30 min);

(4) inhibition of starter in milk autoclaved at 120 °C for longer than 30 min.

According to Greene and Jezeski (1975*b*), the effect of the heat treatment of milk on starter behaviour, i.e. the apparent stimulation/inhibition/stimulation/inhibition cycles (confirmed by Swartling and Mukherji, 1966; Bracquart *et al.*, 1978), could be duplicated by the addition of denatured serum protein or cysteine hydrochloride. The transition from one phase of the cycle to another, in response to different heat treatment exposures, occurs as a result of the release of denatured serum protein nitrogen at concentrations of 0.15–$0.20 \, \text{mg ml}^{-1}$ or cysteine (10–$20 \, \mu\text{g ml}^{-1}$). Furthermore, the same workers found that when cysteine was added artificially, it augmented the sulphydryl groups made available by heating. Thus the cysteine became stimulatory in raw and slightly heated milks, but was inhibitory in highly heated milks. These changes in the amino-nitrogen levels are obviously only part of the picture, and Greene and Jezeski (195*/c*) have dealt in length with the possible interactions that could be involved. The formation of formic acid may also be important.

The temperature of heat treatment of the yoghurt milk can vary from as low as ordinary pasteurisation ($72 \, °C$ for $15 \, \text{s}$) to as high as $133 \, °C$ for $1 \, \text{s}$ (UHT). However, convenient industrial practice involves preheating the milk to $85 \, °C$ for $30 \, \text{min}$ (batch process) or 90–$95 \, °C$ for 5–$10 \, \text{min}$ (continuous process). In the batch system, heating of the milk is carried out in a vessel known as a 'multi-purpose processing tank', where all the stages of yoghurt manufacture are carried out in the one holding unit. These tanks are normally jacketed, and a typical cycle of operations might be as follows:

(1) fill the tank with milk and add the necessary ingredients;
(2) circulate hot water in the jacket to raise the temperature of the basic mix;
(3) hold the milk at the desired temperature for a standard period of time;
(4) circulate chilled water to cool the milk to the incubation temperature, i.e. $42 \, °C$;
(5) maintain this temperature during the entire period of fermentation/coagulation;
(6) circulate chilled water just before the yoghurt reaches the desired acidity, and continue the cooling stage until the temperature of the yoghurt is less than $10 \, °C$.

A hypothetical processing line based on this approach is illustrated in Fig. 3, and it should be noted that while the quality of the end-product

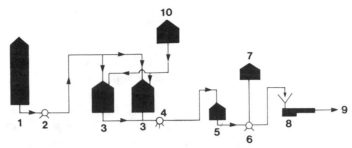

FIG. 3. Diagrammatic representation of a small-scale processing line employed for the production of stirred yoghurt. (1) Silo for bulk storage of milk; (2) centrifugal pump; (3) multi-purpose processing tank; (4) positive pump; (5) intermediate tank; (6) metering pump; (7) fruit tank; (8) packaging/filling machine; (8) cold store; (10) bulk starter tank.

can be extremely good, the time required to manufacture, say 5000 litres of yoghurt, can be as long as 12 h.

The use of multi-purpose processing tanks for the manufacture of yoghurt can, therefore, prove inconvenient, and it is for this reason that plate or tubular heat exchangers are often used for the continuous heat treatment of the basic mix. Norling (1979) has discussed in detail the use of such systems for the manufacture of yoghurt, and the advantages of a continuous process (see Fig. 4) as compared with the batch system are:

(1)   a smaller floor area is required for a given volume of production;
(2)   less energy is required;
(3)   productivity can be increased by utilising the fermentation tanks more than once per day;
(4)   the basic mix can be homogenised more easily; and
(5)   the yoghurt is cooled outside the fermentation tanks.

These two systems of manufacture are mainly used in medium- to large-scale operations, but it should be noted that one of the virtues of yoghurt, particularly as a means of preserving valuable milk protein, is that it is amenable to production in limited volumes. A typical, small-scale system for the manufacture of yoghurt is illustrated in Fig. 5, and this method of production is referred to as the 'churn system'. The basic mix is best prepared and heat treated in a batch pasteuriser, and after cooling to incubation temperature, the milk is inoculated with the starter culture prior to filling the churns. These churns can be left undisturbed in the processing area until coagulation is visible, but in some plants, the churns are sprayed with hot water

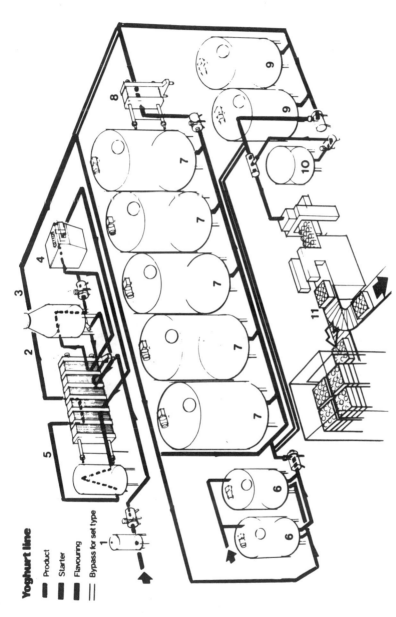

FIG. 4. Flow diagram showing a process line for the production of set and/or stirred yoghurt. (1) Balance tank; (2) plate heat exchanger; (3) vacuum chamber; (4) homogeniser; (5) holding tube; (6) bulk starter tanks; (7) incubation tanks; (8) plate cooler; (9) intermediate tanks; (10) fruit addition; (11) packaging machine. Reproduced by courtesy of Alfa-La·al A/B, Lund, Sweden.

Yoghurt line

■ Product
■ Starter
■ Flavouring
= Bypass for set type

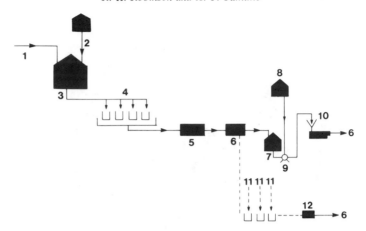

FIG. 5. Diagrammatic representation of the system for producing stirred yoghurt in milk churns. (1) Milk from silo; (2) bulk starter tank; (3) mixing tank; (4) milk churns; (5) incubation chamber; (6) cold store; (7) intermediate tank; (8) fruit tank; (9) metering pump; and (10) packaging/filling machine; or (11) fruit added to churn manually; (12) small-scale filling machine.

in order to maintain the temperature as near to 42°C as possible. The churns are then transferred to the cold store for cooling. Such a method gives rise to a high degree of variability between churns, so that it is advisable, where circumstances permit, for the churns to be incubated in thermostatically controlled cabinets or water baths. In addition, care must be practised during the movement of these churns, especially at the end of the incubation period, otherwise the separation of whey can be encountered in the retail yoghurt.

Incubators/Fermentation Tanks
The acidification of milk during the manufacture of yoghurt is a biological process which must be carried out under controlled conditions in special incubators and/or fermentation tanks. In the case of set yoghurt, the incubation system may be selected from among the following.

Water Baths
The inoculated milk is bottled in glass, retail containers and, after packing in metal trays, the bottles are immersed in shallow water baths or tanks. The temperature of the water is maintained at 40–45°C until the desired acidity is reached, at which point, the warm water is replaced with cold water in order to reduce the metabolic activity

of the starter culture. Final cooling takes place in the cold store. It is, perhaps, worth pointing out that these types of incubator are labour intensive, their energy consumption is high, and they require a large floor area.

Cabinets

The cabinet comprises a small insulated room which is divided into compartments, and most incubators of this type are multi-purpose chambers capable of circulating hot or cold air. In practice, the retail cartons are 'palletised' inside the cabinet, and hot air is circulated during the fermentation period followed by cold air during the cooling stage. Sometimes these cabinets are used as incubators only, and the yoghurt is cooled in a refrigerated cold store; however, the warm coagulum may suffer structural damage during transfer, and hence the latter procedure is avoided if possible.

Tunnel

In the previous two types of incubator, set yoghurt is produced in batches, but in the tunnel system, production can be continuous. The tunnel is divided into two sections, with the first part as a heated chamber, and the second part as a 'cooler'. The trays of yoghurt cartons move through the tunnel on a conveyor belt, and the speed and length of the conveyor are governed by the temperature of incubation, the percentage of starter culture used and the activity of the inoculum; the lower the temperature of incubation, e.g. less than 35 °C, and the smaller the inoculation rate (e.g. 1 %), the longer and slower the conveyor belt. Although the warm coagulum is in motion during the fermentation period, minimum structural damage is achieved by fitting smooth rollers beneath the pallets. Incidentally, any type or size of retail container can be used in the case of cabinet or tunnel incubators.

During the manufacture of stirred yoghurt, the milk is fermented in bulk in a special incubation tank. The use of these tanks in the manufacture of yoghurt has been recently reviewed by Tamime and Greig (1979) and they are divided into two main types.

Fermentation Tanks

These tanks are used as incubators only, and they are usually insulated in order to maintain the appropriate temperature. The processing of the milk and the cooling of the yoghurt is carried out in other equipment on the production line.

Multi-Purpose Tanks

As mentioned earlier, this type of tank is jacketed, and can be used for all stages of yoghurt production.

Cooling

After the incubation period, the yoghurt is cooled in order to control the level of lactic acid in the product. There are a number of different methods which can be used to cool stirred yoghurt, but it is of note that the rate of cooling can affect the structure of the coagulum. Thus, very rapid cooling can lead to whey separation, and this is due to a too rapid contraction of the protein filaments which, in turn, affects their hydrophilic properties (Rasic and Kurmann, 1978).

The normal industrial practice is, therefore, to cool the yoghurt to 15–20 °C before mixing it with fruit/flavours prior to packaging. The final cooling to <5 °C takes place in a refrigerated cold store. The methods used for cooling yoghurt are as follows.

In-Tank Cooling

The cooling of yoghurt in a multi-purpose tank requires the circulation of chilled water through the jacket, and a tank of 2500–5000 litres of yoghurt can take up to 4 h to cool from 45 °C to <10 °C (Jay, 1979).

Plate or Tubular Cooler

An efficient system of cooling yoghurt quickly and continuously involves the use of a plate or tubular cooler, although Steenbergen (1971) and Piersma and Steenbergen (1973) concluded that tubular coolers caused the least structural damage to the coagulum.

Although the usual procedure is to cool the yoghurt to <20 °C, investigations carried out at the Danish Research Institute (Anon., 1977) suggested that the following system might improve the quality of the end-product:

(1) stir the yoghurt in the fermentation tank to obtain a homogeneous mixture;
(2) precool to 24 °C and package;
(3) final cooling: stage 1—at an air temperature of 7–10 °C for 5–6 h; stage 2—at an air temperature of 1–2 °C for the remainder of the cooling period.

However, industry does not appear to have adopted these proposals to any marked extent.

Packaging

Yoghurt packaging machines are based on one of the following principles, namely volumetric level filling, e.g. when fluid yoghurt is poured into glass bottles; or volumetric piston filling, as applied to the packaging of stirred yoghurt in plastic containers. The latter type is, of course, more widely used, but the piston pump can cause some shearing of the coagulum. To minimise this reduction in viscosity, Steenbergen (1971) recommended a low speed of filling, and the use of a filling nozzle with a large orifice. It is also extremely important that the design of the filling head should allow for a high standard of hygiene.

PLANT HYGIENE

The production of food for human consumption must be carried out under hygienic conditions, and the cleaning and sanitisation of processing equipment is an integral part of production and management.

It can be observed from Figs. 3 and 5 that a yoghurt production line consists of various processing components, tanks, pipe-lines and pumps, and hence proper plant hygiene is best achieved with an enclosed system, i.e. cleaning-in-place (CIP). Alternatively, the production line can be dismantled for cleaning purposes, but such an approach is time consuming, and, in many cases, not very effective.

The cleaning cycle of any CIP system, as reported by Charlish and Warman (1980) and Tamplin (1980), is as follows.

1. *Preliminary rinse.* The plant is rinsed with water to remove the bulk of the milk and yoghurt residues from the equipment.
2. *Detergent wash.* Strong alkali is used, e.g. sodium hydroxide, trisodium phosphate or sodium silicate. These chemical compounds break up the organic constituents of milk (proteins), saponify the fats and dissolve precipitates due to the hardness of water. Acid detergents are used to remove milk deposits in plate heat exchangers or multi-purpose tanks, and, of course, hard water scale. Nitric acid can also be used for sanitisation purposes (Zall, 1981).
3. *Water rinse.* The aim is to remove detergent residues from the processing plant prior to sanitisation.
4. *Sanitisation treatment.* The destruction of micro-organisms is achieved by using one of the following sanitising agents: nitric acid,

chloramine, quaternary ammonium compounds or sodium hydroxide. However, while the silos, fermentation tanks and packaging machine can be sanitised by one of the above compounds, hot water circulation (85–90 °C for 15–30 min) is used to sanitise plate heat exchangers.

5. *Final rinse*. Good quality water is used as a final rinse to remove the sterilant residues from the processing plant.

The design of a CIP system varies in relation to the equipment on the production line, and a comprehensive report on the cleaning and sanitisation of dairy plants has been published by the BSI (1977).

MICROBIOLOGY OF NATURAL YOGHURT

The crude inoculum that was employed in the production of traditional yoghurt consisted of a whole range of lactic acid bacteria belonging to the families *Lactobacillaceae* and *Streptococcaceae*, and it was Metchnikoff (1910) who postulated that the derivation of yoghurt depended on the presence of one particular bacterium, *Bulgarian bacillus*. This organism, renamed *Thermobacterium bulgaricum* by Orla-Jensen (1931), is now recognised as being *Lactobacillus bulgaricus*, and its inclusion in starter cultures for yoghurt is now almost obligatory.

Nevertheless, it was clear that other bacteria were also involved in the fermentation, and these were identified as *Thermobacterium jugurt* (*Lactobacillus jugurti*) and *Streptococcus thermophilus* (Orla-Jensen, 1931). *Lactobacillus acidophilus* was reported by Rettger and Cheplin (1921) to be an additional member of the yoghurt microflora, but it is regarded by most authorities as a casual contaminant. The occasional employment of *Lac. jugurti, Lac. lactis* or *Lac. helveticus* in starter cultures is also a distinct possibility, because it is only recently that the classification of this group of bacteria has received close attention. Thus, as discussed by Gilmour and Rowe (1981), the relationship between these associated organisms is fairly close, and the precise identification of any given isolate necessitates a whole range of biochemical tests. However, despite some difficulty with variable biotypes, the organisms associated with the production of yoghurt are normally restricted to *Lac. bulgaricus* and *Str. thermophilus* (Kon, 1972). The regulations in some countries such as New Zealand (Anon., 1973) or Switzerland

(Anon., 1972), do allow other lactic acid bacteria to be included in a starter culture, but, in general, only the two important organisms are employed.

This trend is valuable in that it helps to give yoghurt a distinctive character *vis-à-vis* other fermented milks, such as Acidophilus milk made with *Lac. acidophilus*. There are also sound microbiological reasons for promoting the association, for it has been widely observed that the production of lactic acid during yoghurt manufacture is greater when mixed cultures of *Str. thermophilus* and *Lac. bulgaricus* are used, as compared with the single species (Tamime, 1977). Furthermore, cell numbers, particularly of *Str. thermophilus*, also increase with unit time in mixed cultures more rapidly than in monocultures, and this observation led Pette and Lolkema (1950) to postulate the existence of a synergistic relationship between compatible strains of the two organisms. The fact that *Str. thermophilus* responded more vigorously in mixed culture was attributed to the proteolytic activity of *Lac. bulgaricus*, notably through the release of valine into the milk. Further work by Bautista *et al.* (1966) confirmed this view, but suggested that the essential amino acids were glycine and histidine. This difference in growth requirements may, of course, be a reflection of the different strains that were being employed, but the general thesis that *Lac. bulgaricus* releases amino acids necessary for active growth of *Str. thermophilus* is widely accepted.

The stimulation of *Lac. bulgaricus* is due to a factor originating from the metabolic activity of *Str. thermophilus*, and one that is similar to, or can be replaced by, formic acid (Galesloot *et al.*, 1968). This proposal was confirmed by Bottazzi *et al.* (1971) who showed that formic acid in milk, at levels of 30–50 μg ml^{-1}, resulted in a detectable response by *Lac bulgaricus*. It was also noticed by this latter group that aroma production, including the aldehyde fraction, was highest in mixed cultures (Bottazzi and Vescovo, 1969), and Lees and Jago (1976) showed that *Lac. bulgaricus* was primarily responsible. Thus, although *Str. thermophilus* does form acetaldehyde as a product of metabolism, the pathway is less active at normal fermentation temperatures, i.e. around 40 °C, than the corresponding synthesis by *Lac. bulgaricus*.

The selection of 40–42 °C as the temperature of incubation represents a compromise between the 'metabolic' optima of the two species, for while 45 °C is optimum temperature for lactic acid production by most strains of *Lac. bulgaricus*, 39 °C is the temperature preferred by *Str. thermophilus*. It is also noticeable that, as shown in Table III, the

TABLE III

SOME TYPICAL FIGURES, OBTAINED BY DIRECT
MICROSCOPIC COUNT, SHOWING THE NUMBER
OF CELLS ml^{-1} OF YOGHURT AFTER 4 h
INCUBATION AT THE TEMPERATURES
INDICATED

Temperature ($°C$)	Cell count $ml^{-1} \times 10^{-6}$	
	Streptococcus thermophilus	Lactobacillus bulgaricus
30	834	66
35	1 042	97
40	1 105	221
42	1 636	604
45	906	636
50	221	275

After: Tamime (1977).

increase in total cell count is highest at 42 °C, and this observation
has two important implications for commercial practice.

1. *Streptococcus thermophilus* is normally most active in releasing
    lactic acid during the early stages of incubation, and hence process
    times can be shortened by encouraging the rapid development of
    this species.

2. Metabolites other than lactic acid are important in the aroma and
    flavour of yoghurt, and as these components are derived from both
    species, it is necessary to ensure a balanced growth of the two
    organisms.

It is important, therefore, that the starter culture (see Chapter 4)
provides not only an abundance of viable organisms, but also a balanced
population. The precise level of inoculation is subject to some variation
(Robinson and Tamime, 1975), but, in general, a rate of 2 % (v/v)
will enable the fermentation stage to be complete within 4 h. The aim
is to give the process milk a bacterial load of 30–40 $\times$ 10^6 lactic organisms
ml^{-1}, and although some authorities suggest that the two organisms
are best grown separately and then combined immediately prior to
use, popular practice is to produce a mixed inoculum with a 1:1 ratio
between *Str. thermophilus* and *Lac. bulgaricus*. The exact basis for

this expected balance is open to debate, for while 'clumps' of cells may be appropriate, Sellars (1975) has suggested a ratio of a 'short chain' of *Streptococcus* against single cells of *Lactobacillus*. Alternatively, a cell to cell basis can be employed, with the anticipated ratio of the organisms then being shifted to accommodate the dominance of *Str. thermophilus*, e.g. around 3:1 *S. thermophilus:Lac. bulgaricus*. A similar ratio may be obtained if the starter culture is examined on the basis of a total colony count, and, as shown in Table IV, there are a number of alternative media which can be employed (see also Anon., 1979). However, although plating methods can provide an accurate picture of the microflora, they are expensive of both time and facilities, so that, in practice, a rapid microscopic assessment based on the presence of 'clumps' may well have to suffice as a guide to the balance between the two species.

Although the ratio between the two organisms begins as a nominal 1:1 balance, it alters rapidly as *Str. thermophilus* enters its logarithmic phase of growth, and only as lactic acid begins to accumulate in the milk does *Lac. bulgaricus* become the dominant partner. The overall result is that by the end of the fermentation, when the acidity has reached around 0·90–0·95 % lactic acid, the populations are again roughly balanced. The total colony count may well exceed 2000×10^6 lactic organisms ml^{-1} at this point, and hence their contribution to the organoleptic quality of the end product is significant. Some of the detectable effects of this starter activity are recorded in Table V, and it is clear why such care has to be taken over both the choice of bacterial strains, and the activity of the actual bulk starter being used. Thus, any faults in the bulk inoculum, whether through antibiotic residues, strain imbalance or incompatibility (Moon and Reinbold, 1976) or even infection by bacteriophage, are bound to be reflected at the fermentation stage, and it is for this reason that quality control in the industry covers both starters and retail products.

Microbiological Quality of Yoghurt

An examination of the microbiological quality of a product is usually concerned with two aspects, namely:

(1) protection of the consumer from exposure to any health hazard; and

(2) ensuring that the material does not suffer microbial deterioration during its anticipated shelf-life.

TABLE IV

SOME CULTURE MEDIA THAT CAN BE USED FOR THE ENUMERATION OF STARTER BACTERIA IN YOGHURT; RESPONSES BASED ON COLONY MORPHOLOGY OR DETECTABLE GROWTH

Organism	TGV	LAB	Lee's agar	LAB + Penicillin	Lactic agar (low pH)	Micro-assay	Streptosel agar
Streptoccocus thermophilus	Smooth Irregular, hairy, or rough	Smooth Irregular, hairy, or rough	Yellow	No growth	No growth	Growth	Growth
Lactobacillus bulgaricus			White	Growth	Growth	No growth	No growth

After: Robinson and Tamime (1976).

TABLE V

Carbohydrate metabolism	Lactose is hydrolysed inside the bacterial cell by the enzyme β-D-galactosidase to glucose and galactose. The former is utilised for the production of D($-$) and L($+$) lactic acid, and the galactose accumulates.
	The calcium–caseinate–phosphate complex is destabilised by the lactic acid, and this leads to the formation of the yoghurt coagulum.
	Production of flavour components in yoghurt due to fermentation of the milk sugar, e.g. acetaldehyde (2·4–41·0 ppm), acetone (1·0–4·0 ppm), acetoin (2·5–4·0 ppm) and diacetyl (0·4–13·0 ppm).
Proteolysis	Yoghurt starter cultures are slightly proteolytic, and the peptides and amino acids produced act as precursors for the enzymic and chemical reactions which produce the 'flavour compounds'.
	Protein degradation is associated mainly with *Lac. bulgaricus*, but peptidase enzymes are produced by both *Str. thermophilus* and *Lac. bulgaricus*; the production of bitter peptides is attributed to the use of incubation temperatures below 38 °C and/or enzymic activity during cold storage; strain differences may also be important.
	The free amino acid content of yoghurt can vary from 23·6–70 mg per 100 ml.
Lipolysis	Some degree of fat degradation by the starter culture does take place, and the end products contribute significantly towards the flavour of yoghurt.
	Lipases from the yoghurt starter culture are especially active against short-chain triglycerides.
Miscellaneous	The volatile acidity of yoghurt can reach 7·5 meq kg^{-1}, mainly as the result of the metabolism of *Lac. bulgaricus*.
	On average, there is an increase in niacin and folic acid during fermentation, but the levels of vitamin B$_{12}$, thiamine, riboflavin and pantothenic acid decline.

After: Tamime and Deeth (1980).

Both these aims are of importance to consumer and producer alike. Thus, over and above any moral responsibility that a company accepts for the integrity of its products, the financial losses that follow a public health incident, or even a high level of consumer complaints, provide a very strong incentive for a manufacturer to give quality assurance a high priority. The type of hazard that may be encountered depends, of course, on the nature of the product, and, in general terms, yoghurt can be regarded as 'hygienically safe'.

The reason for this confidence stems from the level of acidity present (around 1 % lactic acid), for in this situation, potential pathogens such as *Salmonella* spp. will be largely inactive (Hobbs, 1972). Similarly, 'coliforms' will be unable to survive the low pH encountered, and this inhibition is reinforced by the production of antibiotic substances by the yoghurt organisms (Lucca, 1975).

The only reported risk comes, therefore, from the possible presence of *Staphylococcus* spp. Thus, Arnott *et al.* (1974) have shown that some species of *Staphylococcus* can survive in samples of commercial yoghurt, but whether this finding implies the existence of a risk from coagulase-positive species is open to debate. Certainly, there are no records of staphylococcal food poisoning being associated with the consumption of yoghurt, at least in the United Kingdom (Gilbert and Wieneke, 1973), and it may be that the chance contamination that was observed is not a portent of risk to the consumer (Rothwell, 1979).

This optimism does not imply that plant hygiene can be given a low priority, because spoilage organisms are often less sensitive to environmental factors than their pathogenic counterparts. Yeasts and moulds, in particular, are little affected by low pH, and with both sucrose and lactose available as energy sources, spoilage can occur rapidly. Yeasts, in particular, are a major concern, and while the lactose-utilising *Kluyveromyces fragilis* can build up on plant surfaces, the main source of contamination for the more popular types of stirred yoghurt is likely to be the fruit. Thus, even pasteurised fruit-purées may not be entirely free from stray contamination by *Saccharomyces cerevisiae*, or some equivalent species, and the sweetened yoghurt provides an ideal medium for growth and metabolism. The 'doming' of aluminium foil caps is a common indication of yeast activity, and Davis *et al.* (1971) have suggested that yoghurt, at the point of sale, should not contain above 100 viable yeast cells ml^{-1}; above 1000 yeast cells ml^{-1} would be 'unsatisfactory'. Mould growth in yoghurts is rather less of a problem mainly because genera like *Mucor*, *Rhizopus*, *Aspergillus*

TABLE VI

SOME OF THE STANDARD METHODS THAT HAVE BEEN PROPOSED FOR EXAMINING THE MICROBIOLOGICAL QUALITY OF YOGHURT

Country	*Colony count*	*Yeast and mould*	*Coliform*
UK	Nutrient agar at 30°C for 72 h	Malt agar or salt dextrose agar at 25–30°C for up to 5 days	MacConkey broth at 30°C for 72 h Presumptive coliforms streaked onto violet red bile agar at 30°C for 24 h
USA	Standard method agar or lactic agar of Elliker *et al.* (1956); parallel count at 7°C for 10 days to detect psychrotrophs	Acidified potato glucose agar—if count is low, distribute 10 ml in 3 plates evenly, and then count total colonies on 3 plates	Violet red bile agar or deoxycholate lactose agar or brilliant green lactose bile broth
IDF	Culture medium free from carbohydrates Incubate twice— (i) 30°C for 48 h (ii) 20°C for 48 h	Agar medium containing yeast extract and glucose plus oxytetracycline at 25°C for 96 h	Brilliant green lactose bile broth at 30°C for 48 h

After: British Standards Institution (1968, 1970); APHA (American Public Health Association) (1967, 1978); IDF (1971*a,b,c*).

or *Penicillium* can only grow satisfactorily at the yoghurt/air interface; agitation during distribution tends, therefore, to suppress development. Nevertheless, the growth of a 'surface' mycelium can occur at retail outlets, and a mould count of $1-10$ cfu ml^{-1} is normally rated as of 'doubtful quality'.

Some of the standard methods employed for the examination of yoghurt are shown in Table VI, and the routine application of these tests can go some way towards the elimination of losses during storage; or at worst, can expedite the identification and eradication of any infections. This aspect of quality control is discussed further in Chapter 7, but it cannot be emphasised too strongly that all these examinations must be carried out to a set pattern. Thus, the real value of microbiological testing comes with the establishment of 'norms', for once an overall 'microbial picture' of a product has been identified, then deleterious changes can be recognised at an early stage. This stricture applies equally to both the desirable activity of starter cultures (see Chapter 4), and the intrusion of contaminants, and it is imperative that the commitment to microbiological quality control should be both conscientious and long-term.

MISCELLANEOUS FERMENTED MILKS

The market for fermented milks is dominated by the various types of yoghurt, but there are many other milk products that enjoy specific national or local appeal. Thus, Kumiss, for example, is virtually unknown outside USSR, and although restricted availability of mare's milk contributes to this limited usage, the final product does represent something of an 'acquired taste'. Similarly, Kefir and Acidophilus milk lack really widespread recognition, and the consumption of Acidophilus milk, in particular, is more often linked with its alleged therapeutic properties, than with its organoleptic appeal. The reason for this belief stems from the fact that the fermentative organism, *Lac. acidophilus*, can become implanted in the human intestine, and its residence is associated with the easing of various forms of intestinal discomfort (Lang and Lang, 1975). Kumiss has also been recommended for its medicinal properties (Berlin, 1962), but it is of note that the possible benefits are associated with an intake of $1-1.5$ litre day^{-1} (Lang and Lang, 1973*a*). Similar consumption patterns must be anticipated in relation to the 'health-giving' properties of other fermented milks, and it would seem likely that only the most dedicated consumers can ever achieve the levels of intake necessary for these products to have any demonstrable effect.

Nevertheless, even if these various milks do little more than provide an attractive vehicle for bringing animal protein to a wider sector of the community, or than offer a valuable means of milk preservation, then the use of fermentation techniques deserves every encouragement.

Production of Fermented Milks

The basic process plant requirements are similar to those already described in relation to yoghurt, and the principal differences arise in respect of the total solids in the milk, and the starter organisms involved. Thus, as shown in Table VII, unfortified skim-milk is the most popular starting point for many fermentations, and the low total solids imparts a minimum of viscosity to the product. The important characteristics are, however, the result of microbial activity, and the basic identity

TABLE VII
SOME ESTABLISHED TYPES OF FERMENTED MILK, AND AN INDICATION OF THE MICRO-ORGANISMS RESPONSIBLE FOR THEIR PRODUCTION

Product	*Raw material*	*Starter organisms*
Acidophilus milk	Whole or skim-milk	*Lac. acidophilus*
Buttermilk (acid)	Skim-milk or buttermilk	*Lac. bulgaricus*
Buttermilk (cultured)	Skim-milk or buttermilk	Commercial butter starter, or mixture of: *Str. lactis, Str. lactis* sub-sp. *diacetylactis, Str. cremoris* and *L. citrovorum*
Cultured cream	Cream with 12–30% fat	*Str. lactis, Str. cremoris, Str. lactis* sub-sp. *diacetylactis*
Kefir	Whole milk or skim-milk	Kefir grains (see text) or mixed inoculum from among: *Str. lactis, Str. cremoris, Str. lactis* sub-sp. *diacetylactis, Candida kefir, Lac. acidophilus, Lac. casei, Kluyveromyces fragilis*
Kumiss	Whole milk (cow) or mare's milk	*Lac. bulgaricus, Lac. acidophilus, Kluyveromyces lactis*

After: Lang and Lang (1973*b*).

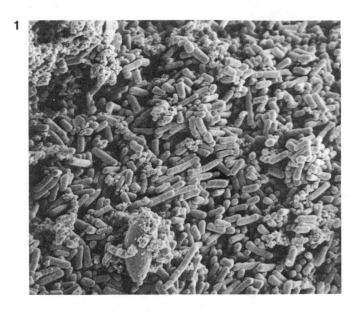

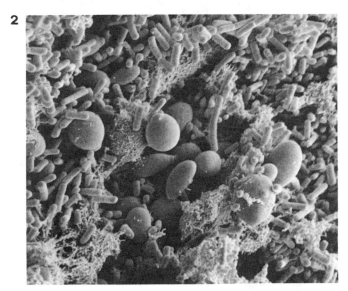

FIG. 6. Micro-organisms observed at various points within a Kefir granule (scanning electron microscope × 3040). (1) rod-shaped bacteria forming peripheral layer; (2) and (3) intermediate zones showing increasing abundance of yeasts and decreasing numbers of bacteria; (4) yeasts embedded in matrix of microbial origin dominating the central region of the granule. (For further details see Bottazzi and Bianchi, 1980.) Reproduced by courtesy of Professor V. Bottazzi.

3

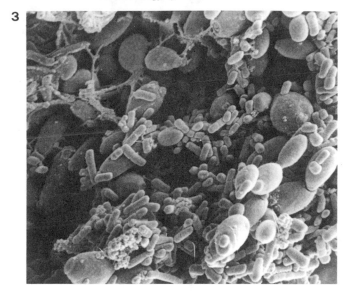

4

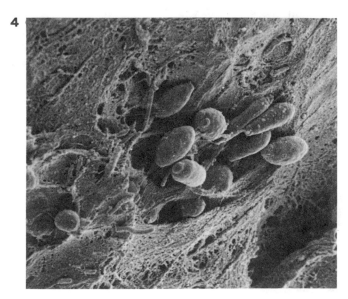

FIG. 6—*contd.*

of each product results from the process conditions providing the correct environment for the bacteria or yeasts concerned. In this way, the basic metabolism of the species or groups of species stamps its unique character on the milk, and although the final flavours will result from a complex range of chemical interactions (cf. Chapter 5), it is the microflora which paves the way. This dependence on microbial activity means that care over starter culture preparation is extremely important, and much of what has been said elsewhere in this volume about starter production and maintenance is equally applicable to the range of fermented milks that are available in various parts of the world.

It must be admitted, however, that with the exception of yoghurt and Acidophilus milk, the microflora of fermented milks has been subject to little systematic investigation. Thus, the role of strain differences and/or incompatibility in relation to the organoleptic properties of the various products is still largely a matter for speculation, and this restraint applies even where interspecific relationships are of major importance. A prime example of this latter situation is the granular starter employed for Kefir, where the nature of the inoculum represents, perhaps, the ultimate complexity.

These granules (Kefir grains) are small, white bodies of some 2–15 mm in diameter, and the outer surfaces are highly contorted. They are composed, as indicated in Table VII, of a mixture of micro-organisms, but the grouping is held together in a highly organised pattern. Thus, the peripheral layers of the granules are dominated by various rod-shaped bacteria (see Fig. 6), but towards the centre, yeasts become the major component of the microflora. What is not known, however, is how these units are built up, or the nature of the relationship between the constituent organisms (Bottazzi and Bianchi, 1980), and yet the granules proliferate freely in milk with no detectable changes in character. This ability of disparate groups of organisms to perpetuate themselves as discrete bodies has also been observed with the inoculum for 'Tibetan yoghurt', and yet the characteristics of this traditional 'health drink' remain constant even after the inoculum has passed through many generations. Nevertheless, in spite of the paucity of detailed information, a number of fermented milks are manufactured on a considerable scale, and a brief outline of some of the more notable ones will serve to indicate how variation between the various products has been achieved.

Acidophilus Milk
Although this product enjoys a marked popularity in Eastern Europe, consumers in the rest of Europe and the United States are more attracted

by its reported therapeutic value than by its flavour. Whatever the merits of this latter claim, the medical profession is sufficiently convinced of its possible virtues to have tacitly encouraged a modest enthusiasm within the dairy industry, and a fairly standard production schedule has now evolved.

The important feature of the system is the care that has to be taken to protect the process milk from contamination, because *Lac. acidophilus* grows only slowly in milk. It is essential, therefore, that the inoculum be maintained in an active state, and daily transfers of the mother culture are desirable to ensure consistent results. Sterilised skim-milk is the usual medium, and, after cooling to 37 °C, it is inoculated with 1 % of an active culture. A controlled incubation to around 0·6–0·7 % lactic acid is required to retain maximum viability of the cells, and this culture can then be employed for producing the bulk starter. This latter culture is handled in exactly the same manner, but with the volumes adjusted to provide a 2–5 % inoculum for the process milk.

The process milk which may be skim-milk or whole milk, must be of good bacteriological quality, but even so, it should receive a heat treatment designed to achieve a state of 'near sterility'. To this end, the milk is usually heated to around 95 °C for up to 1 h and then cooled to 37 °C. The milk is held at this latter temperature for 3–4 h to encourage spore germination, and is then reheated to 90–95 °C to kill any new vegetative cells. The cooled milk (37 °C) is then inoculated with the pure culture of *Lac. acidophilus*, and incubated until coagulation has taken place. In some countries, further fermentation takes the titratable acidity to around 1·0 % lactic acid prior to chilling and bottling, but, for 'medicinal' usage, the acidity is restrained at 0·6–0·7 % lactic acid; in both cases, storage of the bottled product should be between 5 and 10 °C. The main reason for this wide variation in acidity is that the living bacteria survive best at the lower level, and, because viable cells are essential for any therapeutic activity (ideally around 2000–3000 million bacteria ml^{-1}), the properties of the product are adjusted with this end in view.

However, even with studious attention to process details, maximum viability of the bacteria can only be sustained for around 1 week and, to date, attempts to select more robust strains have met with little success.

Cultured Buttermilk

Although a limited quantity of acid buttermilk is still produced employing

Lac. bulgaricus as the starter organism, most buttermilks employ the mixed culture shown in Table VII.

True buttermilk is, of course, the liquid remaining after cream is churned into butter, but skim-milk or whole milk is often used to give an equally acceptable product. In some countries, the fat content is standardised within a range of 0·5–3·0%, and Scandinavian 'sour milk' (Filmjolk) is a typical European variant. However, whatever the national differences in respect of the end-product, manufacturing procedures have much in common. In the most direct process, the milk is simply heated to 82–88 °C, cooled to 21 °C and inoculated, but more modern plants employ two additional stages of de-aeration and homogenisation. In the first stage, the process milk is heated to 78 °C and then treated under vacuum, a step that is regarded as desirable to impart a smooth consistency to the buttermilk. After homogenisation, the milk is finally heated to 90–95 °C in a plate heat exchanger, before cooling to 21 °C.

The bulk starter is usually employed at a rate of 1–2%, and it is important that the culture is both balanced and active. Thus, while *Str. cremoris* and *Str. lactis* are mainly responsible for acid production, *Str. lactis* sub-sp. *diacetylactis* and *Leuconostoc citrovorum* are the primary sources of flavour and aroma. Routine checks on the balance between these organisms is essential, but, in practice, the employment of cultures with an acidity of 0·8–0·85% lactic acid carries some assurance that all the desired organisms will be present and in an active condition.

If these requisites are met, then a fermentation time of 16–20 h at 20 °C will give rise to an excellent product of around 0·8–0·9% lactic acid. It is then cooled, mixed and bottled, and the retail material should be a viscous drink with a pleasing aroma and flavour. Insufficient flavour is often associated with deficiencies in the starter culture, but the most frequent complaint concerns physical separation. Some manufacturers avoid this latter problem by incorporating gelatin or additional fat into the process milk, but the danger of consumer complaints can usually be avoided by close attention to process details.

Cultured Cream

Sour cream is an extremely viscous product with the flavour and aroma of buttermilk, but with a fat content of 12–30%, its method of consumption is more akin to that of normal cream.

A typical schedule for production would involve fortifying whole milk with cream to give the desired fat content, and then heating the mix to around 80 °C for 30 min. Homogenisation at 2000 lb in^{-2} (minimum) and 60–80 °C follows, and, in general, increases in pressure or temperature above the minimum tend to improve the consistency of the retail material. After cooling to the inoculation temperature of around 21 °C, the cream is mixed with 1–2% of active starter, and further incubated at 21 °C until an acidity of 0·6% lactic acid has been reached (18–20 h).

On cooling, the sour cream is then ready for packaging in cartons prior to dispatch, but care must be taken at this stage to avoid any serious deterioration in viscosity. It is for this reason that some manufacturers incubate the cream in the retail cartons, and certainly this process can give rise to a markedly thicker material.

Kefir

This foamy, effervescent drink containing 0·9–1·1% lactic acid and 0·5–1·0% alcohol has achieved great popularity in Eastern Europe, but elsewhere, consumption is limited to a minority market.

The raw material for Kefir is usually whole milk, which is severely heat treated (95 °C for 5 min) to denature the whey proteins. The hydrophilic properties of these denatured proteins improves the viscosity of the end-product, as does the frequently employed process of homogenisation. A portion of this process milk is also employed to prepare the inoculum, and because of the nature of the 'starter culture', strict levels of hygiene are essential.

Thus, the initial stage of culture preparation involves inoculating the pretreated milk with the Kefir grains, and then incubating the mixture at around 23 °C. After some 20 h, the grains are sieved out of the milk and carefully washed in cold water prior to re-use. The remaining milk provides the bulk starter for the commercial-scale fermentation, and it is added to the process milk at the rate of 3–5%. The final incubation at 23 °C will again last around 20 h, and, after cooling, the Kefir is often held for several hours to 'ripen'. This latter stage allows for maximum stability of the coagulum, and the final packaging stage must be designed in such a way that mechanical damage to the product is kept to a minimum. Thus, while Kefir is envisaged as a refreshing drink it has a quite distinct viscosity that is regarded by devotees as an essential feature of a good quality product.

Kumiss

Kumiss is a traditional drink of Central Asia, and although manufactured originally from mare's milk, its modern counterpart is based on cow's milk or skim-milk with added sucrose. After heating the basic milk for 5 min at 90–92 °C and cooling to 26–28 °C, the starter culture is added at a rate of 10–30%. The aim is to give the process milk an acidity of 0·45% lactic acid, and hence the inoculation rate is adjusted to this end. Incubation is then carried out to give an end-product with:

(1)   0·6% lactic acid and 0·7% alcohol;
(2)   0·8% lactic acid and 1·1–1·7% alcohol,
(3)   1·0% lactic acid and 1·7–2·5% alcohol,

with the final choice being geared to consumer demand.

On completion of the fermentation, the coagulum is cooled to 15–16 °C, and vigorously agitated to both provide a smooth consistency, and to lightly aerate the milk. This latter point is important in relation to yeast activity for, after bottling, the product is normally held at ambient temperature for a few hours to accumulate carbon dioxide. The result is a refreshing, effervescent drink with a clean, lactic flavour, and, as such, is widely consumed in Eastern Europe.

Although this brief review has covered the major types of fermented milk, numerous variations can be found around the world. Many of these products are manufactured on a basis of limited microbiological knowledge, and it is fortuitous that, although no two products can ever be handled/monitored in exactly the same way, the general principles remain immutable. The intelligent application of basic techniques can, therefore, go a long way towards avoiding serious problems in relation to starter production, and/or microbial contamination and spoilage of the end-products.

REFERENCES

ANON. (1972) *Swiss Food Act*. Chancelkrie Federal, Berne, Switzerland.
ANON. (1973) *New Zealand Food and Drug Act*. Government Building, Wellington, N.Z.
ANON. (1977) *The Influence of the Cooling Rate on the Quality of Stirred Yoghurt*, Report No. 225. Danish Dairy Research Institute, Hillerod, Denmark.
ANON. (1979) *The Oxoid Manual*, 4th edn. Oxoid Ltd, Basingstoke, UK.
APHA (1967) *Standard Methods for the Examination of Dairy Products*, Ed. Walter, W. G. American Public Health Association, Washington, DC, USA.
APHA (1972) *Standard Methods for the Examination of Dairy Products*, 13th edn, Ed. Hauser, W. J. American Public Health Association, Washington, DC, USA.

APHA (1978) *Standard Methods for the Examination of Dairy Products*, 14th edn, Ed. Marth, E. H. American Public Health Association, Washington, DC, USA.

ARNOTT, D. R., DUITSCHAEVER, C. L. and BULLOCK, D. H. (1974) *J. Milk Fd Technol.*, **37**, 11.

BANKS, W., DALGLEISH, D. G. and ROOK, J. A. F. (1981) In: *Dairy Microbiology*, Vol. 1, Ed. Robinson, R. K. Applied Science Publishers Ltd, London.

BAUTISTA, E. S., DAHIYA, R. S. and SPECK, M. L. (1966) *J. Dairy Res.*, **33**, 299.

BERLIN, P. J. (1962) *Ann. Bull. (Part IV A)*. International Dairy Federation, Bruxelles, Belgium.

BOTTAZZI, V. and BIANCHI, F. (1980) *J. Appl. Bacteriol.*, **48**, 265.

BOTTAZZI, V. and VESCOVO, M. (1969) *Netherlands Milk and Dairy J.*, **23**, 71.

BOTTAZZI, V., BATTISTOTTI, B. and VESCOVO, M. (1971) *Milchwissenschaft*, **26**, 214.

BOYLE, J. L. (1972) *Stabilisation of Cultured Milk Products by Alginates*. Alginate Industries, London.

BRACQUART, P., LORIENT, D. and ALAIS, C. (1978) *Milchwissenschaft*, **33**, 341.

BSI (1968) *BS-4285, Methods of Microbiological Examination for Dairy Purposes*. British Standards Institution, London.

BSI (1970) *Supplement No. 1 to BS-4285, Methods of Microbiological Examination of Milk Products*. British Standards Institution, London.

BSI (1977) *BS-5035, Recommendations for Sterilisation of Plant and Equipment Used in the Dairy Industry*. British Standards Institution, London.

BUNDGAARD, A. G., OLSON, D. J. and MADSEN, R. F. (1972) *Dairy Ind.*, **37**, 539.

CHARLISH, V. R. and WARMAN, K. G. (1980) In: *Hygiene, Design and Operation of Food Plant*, Ed. Jowitt, R. Ellis Horwood Ltd, West Sussex, UK.

DAVIS, J. G. (1967) *Dairy and Ice Cream Industries Directory*, 5. Union Trade Press Ltd, London.

DAVIS, J. G., ASHTON, T. R. and MCCASKILL, M. (1971) *Dairy Ind. Internat.*, **36**, 569.

ELLIKER, P. R., ANDERSON, A. N. and HANNESSON, G. (1950) *J. Dairy Sci.*, **39**, 1611.

FAO/WHO (1973) *Food Standards Programme*, CX 5/70, 16th Session, Rome, Italy.

GALESLOOT, TH. E. and HASSING, F. (1969) *Dairy Sci. Abstr.*, **31**, 63.

GALESLOOT, TH. E., HASSING, F. and VERINGA, H. A. (1968) *Netherlands Milk and Dairy J.*, **22**, 50.

GILBERT, R. J. and WIENEKE, A. A. (1973) In: *The Microbiological Safety of Food*, Eds. Hobbs, B. C. and Christian, J. H. B. Academic Press, London.

GILMOUR, A. and ROWE, M. (1981) In: *Dairy Microbiology*, Vol. 1, Ed. Robinson, R. K. Applied Science Publishers Ltd, London.

GREENE, V. W. and JEZESKI, J. J. (1957*a*) *J. Dairy Sci.*, **40**, 1046.

GREENE, V. W. and JEZESKI, J. J. (1957*b*) *J. Dairy Sci.*, **40**, 1053.

GREENE, V. W. and JEZESKI, J. J. (1957*c*) *J. Dairy Sci.*, **40**, 1056.

HOBBS, B. C. (1972) *J. Soc. Dairy Technol.*, **25**, 47.

IDF (1969) *IDF/47 Compositional Standards for Fermented Milks*. International Dairy Federation, Bruxelles, Belgium.

IDF (1971*a*) *IDF/65 Fermented Milks Count of Coliform*. International Dairy Federation, Bruxelles, Belgium.

IDF (1971*b*) *IDF/66 Fermented Milks Count of Microbial Contaminants*. International Dairy Federation, Bruxelles, Belgium.

IDF (1971*c*) *IDF/67 Fermented Milk Count of Yeasts and Moulds.* International Dairy Federation, Bruxelles, Belgium.

JAY, J. L. (1979) Personal Communication. Wincanton Engineering Ltd. Sherborne, UK.

JENNESS, R. and PATTON, S. (1959) *Principles of Dairy Chemistry.* John Wiley & Sons Inc., London.

KON, S. K. (1972) *FAO Nutrition Study No. 27.* Information Centre, FAO, Rome.

KROGER, M. (1978) *Cultured Dairy Products J.*, **13**, 26.

LANG, F. and LANG, A. (1973*a*) *Milk Ind.*, **73**, 17.

LANG, F. and LANG, A. (1973*b*) *Fd Manufacture*, **48**, 23.

LANG, F. and LANG, A. (1975) *Milk Ind.*, **77**, 4.

LEES, G. J. and JAGO, G. R. (1976) *Australian J. Dairy Technol.*, **24**, 181.

LING, E. R. (1963) *A Text Book of Dairy Chemistry*, 3rd edn. Chapman & Hall Ltd, London.

Lucca, I. (1975) *Il Latte*, **4**, 232.

METCHNIKOFF, E. (1910) *The Prolongation of Life.* Heinemann, London.

MOON, N. J. and REINBOLD, G. W. (1976) *J. Milk Fd Technol.*, **39**, 337.

NIELSEN, V. H. (1972) *Am. Dairy Rev.*, **34**, 26.

NORLING, A. (1979) *Cultured Dairy Products J.*, **14**, 24.

ORLA-JENSEN, S. (1931) *Dairy Bacteriology*, 2nd edn. J. & A. Churchill, London.

PETTE, J. W. and LOLKEMA, H. (1950) *Netherlands Milk and Dairy J.*, **4**, 197.

PIERSMA, H. and STEENBERGEN, A. E. (1973) *Officieel Orgaan FNZ*, **65**, 94.

RASIC, J. LJ. and KURMANN, J. A. (1978) *Yoghurt.* Technical Dairy Publishing House, Copenhagen, Denmark.

RETTGER, L. F. and CHEPLIN, H. A. (1921) *Intestinal Flora*, Ph.D. Thesis, Yale University, USA.

ROBINSON, R. K. (1977*a*) *S. African J. Dairy Technol.*, **9**, 59.

ROBINSON, R. K. (1977*b*) *Nutrition Bull.*, **4**, 191.

ROBINSON, R. K. (1980) *IXth Marschall Internat. Dairy Symp.*, London.

ROBINSON, R. K. and CADENA, M. A. (1978) *Ecol. Fd Nutrit.*, **7**, 131.

ROBINSON, R. K. and TAMIME, A. Y. (1975) *J. Soc. Dairy Technol.*, **28**, 149.

ROBINSON, R. K. and TAMIME, A. Y. (1976) *J. Soc. Dairy Technol.*, **29**, 147.

ROTHWELL, J. (1979) *Health and Hygiene*, **3**, 1.

ROUSSEAU, M. J. (1974) *Dairy Sci. Abstr.*, **36**, 570.

SELLARS, R. L. (1975) Personal Communication. Chr. Hansen's Lab. Inc., Wisconsin, USA.

SPECK, M. L. (1977) *J. Fd Protection*, **40**, 863.

STEENBERGEN, A. E. (1971) *Officieel Orgaan FNZ*, **63**, 996.

SWARTLING, P. and MUKHERJI, S. (1966) *XVIIth Internat. Dairy Congr.*, C2, 337.

TAMIME, A. Y. (1977) *Some Aspects of the Production of Yoghurt and Condensed Yoghurt*, Ph.D. Thesis, University of Reading, England.

TAMIME, A. Y. (1978) *Cultured Dairy Products J.*, **13**, 16.

TAMIME, A. Y. and DEETH, H. C. (1980) *J. Fd Protection*, **43**, 939.

TAMIME, A. Y. and GREIG, R. I. W. (1979) *Dairy Ind. Internat.*, **44**, 8.

TAMPLIN, T. C. (1980) In: *Hygiene, Design and Operation of Food Plant*, Ed. Jowitt, R. Ellis Horwood Ltd, West Sussex, UK.

ZALL, R. R. (1981) In: *Dairy Microbiology*, Vol. 1, Ed. Robinson, R. K. Applied Science Publishers Ltd, London.

7

Quality Control in the Dairy Industry

H. LÜCK

*Animal and Dairy Science Research Institute, Private Bag X2,
Irene, Republic of South Africa*

Modern dairy factories have very sanitary and stream-lined equipment turning out fluid milk, yoghurt, Cottage cheese, ice cream and other products in a highly integrated operation. At the dock area, several milk tankers can be unloaded at the same time and washed.

The raw milk is pumped into silos for further processing. These processes are controlled from a control panel, and the operator is assisted by 'malfunction controls' and other visual or audible alarms. The quality of the product manufactured can only be maintained by adhering to very strict hygienic requirements for production, transport and storage.

Quality, which embraces many factors, is difficult to define. The term 'quality' covers, for instance, chemical, physical, technological, economic, bacteriological and aesthetic characteristics, but quality can also be defined as that which the public likes best. Hence, quality is judged by subjective and objective tests. The results of subjective tests (sensory tests), such as odour, taste, texture (rheological properties) depend on the attitude of the investigator. Objective tests, which include recognised scientific tests, are used more and more in control laboratories. In this chapter, the bacteriological tests used for the quality control of milk and dairy products are considered.

Quality control includes a programme for testing plant hygiene and product quality. To measure plant hygiene, the suitability of the air supply to the processing rooms, and the bacteriological condition of the equipment (sanitation tests) can be checked. A focal point of the process area is the aseptic filling room. The dairy industry must continually conform to public health regulations, and this can only be achieved by regular quality assessments.

MICROBIOLOGICAL CONTROL OF THE AIR OF PROCESSING ROOMS

To ensure longer shelf-life and to prevent disease-causing and toxin-producing micro-organisms from getting to dairy products, every precaution should be taken to avoid air-borne contamination of the product during, and after, processing. The main sources of air-borne contamination are people and packaging materials. To protect the product from non-viable dust and air-borne microbial contamination, mechanical ventilation and air filtration is now considered essential even in the smallest dairy plants. In cheese factories, the problem of phage contamination can be reduced in this way.

General Processing Rooms

Ventilation

Good ventilation will remove the moisture released during processing. It will protect the finish of the building and equipment, and prevent condensation and subsequent mould growth on surfaces. Humidity in processing rooms should not exceed 95 % relative humidity. The ventilation necessary to cope with moisture and heat release in a dairy factory can be calculated (Vickers and McRobert, 1977). Ventilating air must be heated to about 18 °C during colder weather to ensure suitable conditions—for instance, during the manufacture of cheese, to prevent the curd from being seriously cooled. When no provision is made to heat the air, cheesemakers may switch off the ventilation and, consequently, mould will grow on walls and ceilings. Ventilation without air filtration increases the probability of air contamination considerably.

Air Filtration

The microbial flora of air consists mainly of *Bacillus* spp., yeasts, moulds and micrococci, but staphylococci and corynebacteria are also found.

Counts of air-borne micro-organisms at various product areas of dairy plants have ranged from $579\,m^{-3}$ to more than $4085\,m^{-3}$ ($16\cdot4\,ft^{-3}$ to more than $115\cdot7\,ft^{-3}$) for the standard plate count, and $297\,m^{-3}$ to $1088\,m^{-3}$ ($8\cdot4\,ft^{-3}$ to $30\cdot8\,ft^{-3}$) for yeast and moulds. The main sources of contamination were shown to be the ventilation system, air movement into the plant, floor drains and personnel (Hedrick and Heldman, 1969; Hedrick, 1975). To reduce the bacterial count, the air has to be filtered through high-efficiency filters which remove 90–99 % of the

1 μm, or larger, particles. The ultra-high-efficiency filters remove at least 99·9% of all particles 0·1–0·2 μm, or larger.

Today, air filters which even remove 99·999 98% of particles 0·01–2 μm in size can be obtained, and this means a removal of practically 100% of the micro-organisms in the air. The filters, consisting of microfibres of borosilicate, are resistant to relatively extreme conditions of steam, temperature and pressure changes. To extend the life of these air filters, a primary filter to remove all gross dirt (5–10 μm) should always be used in front of them.

TABLE I
PROPOSED TENTATIVE STANDARDS FOR THE BACTERIAL COUNT OF AIR IN VARIOUS PROCESSING AREAS

Product area	Plate count $m^{-3\,a}$		Yeast and mould count $m^{-3\,a}$	
	Good	Poor	Good	Poor
Milk, cream, cultured milk, Cottage cheese	150	1 500	50	1 200
Butter	300	2 000	100	800
Dried milk	250	1 500	100	1 200
Ripened cheeses	300	2 000	400	2 000

^{a} Count $m^{-3} = 35·3 \times$ count ft^{-3} (1 $ft^3 = 28$ litre $= 0·028 \, m^3$).

Despite the use of ventilation, with air filtration and disinfection, a relatively high air-borne count can be measured when there is a considerable movement of personnel, or a high staff density, and when packaging material is stored in the processing room.

Proposed tentative standards for air-borne counts are presented in Table I, but often air counts exceed 5000 m^{-3}. In milk powder factories, even 5 000 000 m^{-3} have been measured. Dalla (1974) determined the contamination of milk bottles by air-borne micro-organisms settling in the empty bottle between the washing machine and the bottling plant. The number ranged from 2–9 bacteria per bottle (bacterial content of the air: 4550 m^{-3}); these bacteria can significantly reduce the shelf-life of pasteurised milk.

Air Count Tests

A consistent relationship exists between the number of viable particles in air and the sedimentation of micro-organisms onto surfaces ($r = 0·797$; standard petri plates/5 min exposure; Cannon and Reddy, 1967). Air counts, by exposure of petri dishes with sterile agar medium for a specified period of time, have been used for many years. The very slow settling rate and the dependence of the results on air currents have led to the development of other methods. Samplers are commercially available that draw a specific volume of air, and deposit the micro-organisms onto the surface of an agar plate or onto a membrane filter. A simpler method consists of bubbling a specified volume of air into sterile water, and making a plate count of the water. These and other methods are described in a monograph published by the US Public Health Service (1959).

Aseptic Filling Lines
The microbiological conditions of the air in general processing rooms, as described in the previous paragraph, are also applicable for rooms with aseptic filling lines. Further precautions must be taken to prevent bacterial contamination. Rooms where dairy products are aseptically bottled or packed should be pressurised and used exclusively for this purpose. A separate room for each machine with a glass partition wall is needed. This allows personnel and visitors to study the operation of the machine without having to enter the room.

The walls and ceiling should be covered with a material that is easy to wash and disinfect. Doors should preferably be equipped with automatic opening/closing devices. Windows are closed during production, but allow good ventilation when the machines are not operating. Openings through walls for conveyors, etc., must be as small as possible to reduce the danger of air-borne infections. The handling of unsterile packaging material in the room has to be avoided.

The room must be provided with a sufficient number of drains which are easy to clean, and with a floor, sloped so that water can drain off without forming puddles. Since floor drains can cause air-borne contamination, sufficient disinfectant should be added to the trap water to destroy the micro-organisms therein; this is done 5 min before the drain is used in the morning.

There are different methods of preventing bacteria which adhere to shoes from contaminating aseptic filling rooms. Door mats moistened with a disinfectant have to be placed at all entrances so that shoes or boots can be wiped clean before entering the room. The mats are to be cleaned at least twice a week.

Washing facilities with hot and cold water, soap, disinfectant and disposable towels should be available. A fundamental rule is to wash the hands with disinfectant soap before every production run starts, and after every visit outside the room. Furthermore, the room should contain a cleaning vat (length approximately 2 m), with hot and cold water connections and feed cocks for disinfectants for cleaning and sanitising filling tubes, product valves, etc.

To ensure troublefree production in aseptic packaging lines, optimal ventilation is essential, and this system includes injection and evacuation posts for the circulation of filtered air (speed: not more than 0.2–$0.3 \, m \, s^{-1}$; over-pressure: 2–4 mm water column). The posts may not be installed within 2–3 m of the machine. To give a guideline: the volume of air (in m^3) which has to be changed five to eight times per hour is calculated by multiplying the floor area of the room (in m^2) by three.

Cheese Curing Rooms and Cheese Stores

Undesirable moulds are normally endemic in cheese factories. Contamination of cheese with these organisms from the atmosphere, walls or other surfaces can be a problem, because it can give rise to defects in the texture and taste of cheese. Since it is known that certain moulds can produce toxic substances, i.e. mycotoxins, mould growth on cheese must not only be considered an economic problem, but also a potential health hazard. The results of mould profile studies have indicated that the main contaminants of mouldy cheeses are penicillia (more than 80 % of the total isolates). *Aspergillus* species are less often found (up to 13 %) and other genera are relatively seldom present (up to 13 %) (Bullerman, 1976; Lück and Wehner, 1979). The reason is probably that penicillia are less dependent on moisture and grow better at low temperature than aspergilli and other fungi.

To combat the presence of mould in cheese curing rooms and cheese stores, close attention should be given to the quality of the air as well as to the presence of mould on surfaces, such as shelves, walls, etc. The use of purified, positive-pressured air in addition to other sanitising measures will help to control mould growth. Gassing with vaporised formaldehyde is very effective in destroying all forms of micro-organisms. After removing all cheese from the room, formalin (38 % formaldehyde solution) is diluted with water (1:1) and potassium permanganate is added (3–5 g of formaldehyde plus 3–5 g of permanganate per m^3 of space). (Vapourised formaldehyde does not destroy cheese mites which are usually killed by fumigating with methylbromide.)

Effective measures to control the air-borne mould count are as follows.

1. The temperature and relative humidity (RH) must be properly supervised. Low temperatures can easily lead to high RH values. RH values of 80 % or slightly lower help to reduce the mould growth. (This is not applicable to curing rooms for mould ripened cheese, where RH values of 95 % are desirable.)

2. Cheese should be coated with plastic emulsion before it comes in contact with mould or mould spores in the curing room. Treatment with plastic coats containing fungistats, such as sorbic acid, sorbate or natamycin (pimaricin) significantly delays mould growth, but does not prevent it completely (Lück and Cheesman, 1978).

3. The elementary rules of hygiene may not be neglected. The shelves after being removed to another room for cleaning must be disinfected, for wooden shelves saturated with whey provide an excellent medium for the growth of mould, and the subsequent contamination of the air. To combat undesirable mould growth they have to be scrubbed from time to time using the following substances: chlorine solution (sodium hypochlorite in trisodium phosphate solution (0·3 %) containing 400–1000 mg kg^{-1} available chlorine), formalin (10 %), quaternary ammonium compounds (800 mg kg^{-1}) or sodium or calcium propionate (10 %).

Other measures to control mould growth, such as ultra-violet light or treatment with ozone are not very effective.

The air should not contain more than 400 yeast and mould cells m^{-3} (Table I). As a routine test for estimating the mould count, although not very accurate, sterile standard petri dishes containing potato dextrose agar can be exposed for 15 min and incubated at 20–25 °C for up to 5 days. The conditions in the room are ideal when there are no mould colonies on these plates; a mould count of 10 is already relatively high.

Milaan and Pulles (1972) have described the construction and layout of air-conditioned cheese stores.

ASSESSMENT OF DAIRY EQUIPMENT HYGIENE

Every precaution should be taken to prevent contaminated equipment surfaces coming into contact with pasteurised products. The enumeration of bacteria on the surfaces of tanks, pipelines and other dairy equipment is important to evaluate the efficiency of cleaning, and to assess the hygienic conditions of a plant. Gram-negative rods, including coliforms, and milk souring streptococci are dominant on poorly cleansed surfaces, but as yet, no bacteriological techniques are available by means of which accurate surface counts can be obtained. Some sampling techniques have good repeatability† or reproducibility,‡ but a poor recovery of contaminants, while other techniques have a good recovery but are only suitable for a limited number of surface types. Different organisms adhere to the same surface with different tenacities. The same bacteria on different types of soil (fat, protein or mineral deposits) are not removed with the same ease. Grease films on surfaces especially influence the recovery of contaminants.

The following sampling techniques are used to assess bacterial contamination on surfaces.

1. Surface sampling methods, i.e. swab contact method, surface rinse method, adhesive tape method, and vacuum method to sample environmental surfaces.

2. Agar contact methods, i.e. replicate organisms direct agar contact (RODAC) plating, agar slice method or direct surface agar plating.

The surface sampling methods and the agar contact methods are widely used in the dairy industry, but these methods are applicable *only* to surfaces that have been previously cleaned. Baldock (1974) has compared the limitations and variability of the different methods.

Surface Sampling Methods
Swab Contact Method
The swab contact method involves rubbing a moistened sterile cotton swab over the test surface, and placing the swab in a rinse solution to be subsequently diluted and examined by the plate count method. Specified surface areas (for instance, two similar $50\,cm^2$ sites designated using thin flexible sheet metal or flexible plastic guide frames) are swabbed with the same swab. When using selective media for plating, the swab method can provide information not only on the number, but also on the type of bacteria on a surface. The method is adaptable

---

† *Repeatability* is the closeness of agreement between successive results obtained with the same method on identical test material and under the same conditions.
‡ *Reproducibility* is the closeness of agreement on identical test materials but under different conditions, such as different operators, different laboratories and/or different times (as defined in Sections 3.18 and 3.19 of ISO/R645-1967).

to a variety of surfaces, but recovery of the surface bacteria is moderate (52–90%, Green and Herman, 1961).

To simplify the bacteriological testing of dairy operations, the swab method was combined with a simple membrane filter technique, (Kristensen, 1977): 1 ml of the swab rinse solution is sucked up by a membrane filter (pore diameter 0·45 μm) containing an absorption layer with the selective medium (swab test kit). After incubation, the colonies on the membrane are counted. A modification of the swab method is the use of calcium alginate swabs to replace the cotton swab. The alginate swab dissolves in the rinse solution and all bacteria entrapped in the swab are freed. This modification has, however, not really improved the performance of the swab method.

Surface Rinse Method

The surface rinse method involves the dislodging of surface contaminants by agitating a sterile fluid over the entire surface, and subsequently enumerating the micro-organisms in the fluid. This method is a suitable technique for the examination of milk bottles, and was subsequently also adopted for milk cans, farm dairy equipment and pipeline milking plants. It is not applicable to surfaces of large equipment. A bibliography of the literature on the rinse method was compiled by Thomas and Thomas (1977). A standard amount, such as 20 ml for milk bottles or 500 ml for 40-litre cans, of sterile quarter-strength Ringer's solution is poured into the container, which is then rolled or shaken a standard number of times, allowed to soak for a standard time, and then rolled again. The colony count of the rinse is determined on nutrient agar (usually at 30°C for 72 h). The rinse method has a moderate recovery (approximately 70%), and fair to good repeatability.

Adhesive Tape Method

Strips of transparent adhesive tape, such as 'Sellotape', free of antimicrobial substances, are pressed for a few seconds onto the surface under examination taking care to avoid air pockets. The strips are then removed and pressed, briefly, onto the surface of the agar medium of prepoured and dried plates. After incubation of the plates, the colonies are counted. The exposed tape can also be stained and viewed, microscopically, but the latter method has the disadvantage of not distinguishing between dead and living cells. The recovery of surface bacteria by the adhesive tape method was only 8–22% (Tamminga and Kampelmacher, 1977).

Vacuum Method

Because of its complexity, the vacuum method has so far hardly been used in the dairy industry. Particles on the surface are removed by vacuum and captured on an agar surface, on a membrane filter, or in a liquid medium. The recovery of the bacteria present is comparable to that of the swab method.

Agar Contact Methods

RODAC Plate Count

The RODAC plate count test is being used more and more as a surface sampling technique for the determination of the number and type of bacteria on surfaces. Plastic plates ($\pm 25 \, cm^2$), which are available commercially, are filled with enough molten agar medium to allow a convex surface to form. When it solidifies, the agar surface is pressed onto the test area, removed, covered with the lid and incubated. The number of colonies developing are counted. The RODAC test does not work on heavily contaminated surfaces, and cannot distinguish between single surface bacteria and clusters of micro-organisms. The maximum possible accurate count is limited to 200 colonies per plate. Furthermore the surface must be smooth, dry and flat or only slightly rounded. Approximately 80% of the bacteria can be recovered. The repeatability is excellent.

Agar Slice Method

What was said about the RODAC test is also applicable to the agar slice method (Walter, 1955). A sterile syringe is filled with molten agar medium. The solidified agar is pushed out to make contact with the test surface. The portion of the agar having made contact is sliced off with a sterile knife and placed in a petri dish for subsequent incubation. The syringe can be replaced by an artificial sausage casing (Cate, 1965). The recovery of the surface bacteria by this agar sausage method is 70–75% (Tamminga and Kampelmacher, 1977).

Direct Surface Agar Plating

The method of direct surface agar plating involves pouring a molten agar medium onto the surface. Due to the direct contact of the surface with the medium during incubation, the recovery of the contaminants on non-porous surfaces is high (approximately 80%). The drawbacks of the method lie in the difficulties encountered in incubating the surface of pipes, tanks, etc.

TABLE II
PROPOSED TENTATIVE STANDARDS FOR THE SURFACE COUNT OF DAIRY EQUIPMENT

Grade	Total bacterial count per $100\,cm^{2}$ [a]	
	For storage of raw products	For storage of pasteurised products
Good	<200	<10
Satisfactory	200–<1 000	10–50
Unsatisfactory	1 000–10 000	>50
Heavily contaminated	>10 000	

[a] Count $ft^{-2} = 9\cdot29 \times$ count per $100\,cm^{2}$; count per $100\,cm^{2} = 0\cdot108 \times$ count ft^{-2}.

Conclusion and Proposed Standards
No one surface sampling technique gives an accurate picture of the surface count. The rinse method is the most accurate for enumerating micro-organisms. The direct surface agar plating technique is the most accurate for counting clumps of viable micro-organisms, provided the contamination is not too great. The convenience of the agar contact method will often be the dominant factor in selecting this technique provided the surface is flat.

The bacterial standards for adequately cleaned surfaces of dairy plants vary from country to country, but to give the reader an idea, some proposed norms are presented in Table II. Dairy equipment is well cleaned when the bacterial surface count is less than 200 per $100\,cm^{2}$. In many cases a surface count of 1000 per $100\,cm^{2}$ is still acceptable. For the surface rinse method, satisfactory colony count standards are based on not more than one viable bacterium per ml capacity of the container, i.e. less than 1000 per litre milk bottle.

These standards are much too lenient for surfaces which come in contact with pasteurised milk products. Even 10 active bacteria per $100\,cm^{2}$ or 10 bacteria per 100 ml can considerably decrease the shelf-life of milk products, and thus is much too high. The shelf-life can only be improved through elimination of post-pasteurisation contamination (see section on Shelf-Life Tests).

ASSESSMENT OF THE HYGIENE OF PACKAGING MATERIAL

The influence of packaging on the contamination of dairy products may be direct, due to the presence of micro-organisms on the material, or indirect, due to the permeability of the material to bacteria. The high temperatures used in the fabrication of plastic articles cause most of them to be sterile at the time of manufacture. The maintenance of hygiene during handling and storage is, therefore, important. In tests, more than 90% of deep-drawn plastic containers which were sealed immediately after discharge were found to be sterile. The same was found for plastic bottles taken from a blow-moulding machine under aseptic conditions (sterile air, immediate closure). Plastic cups manufactured by means of the extrusion technique from granulated plastic at a temperature of 120–220 °C had surface counts of 2–8 per 100 cm² (Voss and Moltzen, 1973). Bacterial counts of plastic and laminated plastic packaging materials were usually not higher than 10 per 100 cm² (Lubiebiecki-V. Schelhorn, 1973).

The microbiological condition of packaging materials should be in keeping with that of the products packed in them. A total bacterial count of not more than 10 per 100 cm² or 10 per 100 ml capacity, and a coliform count of 0 per 100 ml or 100 cm² are proposed. The surface of containers used for aseptic packaging of sterile products must, however, be sterile. This can be achieved by the application of high temperatures during manufacture of the pack, or by passing through a bath of H_2O_2 (15 to 35 % plus wetting agent).

The surface count of non-absorbent packing materials based on paper, cardboard, plastics, aluminium foil, etc., can be determined by the direct surface agar plating method. A specified area of the packing material is aseptically placed on the solid agar medium of a petri dish, and overlaid with the same medium. After incubation, the colonies on both sides of the material can be counted. The use of selective media allows the counting of coliforms, yeasts and moulds, or other organisms.

The efficiency of heat-sealing and the susceptibility to 'microbiological pin-holing', particularly after handling and transporting, of aseptically packaged sterile dairy products, can be assessed by immersing the packs for 24 h in a chilled suspension of *Bacillus polymyxa* (Dicker and Wiles, 1974). The slight vacuum induced in the pack facilitates the penetration of the bacteria which cause spoilage after incubation (identified microscopically).

SAMPLING OF PRODUCTS FOR MICROBIOLOGICAL TESTS

Methods of Sampling
Sampling for bacteriological purposes must be undertaken only by experienced persons, because aseptic precautions have to be taken.

To avoid confusion and distrust, it is imperative that the sample drawn gives a true reflection of the compositional and bacteriological quality of the product from which it has been selected. The general administrative requirement, e.g. labelling of samples, replicate samples, test report, and the technical instructions, e.g. sterilising of equipment and sample containers, are described in national and international standards (see International Dairy Federation, 1969; American Public Health Association, 1978).

Sampling Raw Milk

Different methods of taking samples are practised. For instance, a stainless steel dipper, carried in a tube containing a disinfectant solution, may be used, and the empty dipper is immersed twice in the milk before the sample is taken. After sampling, the dipper is washed and replaced in the disinfectant solution. Another device consists of a metal rod with a stationary clamp holding the sample bottle, and the lower half of the rod and the open sample bottle are immersed in a disinfectant solution (for at least 5 min) before use. This solution can contain chlorine ($100 \, mg \, kg^{-1}$), or an idophor solution (active iodine concentration 50–$60 \, mg \, kg^{-1}$). Before taking the sample, the milk must be agitated for 5 min when the tank capacity is 500 to 4000 litre, and for at least 10 min when the quantity of milk exceeds 4000 litre.

The collection of a representative sample from road tankers may present problems. Proportionate sampling during the process of unloading, by means of drip samples or by automatic sampling devices, is recommended. If the sample has, however, to be taken before the tanker is unloaded, an agitating device is lowered into the manhole and the milk is stirred for at least 10 min. Another method is to circulate the milk for at least 15 min, and in this case, the pick-up hose of the tanker is connected to a clean pipe in the manhole, allowing the milk to be sucked in and pumped out again. Milk from large, factory storage tanks must be agitated (mechanically or by using air) for 30–60 min. The distribution of micro-organisms in raw milk closely follows that of the fat, and is widely independent of the total bacterial count, at least within the range of 10^4–10^7 bacteria ml^{-1} (Reuter and Quente, 1977).

In-Line Sampling

This requires the installation of stainless steel nipples fitted with sterile, rubber serum caps. The samples are collected with syringes inserted through the rubber caps.

Automatic Sampling Devices

If the milk stored in refrigerated farm tanks is not properly mixed, the proportionate sampling by means of automatic devices minimises the effect of inadequate mixing. Nevertheless, the milk should be agitated while being pumped into the tanker.

When proportionate milk samples collected by automatic sampling devices are used for bacteriological tests, the following conditions should be fulfilled (Lück *et al.*, 1978*b*).

1. The relative carry-over effect (ratio of the volume of the residual milk from previous delivery remaining in the system up to the sampling point to the volume of the present milk intake) should not exceed 0·2 %.
2. The variation in the bacterial count of two successive deliveries should not be larger than 100:1.
3. During the summer, the period of time elapsing between two milk intakes should not be longer than $\frac{1}{2}$–1 h to prevent the multiplication of bacteria in the residual milk.

When these precautions are not taken, automatic sampling devices are not suitable for collecting samples for bacteriological tests, unless the residual milk is removed by allowing a certain quantity of new milk (10 litre) to run through the system before sampling commences.

Storage and Transport of Samples

For cheese, butter, liquid non-sterile products, yoghurt or custard, insulated transport containers capable of maintaining a temperature of 0–5 °C should be used. For shelf-stable products (UHT products, sterilised milk, evaporated milk, sweetened condensed milk and milk powder or other dried milk products) the transport and storage temperature of the samples must not exceed 25 °C. For ice cream, the temperature must be − 15 °C, or lower.

Samples must be transported to the laboratory as quickly as possible, and should reach the laboratory within 24 h. In several countries, testing must be completed within 36 h. Under special circumstances, a maximum period of time between sampling and testing of 48 h may be acceptable. When the period of time between sampling and testing is longer than 24 h, the sample must be stored at between 0 and 2 °C.

In many countries, the addition of preservatives to samples used for bacteriological tests is not permitted. In other countries, milk samples taken at farms or factories may be preserved for the determination of the bacterial count and the leucocyte count. For this, a readily soluble, freeze-dried preparation of 1·18 ml of an aqueous solution containing 5% *ortho*-boric acid, 0·75% potassium sorbate, 1% glycerol and small amounts of methylene blue as marker is added to 10 ml of milk (Heeschen *et al.*, 1969). It is claimed that, in circumstances representative of those involved in the delivery of a sample to a testing laboratory (24 h at between 4 and 20 °C followed by 24 h at 4–6 °C), the bacterial count agrees closely with the initial count, and that the somatic cell count remains unchanged. The effectiveness of other preservative mixtures has also been tested. Aqueous solutions consisting of 1% boric acid and 0·2% sodium azide are suitable for maintaining the bacterial count almost unchanged for a period of 27 h at room temperature followed by 24 h at 6 °C (Leesment, 1971). Preservatives may not be used when dye reduction tests, i.e. methylene blue test or resazurin test, are carried out to determine the bacteriological quality of milk.

Usually sterile sample containers are required, but the use of non-sterile containers may, however, be permitted when, in comparison with the bacterial count of the sample, the number of viable bacteria of the container is low (not exceeding 1 bacteria per ml of capacity).

Preparation of Samples and Proposed Tests
Prior to mixing and opening, the container should be examined for signs of deterioration or defects, such as blowing, leakage, faulty closure, etc. Liquid milk is mixed by inverting the container 25 times within about 12 s. If the headspace is insufficient for proper mixing, the contents are aseptically poured into a larger sterile container for mixing. Samples of liquid cream are warmed to about 45 °C immediately before mixing and preparing the dilutions. Thick, coagulated cream is stirred with a sterile spatula and then weighed into a dilution bottle, and a 2% w/v solution of sodium citrate (warmed to 45 °C) is added. Butter and butter products, after being aseptically transferred to a sample container, are melted by standing in a water bath at 45 °C. For the preparation of the dilutions, the melted product is properly mixed, and a 10 g aliquot is diluted with 90 ml of 0·1% peptone water containing 0·1% w/v agar at 45 °C. The diluent for further dilutions is also held at 45 °C.

Surface layers of cheese should only be sampled in order to investigate the surface flora, but otherwise, the surface and/or other abnormal areas of cheese should be avoided. As an emulsifying agent for the preparation of a cheese suspension, a 2% w/v sterile solution of sodium citrate is used in a blender. For dissolving dried dairy products, a 1·25% sterile sodium citrate solution is recommended.

Before opening, cans and tubes must be washed thoroughly with warm water and soap. The part to be opened (lid, cap) is dipped into 95% v/v ethanol, ignited, and then opened using a sterile can-opener or scissors. The screw-caps of sample containers, the tops of retail bottles, cartons or cheeses wrapped in film are cleaned outside, and then treated with 70% ethanol and allowed to dry. The closure is removed aseptically From cartons, one of the top corners is cut using sterile scissors. A sterile knife is used to transfer cheese samples to a sample bottle.

The proposed bacteriological tests to be carried out on different dairy products are summarised in Table III.

Numerical Selection of Samples

In many situations, seller and buyer come to an agreement as to what the quality of the product should be. It is then customary to devise a scheme whereby the product will be tested from time to time to check whether it is of the agreed quality. The test on any one sample may suggest that the product is better or worse than it actually is. Hence, an element of risk is introduced, for only 100% sampling will give 100% certainty. In practice this is not possible, and for normal control purposes, the seller and the buyer accept a certain range of error, e.g. 5 and 10%, respectively. This means that, for legal purposes, more units have to be sampled and tested than for normal quality control. Most milk-producing countries have a dairy policy that makes provision for quality payments on raw milk supplied to dairy factories, and such a scheme will cover both compositional and bacteriological quality. Usually, three samples of milk are tested per month, per supplier. This is too few to obtain an accurate estimate of mean monthly quality, because the day-to-day variations in compositional and bacteriological quality, both within and between herds, are relatively large. It must, therefore, be accepted that a farmer can be overpaid one month and underpaid another, but the situation should even out after a period of several months (Lück *et al.*, 1975).

When the quality of dairy products is tested, the number of 'units'

TABLE III

PROPOSED TESTS FOR DETERMINING THE BACTERIOLOGICAL QUALITY OF DIFFERENT DAIRY PRODUCTS

Sample	Sample size	When carried out	Test or count / Type of test
Raw milk	Not less than 200 ml or unopened retail container	On receipt (held at 0–5°C until tested, but not longer than 24 h) and/or after storage at specified temperatures, e.g. 7°C When testing for antibiotics maintain sample at −30 to −15°C to minimise inactivation of penicillin	Total bacteria Direct microscopic count or plate count Coliforms (*E. coli*) Psychrotrophs Proteolytic or lipolytic organisms Thermoduric bacteria Spores (*B. cereus*) Dye reduction (e.g. methylene blue, resazurin test) Antibiotics Pathogens (if required) Shelf-life
Cream (raw or pasteurised)	See 'milk'	See 'milk'	See 'milk'
Cultured dairy products, e.g. yoghurts, cultured milks, sour creams	Unopened consumer pack to give a minimum sample of 100 g (bulk containers 'are-sampled in the field)	On receipt (held at 0–5°C until tested, but not longer than 24 h)	Direct microscopic examination of the types of bacteria in reference to the culture expected Coliforms *E. coli* Psychrotrophs Yeasts and moulds Staphylococci (coagulase-positive) Viable culture organisms pH Shelf-life

Product	Sample	When tested	Tests
Sweetened condensed milk	Not less than 200 g or unopened retail container	One set: on receipt Another set: after 1 month at 25°C	Total bacteria (plate count or direct microscopic count) Yeasts (low level contamination) Moulds (microscopic examination of lumps for mould mycelia) Micrococci Staphylococci (coagulase-positive) Thermoduric organisms
Canned dairy products (sterile), e.g. evaporated milk, flavoured milk, cream in cans	Intact retail can	One set: on receipt Another set: after 1 month at 25°C	Total bacteria (plate count or direct microscopic count) Anaerobic spores Sterility test pH
UHT milk UHT cream UHT products Sterilised milk	Two or three sets of containers (A and B or A, B and C)	A: on receipt B: after 5–7 days at 35–37°C *or* A: on receipt B: after 7 days at 33°C C: after 7 days at 55°C	Direct microscopic examination (if suspected of being non-sterile) Sterility test
Cheese Cheese spreads	50–200 g or small wrapped domestic portion	On receipt (held at 0–5°C until tested)	Coliforms *E. coli*

(continued)

TABLE III—*contd.*

Sample	Sample size	When carried out	Test or count Type of test
Cheese powders			Yeasts and moulds Staphylococci (coagulase-positive) Contaminating organisms Other pathogens (as required)
Butter Butter products	50–200 g or small packs to yield at least 50 g	On receipt (held at 0–5°C until tested)	Total bacteria (plate count) Contaminating organisms Yeasts and moulds Lipolytic bacteria Proteolytic bacteria Coliforms (*E. coli*) Psychrotrophs Staphylococci (coagulase-positive)

Product	Sample	Condition	Tests
Dried milk Dried dairy products, i.e. buttermilk powder, whey powder, ice cream mix powder	50–500 g	On receipt (held at 0–5°C until tested)	Total bacteria (direct microscopic count and plate count) Coliforms Thermophilic bacteria Spores Staphylococci (coagulase-positive) Salmonellae Antibiotics
Casein Caseinates Coprecipitates	50–500 g	On receipt (held at temperatures below 25°C until tested)	Total bacteria (plate count) Coliforms Thermophilic bacteria Yeasts and moulds Staphylococci (coagulase-positive) Salmonellae
Ice cream Frozen milk products	At least 50 g	Stored at a temperature not exceeding −15°C until tested	Total bacteria (plate count) Psychrotrophs Coliforms *E. coli* Yeasts and moulds Staphylococci (coagulase-positive) Salmonellae

to be sampled from a bulk consignment or 'batch' depends on the size of the 'unit' (large containers or small retail units), and the purpose of the test (determination of the general quality, or the detection of isolated failures). A 'batch' is defined as a fairly uniform quantity (tank, churn) of the product from which the 'units' are made; a 'unit' is a single item of the product in the form in which it is sold. The batch is accepted or rejected according to the sampling plan.

Tables are available which show how sample taking should be carried out in order to obtain a statistically certain basis for the assessment of quality, i.e. the number of containers (units) that should be taken out from a bulk consignment (batch) of a certain size, and because of the large variation in microbiological properties, more units (minimum number of units) have to be sampled to determine the bacteriological quality than to determine chemical properties. The minimum number of defective units per 100 which can be detected by sampling different numbers of units from a consignment is as follows.

No. of units sampled	*Level of detectable contamination*
10	37% or more
20	21% or more
30	16% or more
50	10% or more
100	4% or more

In other words, when 30 milk bottles from a batch of sterilised milk are collected and tested for sterility, there is no guarantee that a non-sterile bottle will be detected when the percentage of contaminated bottles is less than 16%. To give a guideline for the general control of the bacteriological quality of a bulk consignment, the following number of samples have to be taken:

Condensed milk, evaporated milk: at least 20 samples per batch

Dried milk, dried milk products: at least 10 samples per batch

The minimum number of samples to be taken from other dairy products is similar. The number of units sampled is usually related to the size of the batch; 0·1–1% of the filled containers are sampled. Theoretical considerations have, however, revealed that control of quality for a product with a low level of defect, e.g. bacterial contamination of sterilised products, depends upon sample size, and not upon the size of the batch (Bockelmann, 1974). A small defect number leads to biased binomial distributions, and hence, the Poisson distribution should be used. This means that a sampling scheme should be based on a fixed number of samples, and not on percentage sampling, for the certainty of the evaluation is independent of the total volume of production, and depends only on the size of the sample.

For the testing of sterility when non-sterility seldom appears, the so-called 'average sample' (taken continuously during production) hardly gives more information than a sample of the same size which has been taken out at random, i.e. not continuously. It must, however, be borne in mind that more than 1000 units per batch will have to be tested in order to ensure that the percentage of non-sterile units is not larger than the accepted standard of say, one in 1000; because of the destructive nature of the tests and the production costs involved, this approach to sampling is unrealistic. It is, therefore, recommended that approximately 50 to 100 packages of UHT product per machine of each production (day) be tested. When faults occur, the number of units will have to be increased.

METHODS OF DETERMINING THE TOTAL BACTERIAL LOAD OF DAIRY PRODUCTS

The bacteriological load is a function of the species of bacteria present as well as the number. A programme for controlling the total bacterial load of milk and dairy products might include metabolic activity tests, tests involving the enumeration of bacteria, or tests for the presence or absence of bacteria. The standard procedures for determining bacteriological quality of food products are based on the counting of bacteria. The procedure of determining the actual bacterial count is, however, time consuming and expensive. For this reason, metabolic activity tests, such as dye reduction tests, nitrate reduction tests, or tests for the determination of different metabolic products, have been selected as quality tests in the dairy industry. These indirect tests are dependent on the relative metabolic rates of the various micro-organisms present.

Metabolic Activity Tests
Dye Reduction Tests
These are still applied in many countries. The principle of these tests is to add dyes, such as methylene blue, resazurin or triphenyl-tetrazolium chloride, to milk or liquid dairy products, and to measure

the colour change after incubation. The colour change is based on the dehydrogenase activity of the bacteria present. Dehydrogenases, i.e. mainly flavine enzymes, transfer hydrogen from a substrate to biological acceptors or to the dyes added. The period of time needed to change or to decolourise the dye is an index of the bacterial load of the product.

Dye reduction tests are, however, of little value as an index of the bacterial count of refrigerated milk, because this relationship is poorly correlated ($r = -0.36$ to -0.62; Lück et al., 1970a; Lück, 1972). The reason is that most of the bacteria of refrigerated milk are in a dormant state. Furthermore, a relatively large proportion of the bacteria present are psychrotrophs. These micro-organisms have, compared with lactic acid bacteria, a low dehydrogenase activity, a characteristic which contributes to the low correlation between bacterial count and methylene blue reduction time or resazurin disc reading. In order to achieve the same reliability of, say the methylene blue test for non-refrigerated milk, approximately twice as many samples of refrigerated milk have to be tested (Lück and Andrew, 1975). Pre-incubation (13–18 °C for 16–24 h) has been shown to be unsuccessful in improving the relationship between bacterial count before incubation, and the results of a metabolic activity test after pre-incubation. Similar results were obtained with the *nitrate reductase test*, which is only suitable as a method of detecting samples with a content of coliforms, psychrotrophs and other polluting bacteria in excess of acceptable standards (Lück and Dunkeld, 1972).

Because of these problems, the suitability of other metabolic activity tests was investigated. The break-down process which takes place in milk products can be classified as glycolysis, proteolysis and lipolysis (according to the three main groups of nutrients), and often the heat-resistant enzymes, which can survive processing, are produced by bacteria. The conclusion is, therefore, drawn—in opposition to other opinions—that the bacterial count does not adequately reflect bacteriological quality, and that a measurement of the general catabolic activity of the bacteria present would give a better index of the total bacterial load than the colony count.

The key substance of most microbial metabolic processes is pyruvate. Thus, the pyruvate content of milk products is a function of the types, numbers and activity of the micro-organisms which are or were present. Tolle et al. (1972) have carried out intensive investigations on the *pyruvate test* for estimating bacteriological quality. Pasteurisation has no effect on the pyruvate content, but it can, so it seems, be affected by peroxides. The natural pyruvate content (of non-bacterial origin) of raw milk is approximately 0.5 mg kg^{-1}, which would correspond to a plate count of 50 000 to 100 000 ml^{-1}.

What is applicable to pyruvate is also applicable to adenosine triphosphate (ATP), which serves as a universal energy carrier in microbial glycolysis, proteolysis and lipolysis. Therefore, the *ATP content* of bacterial cells, which is fairly constant, has been used as an index of bacterial load of milk (Howard and Westhoff, 1974). Certain limitations were, however, found in the application of this method. The presence of micrococci or yeasts gave erroneously high results, and due to thermal lysis, bacteria can give a higher reading after heating than before. Furthermore, heat-killed bacteria can also contain ATP (Williams, 1971).

Another interesting metabolic activity test is the *oxygen (O_2) test* carried out by means of an O_2 electrode. The magnitude of the decrease in O_2 content increases in relation to the initial bacterial count. However, due to the different procedures of handling milk and liquid dairy products, such as different time lapses until testing, and/or the time and speed of stirring, a direct measure of the O_2 content does not give a clear indication of the bacterial count. The samples have to be saturated with air (e.g. by shaking) followed by a specified period of incubation to obtain a close relationship between O_2 content and bacterial count. The multiple regression relationship between O_2 content and bacterial counts was consistent enough ($r = 0.70$) to recommend the 6 h O_2 test as an index of the total mesophilic and psychrotrophic count (Lück *et al.*, 1970*b*).

In conclusion, it may be stated that metabolic activity tests give an indication of the relative metabolic rates of the various organisms, while the bacterial count gives information about the effectiveness of production hygiene. Metabolic reactions caused by bacteria can be measured when the bacterial count exceeds 50 000–100 000 ml^{-1}, but they are difficult to measure in high quality products.

Pre-incubation of the samples will increase the activity of the bacteria present, but this increase is, however, not very consistently related to the initial bacterial count. When the bacteriological quality of the product has reached a high level, the counting of bacteria or the determination of specific groups of bacteria (psychrotrophs, coliforms, thermoduric bacteria, etc.) is recommended in preference to metabolic activity tests.

Tests for the Enumeration of Total Bacteria

Direct Microscopic Count

The microscopic examination of a sample provides a rapid indication of the quality of liquid milk products and, at the same time, supplies a permanent record of the quality, but it does impose considerable eye-strain on the operator. A small amount (0·01 ml) of the sample is spread over an outlined area of 1 cm^2 of a fat-free slide using a metal rake. After fixing and staining, the smear is placed under the oil immersion lens of a microscope of which the microscopic factor (MF) is known:

$$MF = \frac{10\,000 \times 4}{31\,416d^2} = \frac{1273}{d^2}$$

where d = diameter of the microscopic field in mm.

The average number of organisms per field multiplied by the MF, yields the number of organisms per ml of the milk product or the dilution thereof. The number of fields to be counted depends on the average number of organisms per field, for instance:

No. of organisms per field	*No of fields to be counted*
0–3	64
4–6	32
7–12	16
13–25	8

By the microscopic method, *clump counts* or *individual counts* can be determined. All bacterial cells which are not further than 5 μm from another are considered to belong to one clump. A single cell which is further than 5 μm from another cell is also a 'clump'. When determining the individual count, all bacterial cells within clumps or in isolation are counted. Dead bacteria can also be stained. However, bacteria killed by heat usually disintegrate soon after, or lose their stainability. In addition, there are stain solutions which permit the recognition of dead cells, and then the microscopic count can be applied to heated milk, or reconstituted samples of powdered milk, to furnish information concerning the past history of the product.

The microscopic method can also be used for counting somatic cells. For milk from individual quarters of cows in normal lactation, a *somatic cell count* of 500 000 cells ml^{-1} has been accepted. For bulk milk, cell counts ranging from less than 300 000 to 1 000 000 ml^{-1} were considered acceptable in various countries (International Dairy Federation,

1967). The recommended methods for somatic cell counting, including the microscopic cell count, have been published by the International Dairy Federation (1979).

Colony Count

Pour plate method. The plate count or pour plate method is often used for estimating the number of viable units of micro-organisms in liquid, reconstituted or suspended dairy products. Due to the wide range of bacterial counts occurring in dairy products, bacteria can often only be counted after diluting the liquid. A number of 10-fold serial dilutions of the sample are prepared. The dilutions are made either by

(1)   pipetting, or
(2)   using the loop method.

A small amount of the dilution is mixed with the liquefied sterile agar medium in a sterile petri dish. After solidification of the agar, the petri dishes are incubated at a specific temperature and for a suitable period of time. The bacterial cells grow to recognisable colonies which can be counted. Plates with 30–300 colonies are selected for counting. To obtain the plate count per ml or g of product, the number of colonies per plate is multiplied by the dilution factor, stating the medium and the conditions of incubation.

Mechanisation of the pour plate system. In the literature, several automatic methods, based on the plate loop method of Thompson *et al.* (1960), have been described (Richard and Auclair, 1969; Grappin and Jeunet, 1974). The results obtained indicate that the loop method requires a completely standardised mode of working, i.e. standardisation of the angle at which the loop is removed from the sample, the depth of immersion, and the speed of movement of the loop. When these precautions are taken, the fear that the very small amount (1 μl) of the sample may not be representative of the product tested is hardly justified, except perhaps for the 0·1 μl hook. To cover a wider range of bacterial counts, instruments were developed where two loops (1 and 10 μl or 0·1 and 1 μl) are dipped into the sample, and the contents are flushed with 1–2 ml Ringer's solution into two different petri dishes. For milk with bacterial counts not exceeding 200 000 ml^{-1}, the process of diluting is not needed when the loop method is used.

Since apparatuses for the mechanical preparation of plates have been designed, the plate count test has been adopted as a routine method in the dairy industry. The evaluation of a prototype automated plating instrument based on the loop method (used throughout by two operators— approximately 300 samples h^{-1}) yielded a variance of differences between duplicate tests of 0·0025 ($\log_{10}$ transformed counts), and the accuracy, expressed as a variance of differences between results obtained by the mechanical and the manual method, was 0·007 (Fleming and O'Connor, 1975). With bacterial counts of less than 200 000 ml^{-1}, the plate loop count approximated the standard plate count very closely, but with higher counts, the agreement was, however, not so close. Wright *et al.* (1970) used the following formula to predict the standard plate count (SPC) from the plate loop count (PLC)

$$SPC = (PLC)^{1·04}$$

Automatic colony counters designed to count colonies on agar plates have been developed by several manufacturers, and the usefulness of these colony counters was confirmed by several authors (Jeunet *et al.*, 1973; Wernery *et al.*, 1973; Fleming and O'Connor, 1975; Fruin and Clark, 1977). Most of these colony counters consist of a television camera to detect the colonies on the illuminated petri dish, and a small electronic computer for detection, counting and control (150 plates h^{-1}). The counting of impurities, lumps, undissolved parts in the media or air bubbles can be avoided, as they only occur due to the carelessness of the personnel. More serious is the problem of counting pinpoint colonies, which are sometimes formed by thermoduric bacteria. Colonies smaller than 0·20 mm are usually not counted, while the human eye can see and count colonies exceeding 0·10 mm. Another disadvantage is that, so far, automatic counters cannot distinguish colours or coloured colonies. Colonies on media that absorb light, e.g. blood agar, are difficult to count, while diffuse media, e.g. agar containing milk, also cause difficulties because the contrast between colonies and background is too limited. When the petri dishes known to be difficult to count by an automatic colony counter are eliminated, the counts obtained are as accurate as those obtained manually.

Another approach to counting colonies is *electronic microcolony counting* using particle counters (coulter principle) (Tolle *et al.*, 1968). The microcolonies grow in a solid nutrient medium containing gelatine (for 20 h at 20–22 °C) which is melted before counting. This method can also be applied to bacteria grown on selective media.

Roll tube method. For special determinations of the bacterial count, the roll tube method may be useful, because less culture medium is needed for this method than for the plate count. Universal bottles of the sterile, liquid agar medium are inoculated with 10-fold serial dilutions of the sample, and the bottles are then rolled mechanically, with simultaneous cooling, until an even solid film on the inner surface of the bottles is formed. After the specified incubation period, only those which contain 20 to 200 visible colonies are counted. The roll tube method has also been mechanised (Jaartsveld and Swinkels, 1974; Posthumus *et al.*, 1974).

Surface count. Surface colonies grow faster and can be counted after 24 h. Where it is desired to produce surface colonies, the spread method or the drop method can be applied. In the *spread method*, 0·1 ml of the 10-fold dilutions are transferred to, and spread over, the dry surface of a solid agar medium. After incubation, plates on which the overall growth has been retarded due to overcrowding of the colonies must be discarded. The rest are counted.

Drop or microtitre method. This is a simplified colony count technique (Fung and La Grange, 1969). Drops of standard volume (e.g. 0·02 ml of the dilutions) are transferred to the dry surface of a solid agar medium. The plates are then dried for 30 min at room temperature, and incubated for 24 h before counting, as far as possible, the individual colonies; the colonies are inclined to merge and cannot be easily differentiated. Nevertheless, the method is suitable to determine, at a glance, whether there are too many bacteria in the product (rapid screening procedure), and the technique does save space and material, because on one plate, two samples in two dilutions can be tested. The results of this method are slightly less reproducible than those of the standard plate count method, but the spot technique is accurate enough ($r = 0·83$) to be used as a routine test for estimating the bacterial count of milk (Lück and Dunkeld, 1974).

Flooding technique. The advantages of the flooding technique is that the poured plates can be stored and used at short notice. It is questionable, however, whether this method is really time and labour saving, as the time taken to pour the plates, and to make dilutions, is the same as that of the standard method. The plates have to be poured, dried, packed into plastic bags and stored. The diluted sample is poured over the surface of the agar of the prepoured plates, and the excess of the sample decanted (for 3 s) followed by a sharp shake. During incubation, the plates are placed upside down in their lids, with a filter paper to absorb remaining moisture.

The flooding technique is no less quantitative than the standard plate count (Lawrence et al., 1970). The correlation coefficients associated with the relationship between standard plate count and flooding plate count are large enough (at 24 h, $r = 0.89$; at 48 h, $r = 0.96$; Lück and Dunkeld, 1974) to recommend the flooding technique as a routine method.

The flooding technique is also the principle of the 'lactocult' method (Antila and Kylä-Siurola, 1976). Commercially available 'lactocult' plates, stored in sterile plastic tubes, are dipped into the liquid sample, and then put back into the plastic tube. After incubation, the bacterial count is estimated by comparing the colony density with that on model plates.

Membrane filtration method. When the bacterial count of a sample is low, and the sample or its dilution can be efficiently filtered (without causing a build-up on the filter), the membrane filtration method is most suitable. Membrane filtration apparatuses are commercially available. After the sample is passed through the membrane, the latter is placed on a solid agar medium, or on a filter pad which has been saturated with liquid medium. After incubation, the bacteria grow into colonies on the membrane. The test is only effective if the number of colonies per membrane is in the range of 10–200 (optimum: 50). When the medium does not contain an indicator, it is advisable to stain the membrane, e.g. by gently flooding the surface with a 0.01% aqueous solution of malachite green-oxalate, to confirm the count.

The porosity of the membrane, the type of medium, the conditions of incubation and the presence of spreaders shall be reported.

Most Probable Number (MPN) Method

The MPN test makes use of a statistical technique to determine low counts of bacteria in dairy products. Three sets of three or five tubes, each containing sterile medium, are prepared and inoculated from each of three consecutive, ten-fold dilutions. Tubes showing bacterial growth after incubation are positive. From the number of positive tubes in each set of three or five tubes, the most probable number of bacteria per unit of sample, as presented in McCrady's Tables, is read. When more than three dilutions (more than three sets of tubes) are made, only the results from three consecutive dilutions are significant. The highest dilution which gives positive results in all of the tubes, and the next two succeeding higher dilutions, should be chosen. When the weight or volume of sample in the first dilution is 10 or 100 times less than the weight or volume listed in McCrady's Tables, then the count tabled will be multiplied by 10 and 100, respectively.

Counting Single Bacteria

A new development in this field is an apparatus for counting the total number of single bacteria in milk. The bacteria, stained with a fluorescent dye (acridine orange), are detected by a microscope. Alternatively, a small quantity of the diluted 'milk' is applied to a rotating disc which spins in front of a narrow slit in the detector. The fluorescent light emission of the bacteria is transformed into pulses of electricity which activate a digital display and print out. In order to avoid interference due to other particles, such as fat globules, protein micelles and somatic cells, these materials may be initially separated by centrifugation; the bacteria in the 'milk' are then clearly visible. The separation, dilution, chemical treatment and staining are performed automatically (90 samples h^{-1}; range 10^5–10^7 bacteria ml^{-1}). The same principle is applied to the counting of somatic cells in milk.

Sterility Test

The aim of the quality control of products, aseptically processed and packed, is to limit the number of containers spoiled by microbiological contamination to a level acceptable to the market. The term 'sterility test' is here based upon stability, and not on sterility. Absolute sterility is sometimes difficult to achieve and, in practice, difficult to prove. There may be a very small number of bacteria in the product, but so long as they do not grow during a commercially acceptable period, then this product is commercially 'sterile'. Thus, commercially, sterility means: absence of all disease-causing bacteria, and absence of all micro-organisms capable of multiplication under the conditions of storage and distribution. The isolation of viable organisms, does not in itself, necessarily mean that the product is not safe.

To prevent the production of undesirable chemical or organoleptic changes in 'sterilised' products, heating must be kept at such a level that a very small number of heat-resistant spores will always survive. The larger the number of heat-resistant bacteria in the raw material, the larger the probability that some will survive. The bigger the package or bottle, the greater the probability of contaminants being in the container.

The spore-forming anaerobes, principally of the genus *Clostridium*, are less thermo-resistant, and therefore of lesser concern, than the spores of mesophilic and thermophilic aerobes of the genus *Bacillus*. Ultra-high temperature (UHT) treatment reduces the number of spores by 8–10 powers of ten. This means that, after treatment, milk with 100 spores ml^{-1} (10^5 spores $litre^{-1}$) will contain 0·001 to 0·000 01 spores per litre pack, and thus the number of non-sterile packs will be approximately one in 1000 to one in 100 000. This is acceptable. With survivors at a level of less than $1\,ml^{-1}$, or sometimes as low as $1\,litre^{-1}$, the use of the plate count method is not feasible, and the most probable number test has to be used.

To detect unsterile packages, different incubation temperatures and periods have been suggested. They depend very much on the type of product manufactured. A 2-week incubation, as suggested by the International Dairy Federation (1972), is commercially not practicable, because the whole production must be stored for 2 weeks before the results of the test are available. Tests carried out on UHT milk revealed that only 18% of unsterile packs were detected after 3 days at 30°C, 41% after 5 days and 68% after 7 days (Lück *et al.*, 1978*a*). The results indicate that an incubation period of 3 days is too short, while an incubation of 5 days does not give sufficient security. Thus an incubation period of 7 days, or 5–7 days, is advisable.

Two incubation temperatures have been recommended for UHT milk: 25–30°C and 55°C. However, most of the obligate thermophiles causing spoilage at 55°C remain dormant under ambient storage temperatures, and only grow when transferred to appropriate culture media and temperatures, so that only one incubation temperature, such as 35–37°C, is really necessary. The facultative thermophiles are capable of slow growth at 35°C, but one has to bear in mind that damaged bacteria more easily recover and grow at 27°C than at 35°C.

The quality of sterilised dairy products, as determined by the sterility test, is not assessed along the continuous scale of bacterial counts, but by attributes, i.e. 'defective or not'. As suggested by the International Dairy Federation (1972), defectiveness or spoilage is recorded when the organoleptic properties are different from those normally obtained by prolonged incubation; when the sample does not pass the ethanol (68% v/v) stability test; when the titratable acidity differs from what it was before incubation by more than 0·02% (expressed as grams of lactic acid per 100 g of milk); or when the colony count exceeds 10 per 0·1 ml of milk.

According to Langeveld *et al.* (1978), the decrease in oxygen content at the end of a 5-day incubation period at 30°C is a reliable method of establishing non-sterility, provided that the milk or custard has been indirectly sterilised; products treated by direct steam injection must be packed in a container which has a head-space, or which is sufficiently permeable to oxygen. The method is not suitable for products which show a rather large ability to bind oxygen, e.g. chocolate milk, caramel custard, etc. To prevent the destruction and wastage of many non-defective containers, a method of checking closed packages for non-sterility has been developed (Moisio and Kreula, 1973). This method is based on the fact that bacteria, if present, change the hydro-dynamic behaviour of the product after incubation, e.g. change of viscosity, separation of phases, etc. This change affects the oscillations of the package when placed on a table and swing a few degrees from side to side (6000–8000 packs h^{-1} can be handled by this technique).

Standard

The accepted percentage of non-sterile units varies from country to country and ranges from 0·02 to 0·1 %. A plant cannot be assumed to be functioning satisfactorily if more than one out of 1000 containers are defective.

Shelf-Life Tests

An important factor today is the open dating of perishable dairy products to indicate when the products were packed, or when they should be removed from market. Many dairy factories, therefore, need to rapidly determine the shelf-life of perishable dairy products, i.e. the period of time for which a product will remain acceptable in quality to the consumer. Thus, shelf-life can be expressed by the number of days or months required to attain a certain bacterial count, or an unsatisfactory flavour, under specified storage conditions.

The key to predicting shelf-life is not the total bacterial count immediately after processing, because this count does not differentiate between contaminants and those bacteria that survived pasteurisation, or between bacteria which can grow and those which cannot grow at the specified storage temperature. In many cases, a significant correlation between shelf-life and total bacterial count after processing does not exist (Randolph *et al.*, 1965; Lück *et al.*, 1980). Attention must, therefore, be given to the quality of dairy products at the retail level. Quality checks should be shifted from 'as processed' to 'as purchased' or even 'as consumed'. The shelf-life is, however, related to the time–temperature history of the product. In doing shelf-life tests, it is necessary to simulate the marketing conditions to which the product will actually be subjected. For this reason, a storage temperature of 5–7 °C is recommended.

The shelf-life is dependent on the number of post-pasteurisation contaminants, and especially on the presence of those contaminants which quickly multiply at refrigeration temperature, namely the psychrotrophs. Contaminant counts as low as one or two bacteria per litre can frustrate any attempt to manufacture a product with an extended shelf-life.

The bacteriological content of the raw product has an influence on the shelf-life of the product. This influence must not, however, be overestimated, especially when compared with the effect of post-pasteurisation contamination. Considerable importance is often attached to the incidence of thermoduric bacteria, but except when there is an excessive number of thermoduric bacteria (in excess of $1000 \, g^{-1}$), or a very long shelf-life is desired, the thermoduric microflora has relatively little influence on the keeping quality of pasteurised dairy products.

To determine the number of post-pasteurisation contaminants, selective media for the growth of psychrotrophic, Gram-negative bacteria at higher incubation temperatures have been developed. The literature published on this subject has been reviewed by Thomas (1969). There are significant correlations between shelf-life and the initial post-pasteurisation contamination count, and between shelf-life and the initial psychrotrophic count. The correlations are, however, not consistent enough to recommend these tests for estimating shelf-life ($r = -0.31$ to -0.37; Lück et al., 1980).

The growth rate of bacteria during storage at low temperature gives a better index of the shelf-life. The Moseley count (i.e. the plate count after 5 days of storage at 7 °C minus the plate count after processing; Moseley, 1958) has been shown to be a good index of post-pasteurisation contamination. Using this test, the determination of the bacterial count after processing is often not necessary, because the bacterial count of the product after 5 days of storage at 7 °C gives a good indication of the shelf-life ($r = -0.63$). Attempts have been made to improve the sensitivity of the Moseley count by extending the storage period at 7 °C to 7 days (Moseley, 1975). The long storage period (plus the incubation time of the plates) is the great disadvantage of these shelf-life tests. Other authors applied, therefore, a 3-day test for determining post-pasteurisation contamination of milk (Schilhabel et al., 1978), but because the contaminants determined by this method are mainly coliforms, the relationship between this count and the psychrotrophic count is not very consistent.

The optimum temperature for lipase and protease activity is often lower than the optimum for bacterial growth, and hence, off-flavours may develop even though bacterial counts are low. The flavour rating after different storage periods at 7 °C, is, although subjective, presently the most useful shelf-life test. In milk, flavour defects occur when the psychrotrophic count reaches approximately $10^7 \, \text{ml}^{-1}$.

To find the source of contamination in a plant, it is necessary to sample aseptically at given points in the post-pasteurisation system (see section on In-Line Sampling).

SPECIFIC TESTS FOR CERTAIN GROUPS OF ORGANISMS

Because the total bacterial count does not provide a complete picture of product quality, other enumeration tests have been recommended that may contribute to a better understanding of the bacterial problems in a dairy factory. As proposed in Table III, a programme for controlling the presence of specific groups of organisms might include the following tests.

Coliforms and *Escherichia coli*

The coliform group includes all aerobic and facultative anaerobic, Gram-negative, non-spore-forming bacteria (of both faecal and non-faecal origin), which are capable of fermenting lactose with the production of acid and gas at 32–35 °C within 24–48 h on solid or in liquid media. For the enumeration of coliforms, selective media are used, e.g. violet red bile agar, desoxycholate lactose agar, Endo agar or brilliant green lactose bile broth. On violet red bile agar, coliforms form typical dark red colonies (at least 0·5 mm in diameter) within 24 h at 32 °C. In brilliant green lactose broth, they show gas formation (in Durham tubes) within 48 h at 32 °C.

On pasteurised milk or dairy products with low coliform counts, the most probable number (MPN) test is carried out, because the plate count method is less determinate. A series of dilutions of the sample are inoculated into 2 % brilliant green lactose bile broth, and samples showing gas formation in the Durham tubes are positive. The gas-positive tubes can easily be tested for the presence of *E. coli* (faecal). A small quantity (0·2 ml) from the positive tubes is transferred into a separate tube of tryptone water for the formation of indole (*Eijkman test*). After incubation at 44 °C for 24 h, 0·5 ml of Kovac's indole reagent is added, and the development of a rose-coloured ring at the interface of the two liquids is an indication of indole. From the number of positive tubes, the MPN of *E. coli* can be determined.

Coliforms do not survive proper pasteurisation. A positive coliform test, after pasteurisation, indicates recontamination and, thus, gives an indication of factory hygiene, e.g. the unsatisfactory cleansing of equipment; improperly cleansed or sanitised equipment is mostly contaminated with coliforms. The coliform test carried out on raw milk gives only a limited indication of hygiene, unless the milk is tested 3 to 4 h after milking. Coliforms do rapidly multiply in milk, but good quality raw milk, produced under proper hygienic conditions and stored at 3–5 °C, should still contain less than 100 coliforms ml^{-1}.

When the lactose in the selective media is replaced by glucose, the total number of Enterobacteriaceae can be determined; this number is usually larger than that of the coliforms.

Psychrotrophs

The term psychrotroph or psychrotrophic bacteria was introduced to indicate bacteria capable of growth at approximately 5 °C, whatever their optimum growth temperature. To this group belong, among others, species of Pseudomonas, Alcaligenes, Flavobacterium, Enterobacteriaceae, *Bacillus*, *Streptococcus*, yeasts and moulds. Psychrotrophs are particularly undesirable, for many produce off-flavours, such as 'rancid', 'unclean' or 'malty', at refrigeration temperatures. Some species produce enzymes which may not be inactivated by pasteurisation or even by ultra-high temperatures, and which may give rise to defects in milk and milk products.

The present method recommended for the enumeration of psychrotrophs is the same as the standard plate count, except that the plates are incubated at 7 °C for 10 days (American Public Health Association, 1978), but because of the long incubation period, the psychrotrophic count is seldom determined routinely; with the length of time involved, the data may be of historical value only. Based on a reasonable assumption that the psychrotrophic flora is predominantly Gram-negative, and that measurement of the Gram-negative flora would give an indication of the number of psychrotrophs, several tests have been devised. The two approaches which have shown the most promise, are:

(1)   incubation at an elevated temperature, and
(2)   the addition of inhibitory substances to restrict the growth of Gram-positive organisms.

Compared with the psychrotrophic counts obtained by the standard method (10 days incubation at 7°C), the following methods can be recommended for the enumeration of psychrotrophs.

Incubation of the plates for 3 days at 15°C ($r = 0.967$). When the results are required promptly, a 2-day count after incubation at 17°C ($r = 0.947$) has proved a satisfactory alternative to the standard method (Lück and Hopkins, 1975). Incubating the plates at 15°C for 24 h and then at 7°C for 72 h (Juffs, 1970) is another alternative.

The differences in counts between the standard method, and the methods whereby inhibitory substances are added to the media to restrict the growth of Gram-positive organisms, are often rather great.

The counting of *oxidase-positive colonies* at elevated temperatures has been proposed as a rapid test to find potential psychrotrophs. This test is based on the ability of certain bacteria, which contain a strong cytochrome C oxidase system, to oxidise chemicals and to form dyes. For estimating the psychrotrophic count, the oxidase-positive count at 27 or 32°C ($r = 0.59$ and 0.57, respectively) showed no advantage over the determination of the total count ($r = 0.61$ and 0.61, respectively; Lück *et al.*, 1971).

Thermoduric Bacteria and Bacterial Spores

Thermoduric bacteria are organisms which survive pasteurisation at 72–74°C for 15–16 s. When present in large numbers, they can reduce the shelf-life of pasteurised products (see section on 'Shelf-Life Tests). For the enumeration of these bacteria, a small quantity (a loop) of the sample is mixed with 10 ml of melted plate count agar at 74°C. After exactly 15 s, the medium is poured out into a petri dish which is incubated at 30–32°C for 48 h.

In place of this high temperature–short time treatment, the *laboratory pasteurisation count* is often used, where the sample is heated to 63°C for 30 min. Both counts are significantly different. For additional information on thermoduric organisms, the review by Thomas *et al.* (1967) may be consulted.

Bacterial spores may survive the heat treatment applied and cause defects in the finished product (see section on 'Sterility Test'). For counting spores, the heat treatments of the product, applied in different laboratories, vary, e.g. 80°C for 10 min, 85°C for 10–15 min, 100°C for 2 min. A temperature of 80°C for 10 min is not sufficient to kill all vegetative cells in milk and hence, a treatment at 85°C for 10 min is recommended. After this treatment, the sample is immediately cooled to about 10 °C, and the plate count technique is applied with incubation of the plates at 30 °C for 72 h (for mesophiles), or at 55 °C for 48 h (for thermophiles). The *Bacillus cereus* content of a sample (spores and vegetable cells) can be counted on mannitol egg yolk polymyxin agar (Mossel *et al.*, 1976). Dry colonies with a distinct violet–red background surrounded by a halo of dense white precipitate are presumptive *B. cereus* colonies. Of the colonies isolated in this way, 26 % were not, in fact, *B. cereus* (Walthew and Lück, 1978).

When the spores from anaerobic bacteria are to be counted, a broth of reinforced clostridial medium, plus a 'seal' of 2 % sterile water agar containing 0·5 g litre^{-1} sodium thioglycolate, is recommended; because of the low number of anaerobic spores usually found in dairy products, the most probable number method is carried out. After inoculation, the tubes are sealed by the addition of a layer of the agar containing sodium thioglycolate (25 mm thick), and are incubated at 30 °C for 14 days (mesophiles) or 55 °C for 5 days (thermophiles). The tubes are examined daily, and those forming gas are positive.

The number of spores from anaerobes which produce hydrogen sulphide can be determined using differential reinforced clostridial medium; tubes which show blackening of their contents are positive and can be used for further identification.

Proteolytic and Lipolytic Organisms

The proteolytic count is determined on plate count agar containing 10 % of sterile milk. After incubation at 23–25 °C for 48 h, the plates are flooded with 10 % acetic acid or 1 % HCl. After 60 s the excess acid is removed, and colonies showing clear zones around them, due to proteolysis, are counted. Flooding of the plates with acid is necessary as some acid-forming bacteria are also capable of producing clear zones.

To determine the lipolytic count, a sugar-free nutrient agar (pH 7·5) containing emulsified butter-fat is used. As an indicator, fat-soluble Victoria blue base, decolourised with 10 % sodium hydroxide, is added to the butter-fat; 0·5 ml of this butter-fat is mixed aseptically with 10 ml of molten nutrient agar (48 °C), and the mixture is then shaken vigorously and poured immediately into petri dishes. A dilution of the sample is spread on the surface of the agar of the prepoured and dried plates. After incubation of the plates at 30 °C for 6 days, any plates showing 6–60 well-spaced blue colonies are counted. The blue colour is caused by the free fatty acids formed as the result of hydrolysis of the butter-fat.

Yeasts and Moulds

To determine the yeast and mould count, the pour plate method is applied using malt extract agar or potato dextrose agar which have been acidified with citric acid or tartaric acid to yield a final pH between 3·3 and 3·7. Plates are incubated at 22–25 °C for 5 days. The addition of 5 μg ml^{-1} of rose Bengal restricts the spread of rapidly growing mould mycelia. At present, yeast extract dextrose chloramphenicol agar is recommended as a reference medium for counting yeasts and moulds. To confirm yeast colonies, a representative number of the different colony types are examined microscopically. In flavoured, cultured and condensed milks containing sucrose, gas-forming yeasts can be a problem. The ability to form gas can be checked by inoculating separate colonies into tubes, containing Durham tubes in sucrose-enriched yeast extract tartaric acid broth. The presence or absence of gas is recorded after incubation for 10 days at 25 °C, The presence of mould mycelia on the lids of cans or on the surface of dairy products (lumps on the surface of sweetened condensed milks) are examined microscopically.

Viable Culture Organisms

To determine whether cultured dairy products contain viable culture organisms, 1 ml of the one in 10 dilution of the cultured product is inoculated into each of two tubes of sterile, antibiotic-free skim-milk. The tubes are incubated for 24 h, one at 30 °C, the other at 37 °C, or at the optimum growth temperature of the culture expected. From each tube, a slide is prepared and examined for the presence of culture organisms characteristic of the product tested.

Thermophilic Organisms

When dairy products are held at elevated temperatures, or are cooled down too slowly after heat treatment, thermophiles may cause defects. These thermophiles are usually bacilli which grow at 55 °C or higher, and they may be counted on pour plates incubated at 55 °C for 48 h. When their number if very small, the MPN test using sterile litmus milk has to be used. The presence of thermophiles in the tubes is detected by changes in the litmus milk after 48 h at 55 °C (change in the normal colour or consistency of the milk). Tubes with no apparent growth have to be checked for viable bacteria by the streak plate method.

Micrococci

Micrococci can cause abnormalities in the viscosity of sweetened condensed milk. To detect their presence, a smear of the milk is examined microscopically for the presence of micrococci. Then a loopful of the milk is transferred to the surface of a nutrient agar slope and, after incubation (30 °C for 72 h), examined microscopically for micrococcal colonies. Low levels of micrococci in non-stored containers may be detected by incubating 100 g of the milk for 5 days at 30 °C, and then examining it microscopically. During incubation, the milk is shaken vigorously each day to stimulate the growth of micrococci.

Staphylococcus aureus and Salmonellae

Dairy products may also have to be tested for pathogenic bacteria, but usually this is not done in routine control. Pathogens may enter the dairy products after pasteurisation by contact with flies or other insects, by contact with the hands of operators, by sneezing or coughing, or via drops of contaminated condensate. Hence, products with low total bacterial counts may contain pathogens or their toxins, but conversely high bacterial counts do not give an indication of the presence of pathogens.

The World Health Organisation has described 16 different bacterial intoxications and seven viral and rickettsial diseases caused by ingestion of contaminated dairy products (Kaplan *et al.*, 1962). To test for pathogens, inhibitors are added to the media to inhibit the growth of non-pathogenic organisms, but these enrichment and/or selective plating procedures only reduce the growth of non-pathogenic bacteria to approximately 70 %. Therefore a further identification of the pathogens has to be carried out. In this section, only tests for *Staph. aureus* and salmonellae are mentioned.

For the detection and counting of *Staph. aureus*, the Baird-Parker's egg yolk tellurite glycine pyruvate agar is often used. Black and shiny colonies with a narrow grey–white margin surrounded by a clear zone extending into the opaque medium are counted as presumptive *Staph. aureus*. These colonies have to be submitted to a coagulase test for confirmation. Two methods can be applied.

1. *Spread method:* 0·1 ml of the one in 10 dilution of the sample is transferred to the surface of a prepared plate of Baird-Parker's medium. The plates are incubated at 37 °C, and examined at 24 and 48 h for colonies of coagulase-positive staphylococci.

2. *Enrichment broth:* 1 ml of the one in 10 dilution of the sample is added to an enrichment broth (trypticose soy broth plus 10% sodium chloride or Giolith and Cantoni's broth), incubated at 37 °C for 24 h, and then a loop thereof is streaked onto Baird-Parker's medium.

Slide Coagulase Test

A small drop of distilled water is transferred to a microscope slide, and mixed with part of a single colony from the Baird-Parker plate giving a thick suspension. A loopful of undiluted or reconstituted plasma is mixed with this. Coagulase-positive strains give rise to obvious clumping within 10 s. Approximately the square root of the number of colonies on the plate should be tested. Black colonies which are not surrounded by a clear zone or opacity should be included in this test because they may also be coagulase-positive.

The test for *Salmonella* in dairy products includes a liquid selective enrichment stage, the use of a solid selective medium for the isolation of presumptive salmonellae, and their subsequent biochemical and serological confirmation. Often prior to the enrichment stage, a pre-enrichment medium containing buffered peptone water is recommended to resuscitate the salmonellae (18–24 h at 37 °C).

For the selective enrichment procedure, either mannitol selenite cystine broth or selenite brilliant green sulphonamide broth (18–24 h at 42 °C) and either tetrathionate broth or strontium chloride broth (18–24 h at 37 °C) are used. Each culture of enrichment broth is plated out onto a number of different selective agar media, such as brilliant green phenol red agar, brilliant green sulphonamide agar, bismuth sulphite agar or *Salmonella–Shigella* agar (24–48 h at 37 °C). Typical colonies of presumptive salmonellae are: on bismuth sulphite agar, brown or black with metallic sheen (some strains produce green colonies); on brilliant green phenol red agar, colourless or pink with red surrounding medium; and on *Salmonella–Shigella* agar, uncoloured to pale pink. From each plate, three suspect salmonellae colonies are subcultured into separate tubes of peptone water (3–4 h at 37 °C) for biochemical confirmation. Typical salmonellae give a lysine decarboxylase-positive reaction (purple colour in lysine decarboxylase broth after 18–24 h at 37 °C) and are β-D-galactosidase-negative (colourless in *o*-nitrophenyl-β-D-galactopyranoside (ONPG) broth after 18–24 h at 37 °C).

Members of the *Arizona* group which usually ferment lactose (ONPG-positive = yellow colour) should be considered as potential pathogens. For complete identification, the cultures should be forwarded to an authorised laboratory.

Detection of Antibiotic Residues

The disc assay method is extensively used to determine the presence of residual antibiotics in fluid milk products. The liquid sample is allowed to wet a filter paper disc (diameter: 12–13 mm) by capillary action using dry forceps, and then the disc is placed on the surface of an inoculated agar medium (in a petri dish) containing approximately 10^6 micro-organisms ml^{-1}.

The sample contains antibiotics or other inhibiting substances when it gives a clear zone of inhibition around the disc. The diameter of the zone is an indication of the inhibitor concentration. For quantitative determinations, reference discs, which contain 0·005, 0·01, 0·025, 0·1, 0·25 and 0·5 units of penicillin, are also placed on the plate. Uniformity of size and absorbance of discs, uniformity of inoculum and depth of medium are, however, very important for quantitative studies. The following test organisms can be used: *B. subtilis* (ATCC 6633), after 3–4 h incubation at 37 °C; *Sarcina lutea* (ATCC 9341), after 16–18 h incubation at 30 °C; and *B. stearothermophilus* sub-sp. *calidolactis* (C953) after 2·5–5 h at 55–65 °C. Ampules seeded with *B. stearothermophilus* are commercially available ('Delvo test'). After removal of the tip of the ampule, a nutrient tablet containing bromcresol purple as indicator and 0·1 ml of the sample are added. The ampules are then incubated at 63–65 °C for 2·5 h. A purple colouration of the entire test medium indicates the presence of inhibiting substances.

RELEVANCE OF TECHNIQUES AND INTERPRETATION OF RESULTS

No single laboratory test can produce the full information that is required. No bacteriological test yet evolved is above criticism. There is no 'best' test, and yet often much money is wasted on elaborate tests for which fictitious accuracy is claimed. It is now recognised that regular testing is of far greater importance, and that any test, that is better than mere haphazard classification, will contribute to an improvement in quality. The real value of a test is whether it can detect products of unsatisfactory quality, and thus can make a contribution to improving production hygiene. It is, therefore, not necessary that a test should give an absolute measure of the quality, nor need it be in complete agreement with the results of other tests.

Since none of the different methods employed to determine a specified bacterial content give exactly the same result, one test cannot be automatically substituted for another. Two different tests can be compared statistically in order to arrive at comparable standards, but not necessarily to determine their value. The only conditions of a suitable test are:

(1)   there must be a significant correlation between the results of the test and the quality required, and

(2)   the operator must be fully informed on how to obtain this quality.

Due to the error inherent in a single bacterial count, i.e. limitations of the particular bacterial count/test, seasonal and local variation of the microflora, day-to-day variation of the bacterial count, and counting only colony forming units or clumps instead of individual bacteria, only approximately five-fold differences in plate counts can be regarded as significant when grading dairy products. Hence the quality categories have to be established in such a way that the differences between bacterial counts are large enough to be significant. The realisation of this fact led to the development of rapid screening procedures to meet the requirements of the dairy industry.

Quality control is planned and introduced by management, but diagrams indicating real or anticipated quality levels should however, not only concern the management but also the operators. One operator should be responsible for a specific machine or a specific process, and the relevant diagrams can take the form of control charts showing, on the horizontal axis, the sequence of sample or the date of manufacture, and on the vertical axis, the quality characteristics (shelf-life, log bacterial count, etc.). A line can represent, for instance, the expected average log bacterial count of a product, and above this line, the upper limit line is drawn, such that only one in 20 of the plotted points should lie outside this line.

For shelf-life tests, a lower limit line is drawn. The number of points outside the limit lines indicates whether the process has altered in some way, and they are, therefore, essentially indicators of a need for corrective action.

TABLE IV

SUGGESTED MICROBIOLOGICAL STANDARDS FOR DIFFERENT DAIRY PRODUCTS (FIGURES IN PARENTHESES MEAN 'AIMED AT')

Product	Test	Count or result
Raw milk for liquid consumption after being pasteurised	Total bacterial count	$<250\,000\ (50\,000)\ \text{ml}^{-1}$
	Coliforms	<100
	E. coli (faecal type)	Absent in 0·01 ml
	Thermoduric count	$<1\,000\ \text{ml}^{-1}$
	Spores	$<10\ \text{ml}^{-1}$
	B. cereus (spores)	$<1\ \text{ml}^{-1}$
	Staph. aureus (coagulase-positive)	$<100\ (10)$
	Methylene blue reduction time (at 37°C)	Not less than 5 h
	3-h Resazurin test (at 37°C, Lovibond disc reading)	Not less than 3 h
	Somatic cell count	$<750\,000\ \text{ml}^{-1}$
Raw milk Raw cream (to be consumed raw)	Total bacterial count	$<50\,000\ (10\,000)\ \text{ml}^{-1}$
	Coliforms	$<10\ \text{ml}^{-1}$
	E. coli (faecal type)	Absent in 1 ml
	Methylene blue reduction time (at 37°C)	Not less than 7 h
	3-h Resazurin test (at 37°C, Lovibond disc reading)	Not less than 4 h
	Staph. aureus (coagulase-positive)	$<10\ \text{ml}^{-1}$
	Somatic cell count	$<500\,000\ \text{ml}^{-1}$
Pasteurised market milk Pasteurised cream	Total bacterial count	$<50\,000\ (5\,000)\ \text{ml}^{-1}$
	Coliforms (after processing)	$<1\ (0·1)\ \text{ml}^{-1}$
	E. coli (faecal type)	Absent in 10 ml
Dried milk	Direct microscopic clump count	$<50\,000\,000\ (10\,000\,000)\ \text{g}^{-1}$

Product	Test	Standard
Dried milk products	Total bacterial count (plate count)	$<100\,000$ ($50\,000$) g^{-1}
	Yeast and moulds	<10 g^{-1}
	Coliforms	Absent in 1 g
	E. coli (faecal type)	Absent in 10 g
	Staph. aureus (coagulase-positive)	Absent in 1 g
	Salmonellae	Absent in 25 (100) g
Ice cream	Total count (plate count)	$<50\,000$ ($5\,000$) g^{-1}
	Coliforms	<10 g^{-1}
	Staph. aureus (coagulase-positive)	<10 g^{-1}
	E. coli (faecal type)	Absent in 1 g
	Salmonellae	Absent in 25 (100) g
Cultured milks / Cultured cream	Yeasts and moulds	<10 (1) g^{-1}
	Coliforms	<10 (1) g^{-1}
Sweetened condensed milk	Total count	<1000 (100) g^{-1}
	Yeasts and moulds	<1 g^{-1}
	Coliforms	<10 (1) g^{-1}
Butter	Contaminating organisms (non-lactic acid bacteria)	$<50\,000$ ($10\,000$) g^{-1}
	Proteolytic organisms	<1000 g^{-1}
	Lipolytic organisms	<1000 g^{-1}
	Yeasts and moulds	<10 g^{-1}
	Coliforms	<10 (1) g^{-1}
	E. coli (faecal type)	Absent in 1 g
Cheese	*E. coli* (faecal type)	Absent in 0·01 g
	Staph. aureus (coagulase-positive) (at 1 month)	Absent in 0·1 g
Cottage cheese	Yeasts and moulds	<10 (1) g^{-1}
	Coliforms	<10 g^{-1}
Casein	Coliforms	<10 g^{-1}

It is often necessary to summarise results or to calculate mean counts. When there is a small variation in the bacterial counts of different samples, or when the counts are low, the arithmetic mean of the bacterial counts may be calculated. When there are, however, large variations between counts, and the counts are high, the geometric mean (logarithmic average) should be used (geometric mean = arithmetic mean of the log count, which is subsequently transformed into a bacterial count again). When the plate count method supplies positive results, more emphasis should be attached to this method than to the most probable number test, because the repeatability of the plate count method is higher than that of MPN values.

MICROBIOLOGICAL STANDARDS FOR DIFFERENT DAIRY PRODUCTS

Limitation of poor quality products and health protection form the basis for food standards. Legal and voluntary bacteriological standards vary widely from country to country. In Table IV, certain limits are proposed which may be a useful tool to improve the quality of dairy products. A product complies with the bacteriological specifications when at least four out of five portions or samples contain less than the maximum bacterial count specified. Legal action should only be taken when more than one out of five samples exceed the upper limit.

During recent years, the variation of counts in different samples has been taken into consideration, and the hygiene requirements are, for instance, often expressed as follows. Examine five (n) samples of dried milk, and allow two (c) samples to exceed 50 000 bacteria g^{-1} (m), but none to exceed 200 000 g^{-1} (M) or abbreviated: $n = 5, c = 2, m = 50 000$, $M = 200 000$.

REFERENCES

AMERICAN PUBLIC HEALTH ASSOCIATION (1978) *Standard Methods for the Examination of Dairy Products*, 14th edn. American Public Health Association, New York.
ANTILA, V. and KYLÄ-SIUROLA, A. L. (1976) *Milchwissenschaft*, **31**, 8.
BALDOCK, J. B. (1974) *J. Milk Fd Technol.*, **37**, 361.
BOCKELMANN, B. VON (1974) *Nordeuropaeisk Mejeri Tidsskrift*, **40**, 292.
BULLERMAN, L. B. (1976) *J. Fd Sci.*, **41**, 26.

CANNON, R. V. and REDDY, K. K. (1967) *J. Dairy Sci.*, **50**, 938.

CATE, L. TEN (1965) *J. Appl. Bacteriol.*, **28**, 221.

DALLA, G. (1974) *Industrie Alimentari*, **13**, 113.

DICKER, R. A. and WILES, R. (1974) *XIXth Internat. Dairy Congr.*, 1E, 799.

FLEMING, M. G. and O'CONNOR, F. (1975) *Irish J. Agric. Res.*, **14**, 21, 27.

FRUIN, J. T. and CLARK, W. S. (1977) *J. Fd Protection*, **40**, 552.

FUNG, D. Y. C. and LA GRANGE, W. S. (1969) *J. Milk Fd Technol.*, **32**, 144.

GRAPPIN, R. and JEUNET, R. (1974) *XIXth Internat. Dairy Congr.*, 1E, 530.

GREEN, V. W. and HERMAN, L. G. (1961) *J. Milk Fd Technol.*, **24**, 262.

HEDRICK, T. I. (1975) *Chem. Ind.*, **20**, 868.

HEDRICK, T. I. and HELDMAN, D. R. (1969) *J. Milk Fd Technol.*, **32**, 265.

HEESCHEN, W., REICHMUT, J., TOLLE, A. and ZEIDLER, H. (1969) *Milchwissenschaft*, **24**, 729.

HOWARD, A. R. and WESTHOFF, D. C. (1974) *IVth Internat. Congr. Fd Sci. Technol.*, **4b**, 30.

INTERNATIONAL DAIRY FEDERATION (1967) *Definition of mastitis*, Document No. 34, Ed. IDF, Bruxelles, pp. 1–5.

INTERNATIONAL DAIRY FEDERATION (1969) *Standard methods for sampling milk and milk products*, International Standard FIL.-IDF 50: 1969, Ed. IDF, Bruxelles, 7 pp.

INTERNATIONAL DAIRY FEDERATION (1972) *IDF Monograph on UHT milk*, Annual Bulletin, Part V, Ed. IDF, Bruxelles.

INTERNATIONAL DAIRY FEDERATION (1979) *Somatic cells in milk. Their significance and recommended methods for counting*, Document 114, Ed. IDF, Bruxelles, 20 pp.

JAARTSVELD, F. H. J. and SWINKELS, R. (1974) *Netherlands Milk and Dairy J.*, **28**, 93.

JEUNET, R., GRAPPIN, R., THIOLLIERE, M. H. and RICHARD, M. (1973) *Revue Laitière Française*, **312**, 647.

JUFFS, H. S. (1970) *Australian J. Dairy Technol.*, **25**, 30.

KAPLAN, M. M., ABDUSSALAM, M. and BIJENGA, G. (1962) *Milk Hygiene*, Ed. World Health Organisation, Geneva, pp. 11–79.

KRISTENSEN, J. M. B. (1977) *Maelkeritidende*, **90**, 119, 552.

LANGEVELD, L. P. M., BOLLE, A. C. and CUPERUS, F. (1978) *Netherlands Milk and Dairy J.*, **32**, 69.

LAWRENCE, R. C., MARTLEY, F. G., TEESE, J. G. and NEWSTEAD, D. F. (1970) *New Zealand J. Dairy Sci. Technol.*, **5**, 22.

LEESMENT, H. (1971) *Svenska Mejeritidningen*, **63**, 55.

LUBIEBIECKI-V. SCHELHORN, M. (1973) *Verpackungs-Rundschau, Technisch-Wissenschaftliche Beilage*, **14**(10), 77.

LÜCK, H. (1972) *Dairy Sci. Abstr.*, **34**, 101.

LÜCK, H. and ANDREW, M. J. A. (1975) *S. African J. Dairy Technol.*, **7**, 39.

LÜCK, H. and CHEESMAN, C. E. (1978) *S. African J. Dairy Technol.*, **10**, 143.

LÜCK, H. and DUNKELD, M. (1972) *S. African J. Dairy Technol.*, **4**, 93, 179.

LÜCK, H. and DUNKELD, M. (1974) *S. African J. Dairy Technol.*, **6**, 135.

LÜCK, H. and HOPKINS, F. (1975) *S. African J. Dairy Technol.*, **7**, 89.

LÜCK, H. and WEHNER, F. C. (1979) *S. African J. Dairy Technol.*, **11**, 169.

LÜCK, H., CLARK, P. C. and GROENEVELD, H. T. (1970*a*) *Agroanimalia*, **2**, 69.

LÜCK, H., CLARK, P. C. and TONDER, J. L. VAN (1970*b*) *Milchwissenschaft*, **25**, 155.

324   *H. Lück*

LÜCK, H., HOLZAPFEL, W. H. and BECKER, P. J. (1971) *Milchwissenschaft*, **26**, 424.

LÜCK, H., KELLER, J. J. and ANDREW, M. J. A. (1975) *S. African J. Dairy Technol.*, **7**, 111.

LÜCK, H., MOSTERT, J. F. and HUSMANN, R. A. (1978a) *S. African J. Dairy Technol.*, **10**, 83.

LÜCK, H., WALTHEW, J. and JOUBERT, B. (1978b) *S. African J. Dairy Technol.*, **10**, 3.

LÜCK, H., DUNKELD, M. and BERG, M. VAN DEN (1980) *S. African J. Dairy Technol.*, **12**, 107.

MILAAN, P. W. VAN and PULLES, P. C. W. (1972) *Officieel Orgaan van de Koninklijke Nederlandse Zuivelbond*, **64**, 344.

MOISIO, T. and KREULA, M. (1973) *Milchwissenschaft*, **28**, 477.

MOSELEY, W. K. (1958) *Proc. 51st Annual Convention Milk Industry Foundation (Laboratory Section)*, p. 27.

MOSELEY, W. K. (1975) *Dairy and Ice Cream Field*, **158**, 44.

MOSSEL, D. A. A., KOOPMAN, M. J. and JONGERIUS, E. (1967) *Appl. Microbiol.*, **15**, 650.

POSTHUMUS, G., KLIJN, C. J. and GIESEN, T. J. J. (1974) *Netherlands Milk and Dairy J.*, **28**, 79.

RANDOLPH, H. F., FREEMAN, T. R. and PETERSON, R. W. (1965) *J. Milk Fd Technol.*, **28**, 92.

REUTER, H. and QUENTE, J. (1977) *Milchwissenschaft*, **32**, 395.

RICHARD, J. and AUCLAIR, J. (1969) *Revue Laitière Française*, **261**, 15.

SCHILHABEL, W., WALLNER, U., KLEEBERGER, A. and BUSSE, M. (1978) *XXth Internat. Dairy Congr.*, E, 334.

TAMMINGA, S. K. and KAMPELMACHER, E. H. (1977) *Zentralblatt für Bakteriologie, Parasitenkunde und Hygiene. I Abteilung Originale*, **B.165**, 423.

THOMAS, S. B. (1969) *J. Appl. Bacteriol.*, **32**, 269.

THOMAS, S. B. and THOMAS, B. F. (1977) *Dairy Ind. Internat.*, **42**, 7.

THOMAS, S. B., DRUCE, R. G., PETERS, G. J. and GRIFFITHS, D. G. (1967) *J. Appl. Bacteriol.*, **30**, 265.

THOMPSON, D. J., DONELLY, C. B. and BLACK, L. A. (1960) *J. Milk Fd Technol.*, **23**, 167.

TOLLE, A., ZEIDLER, H. and HEESCHEN, W. (1968) *Milchwissenschaft*, **23**, 65.

TOLLE, A., HEESCHEN, W., WERNEY, H., REICHMUT, J. and SUHREN, G. (1972) *Milchwissenschaft*, **27**, 343.

US PUBLIC HEALTH SERVICE (1959) *Sampling microbiological aerosols*, Public Health Monograph No. 60, Washington, DC, Public Health Service.

VICKERS, V. T. and MCROBERT, A. G. (1977) *New Zealand J. Dairy Sci. Technol.*, **12**, 5.

VOSS, E. and MOLTZEN, B. (1973) *Milchwissenschaft*, **28**, 479.

WALTER, W. G. (1955) *Bacteriol. Rev.*, **19**, 284.

WALTHEW, J. and LÜCK, H. (1978) *S. African J. Dairy Technol.*, **10**, 47.

WERNERY, H., REICHMUT, J., HEESCHEN, W. and TOLLE, A. (1973) *Deutsche Molkerei-Zeitung*, **94**, 524.

WILLIAMS, M. L. B. (1971) *Canadian Inst. Fd Technol. J.*, **4**, 187.

WRIGHT, E. O., REINBOLD, G. W., BURMEISTER, L. and MELLON, J. (1970) *J. Milk Fd Technol.*, **33**, 168.

Index